TROUBLESHOOTING ELECTRICAL/ELECTRONIC SYSTEMS

AMERICAN TECHNICAL PUBLISHERS, INC.
HOMEWOOD, ILLINOIS 60430

Glen A. Mazur
Thomas E. Proctor

© 1994 by American Technical Publishers, Inc.
All rights reserved

1 2 3 4 5 6 7 8 9 – 94 – 9 8 7 6 5 4 3 2 1

Printed in the United States of America

ISBN 0-8269-1775-5

CONTENTS

Troubleshooting	**1**	Troubleshooting Levels • Troubleshooting Methods • Troubleshooting Procedure • Equipment Reliability • Troubleshooter Responsibilities • Electrical Shock • Electrical Safety	... 1 ... 17
Basic Electrical Theory	**2**	Electrical Prefixes • Using Ohm's Law • Using the Power Formula • Temperature Conversion • AC Sine Wave • Drawings and Diagrams	... 25 ... 43
Symbols and Circuits	**3**	Electrical Symbols and Abbreviations • Series, Parallel, and Series/Parallel Connections • Circuit Polarity • Batteries and Photocells • Resistance In a Series Circuit • Resistance In a Parallel Circuit • Resistance In a Series/Parallel Circuit • Capacitor Connection	... 55 ... 63
Meter Symbols and Terminology	**4**	Measurement Precautions • Meter Abbreviations and Symbols • Meter Terminology • Reading Analog Displays • Reading Digital Displays	... 73 ... 81
Meters	**5**	Multimeters • Test Lights • Voltage Testers • Voltmeters • In-Line Ammeters • Clamp-On Ammeters • Ohmmeters • Continuity Checkers • Wattmeters	... 87 ... 99
Special Meters	**6**	Contact Thermometers • Infrared Meters • Contact Tachometers • Photo Tachometers • Strobe Tachometers • Megohmmeters • Digital Logic Probes • Oscilloscopes/Scopemeters	...109 ...127
Basic Circuit Measurements	**7**	Branch Circuit Voltage Drop Measurement • Recording Minimum and Maximum Readings • Meter Loading Effect • Voltmeter Circuit Connections • Harmonic Distortion • Troubleshooting Fuses • Troubleshooting Circuit Breakers	...135 ...143
Relays and Motor Starters	**8**	Relays • Contactors and Motor Starters	...149 ...159
Troubleshooting Relays and Starters	**9**	Troubleshooting Electromechanical Relays • Troubleshooting Solid-State Relays • Troubleshooting Contactors and Motor Starters	...165 ...173
Motor Electrical Problems	**10**	Phase Unbalance • Voltage Unbalance • Single Phasing • Improper Phase Sequence • Voltage Surges • AC Voltage and Frequency Variations • DC Voltage Variations • Allowable Motor Starting Time • Megohmmeter Tests • Insulation Spot Test • Dielectric Absorption Test • Insulation Step Voltage Test	...179 ...189
Motor Mechanical Problems	**11**	Heat Problems • Altitude Correction • Motor Mounting and Positioning • Loose Connections • Motor Defect	...197 ...203
Troubleshooting DC Motors	**12**	DC Motors • Troubleshooting DC Motors	...209 ...217

Chapter	Title	Topics	Pages
13	Troubleshooting AC Motors	AC Motors • Troubleshooting AC Motors	☐...221 ○...239
14	Power Distribution	Power Conditioning and Protection • Isolated Ground Receptacles • Transformer Load Cycle • Wye and Delta Transformer Connections • Transformer Tap Connections • Transformer Sizing and Balancing	☐...245 ○...259
15	Troubleshooting Motor Control Circuits	Power Loss • Troubleshooting Power Circuits • Troubleshooting Control Circuits	☐...269 ○...277
16	Lighting Circuits	Lighting Terminology • Incandescent Lamps • Fluorescent Lamps • High-Intensity Discharge Lamps	☐...283 ○...303
17	Troubleshooting Lighting Circuits	Troubleshooting Incandescent Lamps • Troubleshooting Fluorescent Lamps • Troubleshooting High-Intensity Discharge Lamps	☐...309 ○...323
18	Solenoids and Heating Elements	Solenoid-Operated Valves • Troubleshooting Solenoid-Operated Valves • Heating Elements • Troubleshooting Heating Elements	☐...331 ○...343
19	Mechanical and Solid-State Switches	Troubleshooting Mechanical Switches • Troubleshooting Solid-State Switches • Protecting Switch Contacts • Proximity Sensor Installation • Protecting Pressure Switches • Flow Switches • Level Control Problems	☐...355 ○...365
20	Diodes, Transistors, and Thyristors	Diodes • Transistors • Thyristors	☐...373 ○...381
21	Testing Diodes, Transistors, and Thyristors	Diode Testing • Transistor Testing • Thyristor Testing • Electronic Heat Problems	☐...387 ○...393
22	Programmable Controllers	Programmable Controllers • Status Indicators and Error Codes • Force and Disable • Startup Procedure • PC Preventive Maintenance	☐...399 ○...417
23	Preventive Maintenance	General Motor Maintenance • Recording Meters • Lockout / Tagout • Enclosure Selection • Conveyor Belt Tracking	☐...427 ○...437

Glossary ...441

Appendix ...453

Index ...473

TROUBLESHOOTING

Applications

Application — Troubleshooting Levels

Troubleshooting is the systematic elimination of the various parts of a system or process to locate a malfunctioning part. A *system* is a combination of components, units, or modules that are connected to perform work or meet a specific need. A *process* is a sequence of operations that accomplishes desired results. A *malfunction* is the failure of a system, equipment, or part to operate properly.

To locate and correct a malfunction quickly, troubleshooting is performed at different levels. The different levels of troubleshooting electrical and electronic systems are the system, equipment/unit, board/module, and component levels. See Troubleshooting Levels.

TROUBLESHOOTING LEVELS

System Level

A system is a combination of interconnected individual units, modules, and components. In a system, the combination of units, modules, and components works together to produce the desired results.

A system may include main or hand-held programming terminals, fixed or modular controllers, operator terminals, displays, and interfaces. See Electrical/Electronic System.

1

2 TROUBLESHOOTING ELECTRICAL/ELECTRONIC SYSTEMS

ELECTRICAL/ELECTRONIC SYSTEM

The programming terminals are usually located at a central location. Hand-held programming terminals for local programming changes and troubleshooting are connected at any point in the system. The controllers are located at individual machines or processes throughout the plant.

Troubleshooting at the system level requires a knowledge of the operation of the hardware, software, and interfaces in the system. *Hardware* is the physical components in a system. Hardware malfunctions occur when two or more pieces of equipment are not properly sending or receiving data.

Software is the computer programs and procedures that operate the hardware. Software malfunctions are problems with the computer program(s).

An *interface* is an electronic device that allows different levels or types of components to be used together in the same circuit. Interface problems occur when two or more pieces of equipment are working properly but are not able to communicate with each other.

A *systems analyst* is an individual who troubleshoots at the system level. *System analysis* is the breakdown of system requirements and components performed when designing, implementing, maintaining, or troubleshooting a system. A systems analyst observes each interface, connecting cable, program, piece of equipment, and language used.

System troubleshooting is performed on-site or off-site. *On-site troubleshooting* is troubleshooting at the location where the hardware is installed. *Off-site troubleshooting* is troubleshooting at a location other than where the hardware is installed. Off-site troubleshooting is performed when an on-site computer sends information about a malfunction to an off-site computer. The off-site computer is programmed to solve malfunctions in systems. When the malfunction is found, the off-site computer sends information to the on-site computer. See Off-site Troubleshooting.

OFF-SITE TROUBLESHOOTING

Equipment/Unit Level

A *unit* is an individual component that performs a specific task by itself. Systems are made when units are connected. For example, a TV, VCR, CD player, and AM-FM receiver each are separate units. When connected, an entertainment system is developed.

Troubleshooting at the equipment/unit level requires identification of the equipment or unit that is malfunctioning. When the malfunctioning equipment or unit is identified, substitution or testing is performed.

Substitution is the replacement of malfunctioning equipment or units. Substitution is performed when the equipment or unit is small, easily accessible, and a replacement is available.

Testing is performed when the equipment or unit is a hard-wired unit or is large in size, or when no replacement is available. A *hard-wired unit* has conductors connected to terminal screws. A programmable controller may have hundreds of hard-wired connections. Testing requires knowledge of using test equipment and ability to correctly interpret the test results.

For example, if a main computer cannot operate a machine connected to a programmable controller, a malfunction exists in either the main programming terminal, interface, connecting cables, or programmable controller. A hand-held programming terminal is plugged into the programmable controller to troubleshoot at the equipment/unit level without disconnecting the equipment. See Hand-held Programming Terminal.

HAND-HELD PROGRAMMING TERMINAL

A hand-held programming terminal monitors information from and sends information to the programmable controller. Monitoring indicates whether information is received from the main computer. If no information is received, a malfunction exists between the programmable controller and the main computer. If information is received, a malfunction exists between the programmable controller and the machine. When a malfunctioning piece of equipment is identified, it is replaced or serviced at the board/module or component level.

Board/Module Level

A *board* is a group of electronic components placed on a printed circuit (PC) board that performs a set task. A *module* is a group of electronic or electrical components housed in an enclosure that performs a set task.

Troubleshooting at the board/module level requires finding the malfunctioning board or module and replacing it. When troubleshooting electronic equipment at the board/module level, the entire board or module, such as a timer, counter, limit switch, photoelectric control, motor starter, or motor, that contains the malfunction is replaced. For example, a plug-in module on a programmable controller connects inputs (pushbuttons, limit switches, etc.) or outputs (lights, motor starters, solenoids, etc.). If a problem develops with the module, the module is removed, serviced, or replaced. See Programmable Controller.

PROGRAMMABLE CONTROLLER

Component Level

A *component* is an individual device used in a board or module. Components include springs, resistors, diodes, contacts, transistors, and capacitors. Troubleshooting at the component level requires finding the exact component that is malfunctioning. Examples of malfunctioning components include bad transistors or capacitors inside a timer or photoelectric control, or bad contacts or springs inside a magnetic motor starter or pushbutton station.

Troubleshooting at the component level is time consuming. It is often economical to replace an entire board or module. A voltmeter tests components or a group of components on a board. See Component Checking.

COMPONENT CHECKING

▪ Application — Troubleshooting Methods

A skilled troubleshooter follows a logical plan that finds a malfunction quickly and efficiently. When a malfunctioning part is found, the part is repaired. Preventive maintenance is then performed to prevent future problems.

Preventive maintenance is maintenance performed to keep machines, assembly lines, production operations, and plant operations running with little or no downtime. Preventative maintenance keeps equipment in good condition. Troubleshooting methods include troubleshooting by knowledge and experience, troubleshooting by plant procedures, troubleshooting by manufacturer's procedures, or a combination of these.

Troubleshooting by Knowledge and Experience

Troubleshooting by knowledge and experience is a method of finding a malfunction in a system or process through applying information learned from past malfunctions. Troubleshooting by knowledge and experience is only temporarily effective because the primary malfunction is not corrected. For example, a fuse may blow on a machine, stopping production as a result. Changing fuses (or resetting breakers) starts production, but the reason the fuse blew is not clear.

Troubleshooting by Plant Procedures

In most plants, plant procedures are followed when troubleshooting equipment. Plant procedures are developed by supervisors or operators to ensure safe and efficient operation of equipment by plant personnel. Plant procedures are specific to the system or process used. A plant procedure for replacing a defective module on a programmable controller includes the steps:

MIDWESTERN MANUFACTURING CO.
Defective Module Replacement

1. Inform the machine operator that power will be turned OFF.
2. Turn OFF, lock out, and tag the disconnect switch feeding power to the machine.
3. Using a voltmeter, test to ensure that no voltage is present at the power terminals of the programmable controller.
4. Remove the conductor connected to each terminal screw. Mark each conductor with the same number as the terminal screw.
5. Pull the module locking lever out and down until it is perpendicular to the face of the module.
6. Slide the module out and away from the receptacle.
7. Insert the replacement module into the receptacle and lock it in place using the locking lever.
8. Reconnect the conductors to the terminal screws.
9. Inform the machine operator that power will be turned ON.
10. Place all machine selector switches in the manual position.
11. Remove the lock and tag from the disconnect switch. Make sure the machine is clear of all employees.
12. Turn power ON.
13. Restart the machine.
14. Cycle the machine through one operation using the manual switches.
15. If the machine operates properly, place the selector switches in the automatic position.
16. Cycle the machine automatically several times.
17. Inform the supervisor if the machine is not operating properly.
18. Fill out a repair report if the machine is operating properly. Place one copy of the report in the maintenance folder inside the machine cabinet and return the other copy to the supervisor.
19. Remain with the operator until the machine is back in production.
20. Clean the area of all tools and any litter resulting from the maintenance call.

Troubleshooting by Manufacturer's Procedures

Troubleshooting by manufacturer's procedures is a method of finding malfunctioning equipment by using the procedures and recommendations provided by the manufacturer. Manufacturer's procedures differ from plant procedures in that manufacturer's procedures are shorter and more general than plant procedures. A manufacturer's procedure for troubleshooting and replacing the power supply on a programmable controller includes the steps:

MIDCO TOOL DESIGN INC.
Programmable Controller Troubleshooting & Replacement

1. If there is no indication of power (status lights OFF) on the programmable controller, measure the voltage at the incoming power terminals on the power supply module.
2. If voltage is present and correct, replace power supply on programmable controller.
3. Remove power from the programmable controller.
4. Disconnect the power lines from the power supply terminals.
5. Disconnect the processor power cable from the power supply output terminal.
6. Remove the four mounting screws on power supply to free power supply from main panel.
7. Grasp the power supply firmly and pull out.
8. Press the replacement power supply into the main panel.
9. Replace and tighten the four mounting screws.
10. Connect the processor power cable.
11. Connect the power lines.
12. Turn power ON.

Plant or manufacturer's troubleshooting procedures may be in the form of a flow chart. A *flow chart* is a diagram that shows a logical sequence of steps for a given set of conditions. Flow charts help a troubleshooter to follow a logical path when trying to solve a problem. Flow charts use symbols and interconnecting lines to provide direction.

Symbols used with flow charts include an ellipse, rectangle, diamond, and arrow. An ellipse indicates the beginning and end of a flow chart or section of a flow chart. A rectangle contains a set of instructions. A diamond contains a question stated so that a yes or no answer is achieved. The yes or no answer determines the direction to follow through the flow chart. An arrow indicates the direction to follow based on the answers. See Motor Troubleshooting Flow Chart.

MOTOR TROUBLESHOOTING FLOW CHART

■ Application — Troubleshooting Procedure

A *troubleshooting procedure* is a logical step-by-step process used to find a malfunction in a system or process as quickly and easily as possible. When troubleshooting a system, unit, or component, the following steps are performed: obtain information, observe system operation, isolate problems, test faults, and document findings.

Obtain Information

Obtain information from the manufacturer about the system, components, and potential malfunctions before a malfunction arises. Information is obtained by:
- Gathering technical information from the manufacturer, including service manuals, prints, and recommended spare parts lists. For example, when servicing a manual motor starter, contacts are often burnt and require replacement. By obtaining service bulletins, the exact part number is easily found. See Service Bulletin.
- Consulting listings of local dealer addresses. This provides several places to obtain help concerning a part or malfunction.
- Asking local dealers and manufacturers about training programs on the equipment or product line. Most major manufacturers offer technical training.

SERVICE BULLETIN

PARTS LIST

Item	Description	Part number	Size M-0 2-pole	Size M-0 3-pole	Size M-1 2-pole	Size M-1 3-pole	Size M-1P 2-pole
A	Contact Block	00401	1	1	1	1	1
B	2-pole Contact actuator	01753	1	—	1	—	1
	3-pole Contact actuator	01751	—	1	—	1	—
K	Internal interlock (NO)	2065	—	—	—	—	—
	Internal interlock (NC)	2066	—	—	—	—	—
L	Return spring	17901	2	2	2	2	2

Observe System Operation

Observe a normally operating system or process during startup, slow operation, fast operation, and shutdown. Determine what part of the system is or is not working, and also the condition of the outputs (energized or de-energized), circuit, equipment, and operator. Look for problems, such as jammed, burnt, or broken material. Check fuses, circuit breakers, and overload resets. Also, check for any recent modifications or unauthorized changes.

Typical questions asked when troubleshooting include:
- What affect does a long downtime have on the process or product?
- Is it economical to repair or replace?
- Can the defective part be replaced with a spare or be fixed in the maintenance shop where time is not important to production?

- Is the repair going to last as long as a new component would?
- Is this a recurring problem?
- Are the required tools, instruments, and prints available to aid in troubleshooting?
- Are spare parts available?

Isolate Problems

Isolate the area or section that is probably malfunctioning. Narrow down the sources of the malfunction based on information obtained from observation and questioning. Look for potential secondary malfunction(s). For example, a motor drives a pump that moves product through a piping system. See Piping System.

PIPING SYSTEM

The pump develops a higher-than-normal pressure if the pipe that moves the product is partially blocked. To develop the higher pressure, the motor draws additional current. The additional current blows a fuse. The blown fuse stops the motor. The primary malfunction is that the blown fuse stopped the motor. The secondary malfunction is the scale buildup in the pipe. Changing the fuse solves the problem only until the fuse blows again.

Test Malfunctions

Use test equipment to test and check the components that may be malfunctioning. Testing components help to determine primary and secondary malfunctions. For example, after changing fuses in a pump application, an ammeter is used to test the current draw of the motor. The motor is disconnected from the pump and rechecked if the current draw is too high. There is a motor problem if the current is still too high. There is a pump and/or piping system problem if the current draw is normal.

Document Findings

Record malfunctions found and service performed. List all findings. List components checked if a malfunction is not corrected. List any suggestions that may help to prevent the malfunction from recurring. Suggestions include replacing parts at a certain time, adding additional fuses, or changing the circuit design or operation. See Motor Repair and Service Record.

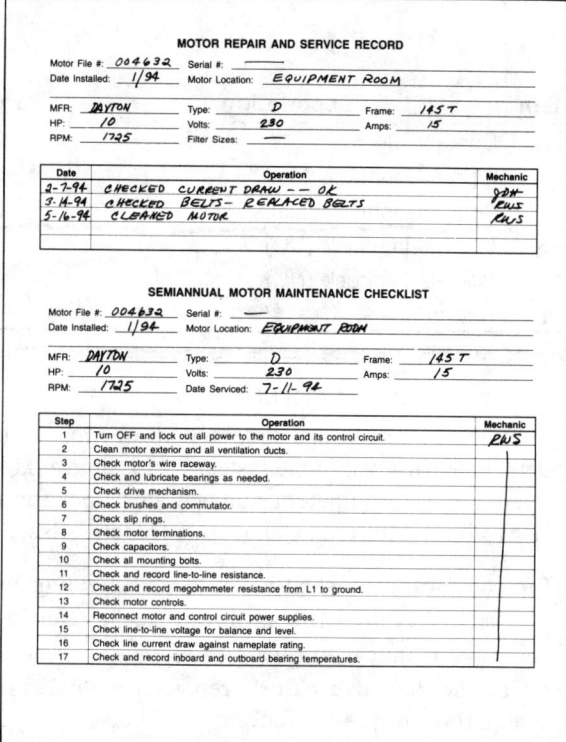

Application — Equipment Reliability

When troubleshooting or servicing a piece of equipment, the age of the equipment should be considered. Most equipment follows a typical life expectancy curve. See Typical Equipment Life Expectancy Curve.

Break-in Period

A *break-in period* is the time just after the installation of a new piece of equipment in which defects resulting from defective parts, poor manufacturing quality, contamination, or environmental stress appear. The highest failure rate for most equipment occurs during the break-in period. Troubleshooting a malfunctioning piece of equipment during the break-in period includes:
- Looking for signs of improper operator use.
- Determining if the equipment is correctly installed. The equipment may be connected to an improper voltage level, improperly grounded, or installed in an environment for which it was not designed.

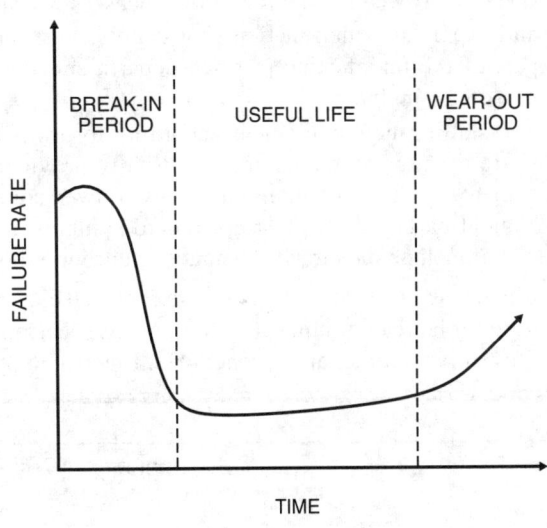

TYPICAL EQUIPMENT
LIFE EXPECTANCY CURVE

Useful Life

Useful life is the period of time after the break-in period when most equipment operates properly. Equipment may fail during its useful life. Troubleshooting a malfunctioning piece of equipment during its useful life includes:
- Checking for signs of improper maintenance, such as dirty or clogged filters. Most equipment needs periodic maintenance. Periodic maintenance usually includes cleaning and lubricating.
- Determining if the equipment is properly protected from improper environmental conditions. Higher-than-normal temperatures, high humidity, salt, corrosive chemicals, and dirt contribute to early failures.
- Looking for signs of damage caused by misuse or accidents. Accidents often occur because of careless operator use or improper environmental conditions. Such environmental conditions include voltage surges and flooding.

Wear-out Period

Wear-out period is the period of time after a piece of equipment's useful life terminates when normal equipment failures occur. Troubleshooting a piece of equipment in the wear-out period includes:
- Determining if the malfunction is worth repairing, and if the equipment will last long enough to justify the expense of the repair. For example, if a key on an old computer keyboard fails, it is likely that more keys will soon fail.
- Determining if the equipment is safe. Deterioration occurs in older equipment. If conductor insulation has deteriorated or if safety guards are not present, the equipment will not be safe even after a malfunction is repaired.

10 TROUBLESHOOTING ELECTRICAL/ELECTRONIC SYSTEMS

Application — Troubleshooter Responsibilities

All industries have systems and/or processes that must operate properly in order to produce products and services. To keep systems and/or processes operating properly, trained personnel are required to troubleshoot and repair any malfunction that occurs. Personnel who troubleshoot and repair systems or processes are part of the maintenance/service department of a company. The responsibilities of the maintenance/service personnel include:

- Installing new equipment according to prints and codes. Prints may be paper or electronic. Paper prints are the most common and are usually provided with each component or machine when purchased. Paper prints include diagrams that show power connections, assembly or connection procedures, and replacement parts. All prints use standard symbols and abbreviations to show circuit operation. See Print. Prints that include the circuit schematic and troubleshooting procedures are also available from most manufacturers.

Electronic prints are prints that are captured electronically and stored on a computer disk. Electronic prints are used for large or complicated equipment. Electronic prints often include setup, operating, troubleshooting, and preventive maintenance procedures. Electronic prints are usually sold as an extra package.

- Following a logical startup procedure for a new system or process. Startup procedures include the following:
    Perform general checks of fluid levels, guard positions, belt tightness, and loose parts before turning power ON.
    Turn ON and test one part of a machine or system at a time.
    Test all manual operations first.
    Determine if all safety equipment and features are working properly.
    Anticipate component failure. Operate a machine long enough to allow pressure to build, belts to slip, and oil to heat up. The majority of component failure on new equipment occurs during the first few hours or days of operation.

- Applying a preventive maintenance program. Preventive maintenance is action taken to maintain equipment in good operating condition to ensure uninterrupted operation. Preventive maintenance includes tightening belts, adding lubricant, cleaning, changing filters, calibrating, general inspecting, and setting control limits. Preventive maintenance is usually performed according to a set schedule. Monthly, semiannual, and annual preventive maintenance schedules are common. See Motor Maintenance Checklists.
- Preventing malfunctions. Every process or operation has potential problems that may cause malfunctions. Preventing malfunctions requires close attention to the entire process. This includes ensuring that the correct size and type of parts are fed into a machine, that the machine is operated properly, and that improvements, modifications, or design changes are made.

- Identifying malfunctions that occur. A skilled troubleshooter tries to identify the cause of a malfunction and correct it before the malfunction occurs again. Note all malfunctions that occur. The true cause of a malfunction may not be clear. For example, a part that jams in a machine may be caused by a part that is not within specifications, a machine that is not within specifications, operator misuse, improper lubrication, or a control malfunction. Most malfunctions are identified after they have appeared several times.
- Repairing equipment in an efficient manner. Repairing equipment in an efficient manner includes the steps:

  Anticipate and have the required tools and test equipment for a troubleshooting call.

  Troubleshoot from the simple to the complex.

  Ask the opinions of operators and other personnel that are familiar with the system.

  Determine if problems can be fixed temporarily to allow production to continue until the end of the operation.

- Maintaining good records of maintenance, equipment condition, and downtime. Provide justification for overtime, new tools and equipment, additional personnel, and improved safety to improve the maintenance operation.
- Handling emergencies. Emergencies that occur in a plant may be handled by the maintenance department until additional help arrives. Emergencies that may require trained maintenance personnel include:

  General emergencies. Examples of general emergencies include fires, floods, accidents, and power outages. Personnel should know plant evacuation procedures, emergency telephone numbers, the meaning of different alarms, and basic first aid procedures.

  Trade-related emergencies. Examples of trade-related emergencies include electrical shock, burns, and exposure to specific hazardous conditions. Maintenance personnel should know power disconnection procedures, shock treatment, CPR, and first aid.

  Operational emergencies. Operational emergencies are caused by hazardous materials, fire hazards, and safety problems. Personnel should know specific chemical, fire, and physical safety procedures.

- Remain informed about the latest equipment, procedures, and technologies. Trade and technical magazines provide information on new equipment, techniques, and technologies. Manufacturers and local suppliers offer training seminars about various product lines. The latest equipment, procedures, and technologies are often taught at local trade schools and in-plant training courses.

Application — Electrical Shock

According to the National Safety Council, over 1000 people are killed each year in the United States from electrical shock. Electricity is the number one cause of fires. More than 100,000 people are killed in electrical fires each year.

Improper electrical wiring or misuse of electricity causes destruction of equipment and fire damage to property. Safe working habits are required when troubleshooting an electrical circuit or component because the electric parts that are normally enclosed are exposed.

An electrical shock results anytime a body becomes part of an electrical circuit. Electrical shock varies from a mild shock to fatal current. The severity of an electrical shock depends on the amount of electric current (in mA) that flows through the body, the length of time the body is exposed to the current flow, the path the current takes through the body, the physical size and condition of the body through which the current passes, and the amount of body area exposed to the electric contact.

The amount of current that passes through a circuit depends on the voltage and resistance of the circuit. During an electrical shock, a person's body becomes part of the circuit. The resistance a body offers to the flow of current varies. Sweaty hands have less resistance than dry hands. A wet floor has less resistance than a dry floor. The lower the resistance, the greater the current flow. The greater the current flow, the greater the severity of shock. See Effect of Electric Current.

EFFECT OF ELECTRIC CURRENT	
Current (in mA)	Effect On Body
8 or less	Sensation of shock but probably not painful
8 to 15	Painful shock Removal from contact point by natural reflexes
15 to 20	Painful shock May be frozen or locked to point of electric contact until circuit is de-energized
over 20	Causes severe muscular contractions, paralysis of breathing, heart convulsions

EFFECT OF ELECTRIC CURRENT

When handling an accident caused by electrical shock, apply the procedure:
1. Break the circuit to free the person immediately and safely. Never touch any part of a victim's body when in contact with the circuit. If the circuit cannot be turned OFF, use any nonconducting device to free the person. Resist the temptation to touch a person if the power is not turned OFF.
2. After the person is free from the circuit, send for help and determine if the person is breathing. If there is no breathing or pulse, start CPR. Always get medical attention for a victim of electrical shock.
3. If a person is breathing and has a pulse, check for burns and cuts. Burns are caused by contact with the live circuit, and are found at the points that the electricity entered and exited the body. Treat the entrance and exit burns as thermal burns and get medical help.

Application — Electrical Safety

A troubleshooter must work safely at all times. Following are some basic safety rules that must be followed when working with electrical equipment.

General Electrical Safety Rules
- Always comply with the NEC®.
- Use UL® approved appliances, components, and equipment.
- Keep electrical grounding circuits in good condition. Ground any conductive component or element that does not have to be energized. The grounding connection must be a low-resistance conductor heavy enough to carry the largest fault current that may occur.
- Turn OFF, lock out, and tag disconnect switches when working on any electrical circuit or equipment. Test all circuits after they are turned OFF. Insulators may not insulate, grounding circuits may not ground, and switches may not open the circuit.
- Use double-insulated power tools or power tools that include a third conductor grounding terminal that provides a path for fault current. Never use a power tool that has the third conductor grounding terminal removed. See Normal Current and Fault Current.
- Always use protective and safety equipment.
- Know what to do in an emergency.
- Check conductors, cords, components, and equipment for signs of wear or damage. Replace any equipment that is not safe.

NORMAL CURRENT   FAULT CURRENT

14 TROUBLESHOOTING ELECTRICAL/ELECTRONIC SYSTEMS

- Never throw water on an electrical fire. Turn OFF the power and use a Class C rated fire extinguisher. See Fire Extinguisher Classes.

    Fires are classified by the combustible material involved. The four classes of fires are Class A, Class B, Class C, and Class D. Class A fires include burning wood, paper, textiles, and other ordinary combustible materials containing carbon. Class B fires include burning oil, gas, grease, paint, and other liquids that convert to a gas when heated. Class C fires include burning electrical devices, motors, and transformers. Class D is a specialized class of fires including burning metals such as zirconium, titanium, magnesium, sodium, and potassium.

FIRE EXTINGUISHER CLASSES

- When working in a dangerous area or with dangerous equipment, work with another individual.
- Learn CPR and first aid.
- Do not work when tired or taking medicine that causes drowsiness.
- Do not work in poorly lighted areas.
- Always use nonconductive ladders. Never use a metal ladder when working on or around electrical equipment.
- Ensure there are no atmospheric hazards such as flammable dust or vapor in the area. A live electrical circuit can emit a spark at any time.
- Use one hand when working on a live circuit to reduce the chance of an electrical shock passing through the heart and lungs.
- Never bypass or disable fuses or circuit breakers.
- Extra care must be taken in an electrical fire because burning insulation produces toxic fumes.
- Always fill out accident forms and report any electrical shock.

Electric Motor Safety

Two areas requiring attention when working with electric motors are the electrical circuit and rotating shaft. Basic electric motor safety rules include:
- Connecting a motor to the correct grounding system.
- Ensuring that guards or housings are connected to the rotating parts of a motor or anything connected to the motor. See Motor Guard.

MOTOR GUARD

- Using the correct motor type for the location. For example, a DC or universal motor shall never be used in a hazardous location that contains flammable materials because the sparking at the brushes can ignite the material.
- Connecting a motor to the correct voltage and power source.
- Providing the motor with the correct overload and overcurrent protection to protect the motor when starting or shorted (overcurrent protection) and when running (overload protection).

Grounding

Electrical circuits are grounded to safeguard equipment and personnel against the hazards of electrical shock. Proper grounding of electrical tools, machines, equipment, and delivery systems is one of the most important factors in preventing hazardous conditions.

Grounding is the connection of all exposed non-current-carrying metal parts to the earth. Grounding provides a direct path for unwanted (fault) current to the earth without causing harm to persons or equipment. Grounding is accomplished by connecting the circuit to a metal underground pipe, a metal frame of a building, a concrete-encased electrode, or a ground ring. See Grounding Methods.

GROUNDING METHODS

Non-current-carrying metal parts that are connected to ground include all metal boxes, raceways, enclosures, and equipment. Unwanted current exists because of insulation failure or if a current-carrying conductor makes contact with a non-current-carrying part of the system. In a properly grounded system, the unwanted current flow trips fuses or circuit breakers. Once the fuse or circuit breaker is tripped, the circuit is opened and no additional current flows.

Codes and Standards

To protect people and property from electrical dangers, national, state, and local codes and standards are used. See Codes and Standards. A *code* is a regulation or minimum requirement. A *standard* is an accepted reference or practice. Codes and standards ensure electrical equipment is built and installed safely and every effort is made to protect people from electrical shock.

National Fire Protection Association (NFPA). NFPA is a national organization that provides guidance in assessing the hazards of the products of combustion. The NFPA publishes the National Electrical Code (NEC®). The NEC's purpose is the practical safeguarding of persons and property from the hazards arising from the use of electricity. The NEC® is updated every three years. Many city, county, state, and federal agencies use the NEC® to set requirements for electrical installations. The NEC® is primarily concerned with protecting lives and property.

CODES AND STANDARDS

NFPA
Batterymarch Park
Quincy, MA 02269

NEMA
2101 L Street, N.W.
Washington, D.C. 20037

ANSI
11 West 42nd Street
New York, NY 10036

UL
333 Pfingsten Rd.
Northbrook, IL 60062

OSHA
230 South Dearborn Street
32nd Floor, Room 3244
Chicago, IL 60604

CSA
178 Rexdale Blvd.
Rexdale, ON M9W 1R3

American National Standards Institute (ANSI). ANSI is a national organization that helps identify industrial and public needs for standards. ANSI coordinates and encourages activities in national standards development.

Occupational Safety and Health Act (OSHA). OSHA is a federal act that requires all employers to provide a safe environment for their employees. All work areas must be free from hazards likely to cause serious harm. The provisions of this act are enforced by federal inspection. With few exceptions, OSHA uses the NEC® to help assure a safe electrical environment.

National Electrical Manufacturers Association (NEMA). NEMA is a national organization that assists with information and standards concerning proper selection, ratings, construction, testing, and performance of electrical equipment. NEMA standards are used as guidelines for the manufacture and use of electrical equipment.

Underwriters' Laboratories, Inc. (UL®). UL® is an independent organization that tests equipment and products to see if they conform to national codes and standards. Equipment tested and approved by UL® carries the UL® label. UL® approved equipment and products are listed in its annual publication.

Canadian Standards Association (CSA). CSA is a Canadian organization similar to UL® that tests equipment and products to ensure they meet national standards.

Customer Relations

Troubleshooting is performed on broken or malfunctioning equipment. When performing a troubleshooting call, the troubleshooter should be prepared to communicate with upset individuals. Communicating with an upset individual can be accomplished by:

• Relaxing

Relaxing takes the tension out of one's voice. Approach the individual as a friend. The upset individual is primarily mad at the situation or equipment.

• Listening

Ask for an explanation of the problem even if it is already known. Let the individual explain without interruption.

• Agreeing

Find some point of agreement. Agreeing demonstrates willingness to help and places the problem on the equipment.

• Finding the problem

Find the problem and its cause.

• Making a decision

Determine if the problem can be fixed and at what cost. State what is required to return the equipment to service.

• Coming to an agreement

Agree to fix the problem, give the individual time to decide, call in a second opinion, or do nothing at the time.

TROUBLESHOOTING

1

Activities

Name _____ Date _____

◯ Activity 1-1. Troubleshooting Levels

State the level at which the malfunction occurred and was serviced (system, equipment, etc.).

1. After a bulk railcar is emptied halfway, no product moves to a storage silo.

    _____ **A.** Malfunction occurred at _____ level.

    _____ **B.** Malfunction serviced at _____ level.

UNLOADER C

TROUBLESHOOTING CHART			**UNLOADER C**	
Date 3-21-94	**Time** 8:30	to 2:30	**Service Person**	G.H.

Symptoms Observed:	The vacuum pressure unloader pump is ON. Hoses from railcar to unloader and unloader to silo intake pipe are not kinked.
Probable Area of Malfunction:	vacuum pressure unloader
Probable Malfunction:	obstruction in hose or vacuum pump
Potential Secondary Malfunction(s):	shaft from pump motor to vacuum pressure unloader broken, hose or silo conveyor clogged
Findings/Service Performed:	Turned vacuum unloader OFF and disconnected hose from silo intake pipe. Placed hose in portable barrel. Turned unloader ON, no product was pumped out. Turned unloader OFF and connected hose to silo. Disconnected hose at railcar. Turned unloader ON. Vacuum was felt at disconnected hose leading to unloader. Turned unloader OFF and removed discharge valve from railcar. Valve was clogged with a paper bag. Removed bag and placed unit back in service.

18 TROUBLESHOOTING ELECTRICAL/ELECTRONIC SYSTEMS

2. The letter *m* on a keyboard sticks and produces a line of *m*'s when pressed.

_____ **A.** Malfunction occurred at _____ level.

_____ **B.** Malfunction serviced at _____ level.

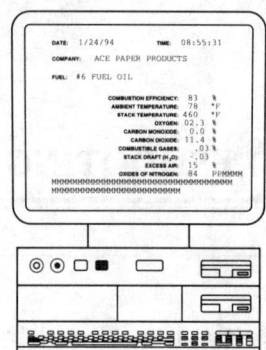

PC TERMINAL 0065

TROUBLESHOOTING CHART PC TERMINAL 0065

Date 1-24-94   Time 10:00 to 10:45   Service Person J.P.

Symptoms Observed: The letters *b*, *n*, and *m* have a sticky feeling when pressed. The letter *m* reproduces several *m*'s at times.

Probable Area of Malfunction: keyboard, letter *m*

Probable Malfunction: defective *m* key

Potential Secondary Malfunction(s): customer misuse

Findings/Service Performed: Found sticky substance in keyboard. Cleaned and tested keyboard. No further problems.

Suggestions: Ask operators not to eat or drink around keyboard. Keep keyboard covered when not in use.

3. One button on a beverage gun does not dispense product when pressed.

_____ **A.** Malfunction occurred at _____ level.

_____ **B.** Malfunction serviced at _____ level.

BEVERAGE GUN 01456

TROUBLESHOOTING CHART BEVERAGE GUN 01456

Date 1-11-94   Time 9:00 to 10:00   Service Person J.R.

Symptoms Observed: All buttons except one dispense product.

Probable Area of Malfunction: beverage gun unit

Probable Malfunction: bad button or malfunctioning component in printed circuit board of beverage gun

Potential Secondary Malfunction(s): cable leading from gun to solenoid valves that dispense product, solenoid valve

Findings/Service Performed: Opened beverage gun and found cracked printed circuit board under malfunctioning button. Replaced entire beverage gun.

Suggestions: Buttons do not have to be pressed hard.

4. A service door does not open when the up pushbutton is pressed.

　　_____　　**A.** Malfunction occurred at _____ level.

　　_____　　**B.** Malfunction serviced at _____ level.

OVERHEAD DOOR B — MOTOR DRIVE UNIT

TROUBLESHOOTING CHART　　　　OVERHEAD DOOR B

Date 2-9-94　　**Time** 8:15 **to** 10:00　　**Service Person** T.R.

Symptoms Observed: Door does not open and motor makes no sound when the up pushbutton is pressed. Door has ice buildup along the bottom. Heat is turned down at night. Ice melts when heat is ON for some time.

Probable Area of Malfunction: fused disconnect unit, motor starter, motor/drive unit

Probable Malfunction: blown disconnect fuses, overloads tripped at motor starter, motor burned out

Potential Secondary Malfunction(s): door jammed from misalignment, ice buildup caused fuses or overloads to trip

Findings/Service Performed: Motor overloads on reversing starter were tripped. Reset overloads and tested door. Door moves freely with no apparent problems. Current draw on motor in operation is within limits.

5. A counter is not counting the parts that are moving down a production line.

　　_____　　**A.** Malfunction occurred at _____ level.

　　_____　　**B.** Malfunction serviced at _____ level.

MAGNETS — LINE 2, COUNTER 4

TROUBLESHOOTING CHART　　　　LINE 2, COUNTER 4

Date 2-23-94　　**Time** 2:30 **to** 3:30　　**Service Person** M.J.

Symptoms Observed: Counter is lit and displaying 0000. No counts are registering as product moves through the sensor.

Probable Area of Malfunction: counter or proximity switch

Probable Malfunction: proximity switch because counter is lit

Potential Secondary Malfunction(s): open or shorted cable from proximity switch to counter

Findings/Service Performed: Cable in good shape. Replaced proximity switch, no counts registering. Replaced counter, new counter working. Left new proximity switch to permit production.

Activity 1-2. Troubleshooting Methods

1. Complete the flow chart by placing the correct letter in the diamond or rectangle.

Problem: No heat, furnace fails to start

A. Is thermostat switch in OFF or cool position?
B. Is thermostat set too low?
C. Are breakers in tripped position?
D. Are fuses blown?
E. Is pilot out?
F. Is 24 V ±2 V present on transformer secondary?
G. Is 115 V ±5 V present on transformer primary?

H. Does furnace start?
I. Set thermostat higher.
J. Close switch.
K. Reset breakers.
L. Replace blown fuses.
M. Turn valve ON and relight pilot.

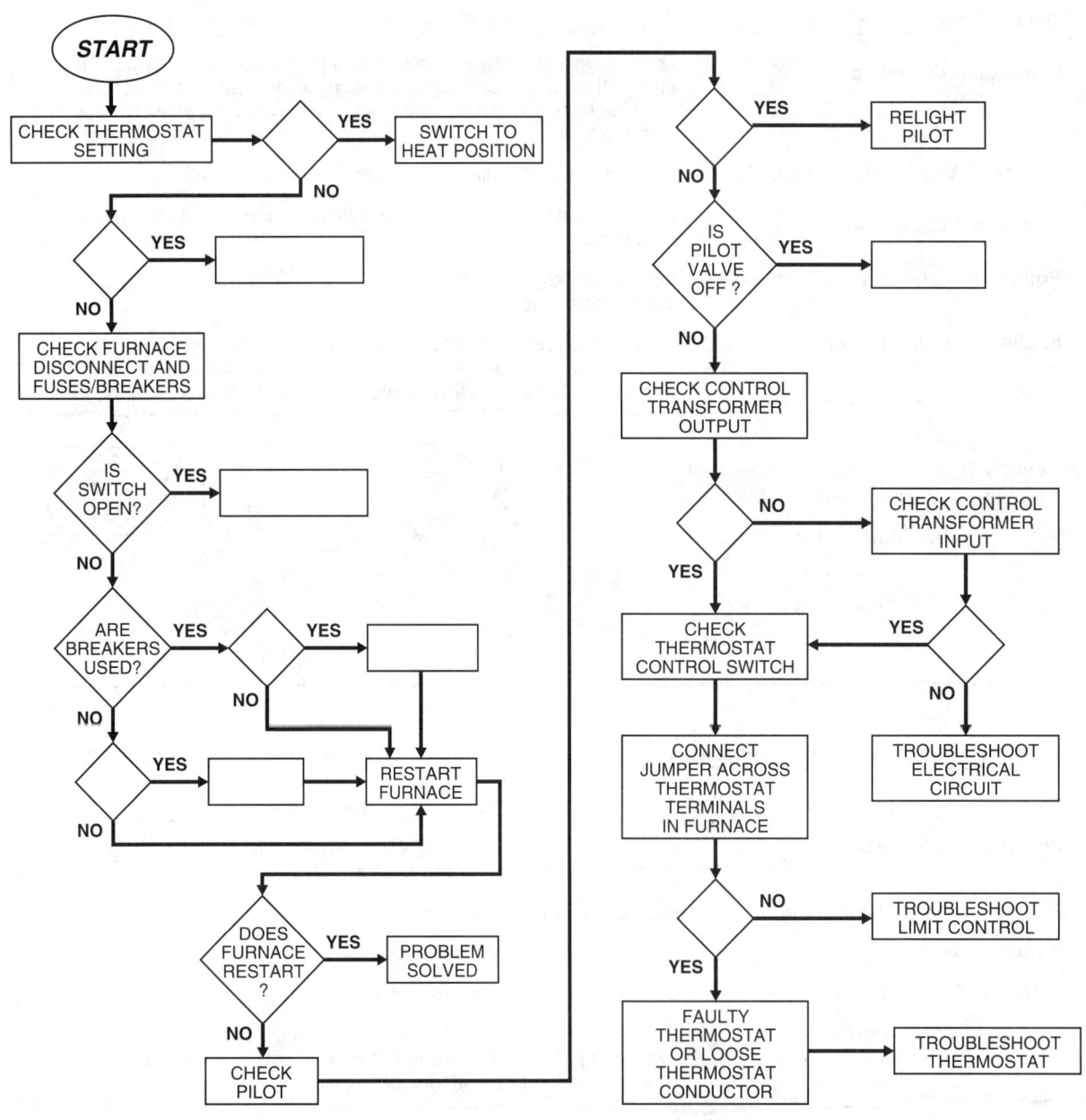

Activity 1-3. Troubleshooting Procedure

Using Annual Motor Maintenance Checklist and Motor Repair and Service Record, answer the following questions.

ANNUAL MOTOR MAINTENANCE CHECKLIST

Motor File #: DC-10-12 Serial #: 1000112B Date Installed: 11-18-92
Motor Location: PLANT 3, BUILDING A, LEVEL 2, FILL LINE 3, INFEED MOTOR MFR: NORTH CO.
Type: DC SHUNT Frame: 256T HP: 10 Volts: 500
Amps: 18 RPM: 650 Date Serviced: 2-1-94 Class: F
Duty: CONT SF: 1.00 Starting Torque (lb-ft): 105

Step	Operation	Mechanic
1	Turn OFF and lock out all power to the motor and its control circuit.	RJ
2	Clean motor exterior and all ventilation ducts. MOTOR VERY DIRTY	
3	Uncouple motor from load and disassemble.	
4	Clean inside of motor.	
5	Check rotors, armatures, and field windings. ROTOR HAS SOME BURNING	
6	Check all peripheral equipment. BELT OK (NO WEAR)	
7	Check bearings. ADDED SAE #10 TO SLEEVE BEARINGS	
8	Check brushes and commutator. BRUSHES ½ WORN. SERVICE SOON.	
9	Reassemble motor and couple to load.	
10	Check motor's wire raceway. OK	
11	Check drive mechanism. OK	
12	Check motor terminations. OK	
13	Check all mounting bolts. OK	
14	Check and record line-to-line resistance. OK	
15	Check and record megohmmeter resistance from T1 to ground. OK	
16	Check and record insulation polarization index. OK	
17	Check motor controls. OK	
18	Reconnect motor and control circuit power supplies. OK	
19	Check line-to-line voltage for balance and level. VOLTAGE LOW (460V)	

MOTOR REPAIR AND SERVICE RECORD

Date	Operation	Mechanic
10-29-93	FUSES BLOWN - CHANGED W/SAME SIZE	MA
11-5-93	FUSES BLOWN - CHANGED 20A TO 25A	DF
2-4-94	ANNUAL SERVICE - SEE CHECKLIST	RJ

_____ 1. The motor was installed on _____.

_____ 2. The motor was in service for _____ months.

_____ 3. The first sign of a problem occurred on _____.

_____ 4. The first problem was _____.

_____ 5. Was the motor serviced before the annual checkup was completed?

6. The current rating of the motor is _____ A.

_____ 7. The measured current draw of the motor during the annual checkup was _____ A.

_____ 8. The starting torque of the motor is _____ lb-ft.

_____ 9. The recommendation made by the maintenance person during the annual checkup was _____.

_____ 10. Is there any record that the recommendations made on the annual checkup report were followed?

_____ 11. The motor failed because of _____.

_____ 12. Was the motor undersized for the application?

_____ 13. Was the motor properly protected from contaminants?

_____ 14. Could a power line voltage lower than the voltage on the motor nameplate cause the problems?

_____ 15. Could the motor that replaced this motor have the same problems if preventive maintenance is not performed?

◯ Activity 1-4. Equipment Reliability

For each maintenance procedure, list each period during which the procedure is generally performed. A = Break-in period, B = Useful life period, and C = Wear-out period.

_____ 1. Add new updates to a system or equipment.

_____ 2. Fill out and return manufacturer's warranty card.

_____ 3. Check the price of replacement equipment.

_____ 4. Read instruction manual.

_____ 5. Check that the incoming voltage level is correct.

_____ 6. Clean equipment components and change filters.

_____ 7. Look for signs of damage caused by misuse or accidents.

_____ 8. Determine if operators need training on the equipment.

_____ 9. Offer a service contract on the equipment.

_____ 10. Check that the equipment is properly grounded and safety features are working.

_____ 11. Check wire insulation and electrical connections for signs of wear.

_____ 12. File equipment manuals so that they are easily referenced.

◯ Activity 1-5. Troubleshooter Responsibilities

Complete the following statements.

_____ 1. Four checks that should be taken before starting new equipment include checking fluid levels, belt tightness, loose parts, and _____.

_____ 2. Four general emergencies that a trained maintenance person should be prepared to handle include fires, floods, accidents, and _____.

_____ 3. Two electrical trade-related emergencies that a trained maintenance person should be prepared to handle are electrical shock and _____.

_____ 4. In addition to fire hazards, two hazards that can cause an operational emergency are safety problems and _____.

_____ 5. Six steps that a troubleshooter should take when communicating with an upset customer include relaxing, agreeing, finding the problem, making a decision, coming to an agreement, and _____.

○ Activity 1-6. Electrical Shock

1. Complete the table and plot the current for each resistance value on the graph. Mark the following points on the graph: A = Sensation of shock, B = Painful shock, able to remove oneself from contact point, C = Painful shock, may be locked to contact point, D = Fatal electrical shock.

$$I = \frac{E}{R} \times 1000 \qquad \begin{array}{l} I = \text{CURRENT (IN mA)} \\ E = \text{VOLTAGE (IN VOLTS)} \\ R = \text{RESISTANCE (IN OHMS)} \end{array}$$

R	48,000	24,000	19,200	12,000	9600	4800	2400	1600
E	120	120	120	120	120	120	120	120
I								

○ Activity 1-7. Electrical Safety

Mark each statement as true or false.

T   F   1. Flammable vapors or dust in the atmosphere can cause an explosion if there is an electrical spark.

T   F   2. Only power tools that are double-insulated or include a third-wire power cord grounding conductor should be used.

T   F   3. The grounding post on a plug can be removed when a three-prong plug does not fit into a socket.

24 TROUBLESHOOTING ELECTRICAL/ELECTRONIC SYSTEMS

T   F   **4.** A wood, metal, or plastic ladder can be used when working on an electrical circuit.

T   F   **5.** When working on a live electrical circuit, only one hand should be used so that the electric current will be prevented from flowing through the body.

T   F   **6.** When troubleshooting a problem, it is safe to bypass or disable a fuse or circuit breaker.

T   F   **7.** Always notify a supervisor and file an accident report after an electrical shock is received.

T   F   **8.** Using bare hands to free a shock victim may result in one's receiving a severe shock.

T   F   **9.** Ground any conductive component that does not have to be energized.

T   F   **10.** Before working on any electrical equipment, the circuit should be tested to make sure power is OFF, even if the disconnect switch is turned OFF and locked out.

T   F   **11.** It is safe to work on an electrical circuit when tired or drowsy?

T   F   **12.** Proper grounding helps prevent electrical shock.

T   F   **13.** To ground a system, all exposed non-current-carrying metal parts are connected to the earth ground.

T   F   **14.** Motors must be protected from overload and overcurrent conditions.

T   F   **15.** Grounding conductors carry only fault current and may be smaller and have a higher resistance than normal current-carrying conductors.

○ Activity 1-8. Electrical Safety

> Answer the following questions.

_____ **1.** Electricity can take multiple paths, but the largest current flows through the path of _____ resistance.

_____ **2.** Always use a Class _____ fire extinguisher on an electrical fire.

_____ **3.** Electrical fires give off _____.

_____ **4.** Always use _____ listed appliances, components, and equipment.

_____ **5.** Disconnect switches should be turned OFF, locked, and _____ before working on equipment.

_____ **6.** A(n) _____ protection device is used to protect a motor when starting.

_____ **7.** A(n) _____ protection device is used to protect a motor when running.

_____ **8.** A grounding system is designed to carry all _____ current to ground.

_____ **9.** A fuse or circuit breaker _____ when an overcurrent condition occurs in a circuit.

_____ **10.** _____ emergencies are caused by hazardous materials, fire hazards, and safety problems.

BASIC ELECTRICAL THEORY

Applications

▢ Application — Electrical Prefixes

Prefixes

Prefixes are used to avoid long expressions of units that are smaller and larger than the base unit. See Common Prefixes. For example, sentences 1 and 2 do not use prefixes. Sentences 3 and 4 use prefixes.
1. A solid-state device draws 0.000001 amperes (A).
2. A generator produces 100,000 watts (W).
3. A solid-state device draws 1 microampere (μA).
4. A generator produces 100 kilowatts (kW).

Converting Units

To convert between different units, the decimal point is moved to the left or right, depending on the unit. See Conversion Table. For example, an electronic circuit has a current flow of .000001 A. The current value is converted to simplest terms by moving the decimal point six places to the right to obtain 1.0 μA (from Conversion Table).

.000001. A = 1.0 μA
— MOVE DECIMAL POINT 6 PLACES TO RIGHT

Common Electrical Quantities

Abbreviations are used to simplify the expression of common electrical quantities. See Common Electrical Quantities. For example, milliwatt is abbreviated mW, kilovolt is abbreviated kV, and ampere is abbreviated A.

COMMON PREFIXES

Symbol	Prefix	Equivalent
G	giga	1,000,000,000
M	mega	1,000,000
k	kilo	1000
base unit	—	1
m	milli	.001
m	micro	.000001
n	nano	.000000001

CONVERSION TABLE

Initial Units	Final Units						
	giga	mega	kilo	base unit	milli	micro	nano
giga		3R	6R	9R	12R	15R	18R
mega	3L		3R	6R	9R	12R	15R
kilo	6L	3L		3R	6R	9R	12R
base unit	9L	6L	3L		3R	6R	9R
milli	12L	9L	6L	3L		3R	6R
micro	15L	12L	9L	6L	3L		3R
nano	18L	15L	12L	9L	6L	3L	

R = move the decimal point to the right.
L = move the decimal point to the left.

COMMON ELECTRICAL QUANTITIES

Variable	Name	Unit of Measure and Abbreviation
E	voltage	volt — V
I	current	ampere — A
R	resistance	ohm — Ω
P	power	watt — W
P	power (apparent)	volt-amp — VA
C	capacitance	farad — F
L	inductance	henry — H
Z	impedance	ohm — Ω
G	conductance	siemens — S
f	frequency	hertz — Hz
T	period	second — s

Application — Using Ohm's Law

Ohm's Law

Ohm's law is the relationship between the voltage, current, and resistance in an electrical circuit. Ohm's law states that current in a circuit is proportional to the voltage and inversely proportional to the resistance. Any value in these relationships is found using Ohm's law, which is written $I = \frac{E}{R}$, $E = R \times I$, and $R = \frac{E}{I}$. A commonly used variation of Ohm's law is $I = \frac{V}{R}$. See Ohm's Law and Power Formula.

P = WATTS
I = AMPS
R = OHMS
E = VOLTS

VALUES IN INNER CIRCLE ARE EQUAL TO VALUES IN CORRESPONDING OUTER CIRCLE

OHM'S LAW AND POWER FORMULA

Current Calculation

Current (I) is the amount of electrons flowing through an electrical circuit. Current is measured in amperes (A). Current may be direct or alternating. *Direct current (DC)* is current that flows in one direction. *Alternating current (AC)* is current that reverses its direction of flow at regular intervals. To calculate current, apply the formula:

$$I = \frac{E}{R}$$

where
I = current (in amps)
E = voltage (in volts)
R = resistance (in ohms)

also used as $I = \frac{V}{R}$

where
I = current (in amps)
V = voltage (in volts)
R = resistance (in ohms)

Example: Calculating Current

An electrical circuit has a resistance of 80 Ω and is connected to a 120 V supply. Calculate the amount of current in the circuit. See Current Calculation.

$$I = \frac{E}{R}$$
$$I = \frac{120}{80}$$
$$I = 1.5 \text{ A}$$

CURRENT CALCULATION

Voltage Calculation

Voltage (E) is the amount of electrical pressure in a circuit. Voltage is measured in volts (V). Voltage is produced by the conversion of chemical energy (battery), light (photocell), electromagnetic energy (generator), heat (thermocouple), pressure (piezoelectricity), or friction (static electricity) to electrical energy. Voltage is produced by generating an excess of electrons at one terminal of a voltage source and a deficiency of electrons at the other terminal. The greater the difference in electrons between the terminals, the higher the voltage. To calculate voltage, apply the formula:

$$E = I \times R$$

where
E = voltage (in volts)
I = current (in amps)
R = resistance (in ohms)

Example: Calculating Voltage

An electrical circuit has 5 A of current and a resistance of 46 Ω. Calculate the voltage in the circuit. See Voltage Calculation.

$E = I \times R$
$E = 5 \times 46$
$E = \mathbf{230\ V}$

VOLTAGE CALCULATION

Resistance Calculation

Resistance (R) is opposition to the flow of electrons. Resistance is measured in ohms (Ω). *Conductors* are materials through which current flows easily. Examples of conductors include copper, aluminum, brass, gold, and other metals. *Insulators* are materials through which current cannot flow easily. Examples of insulators include rubber, glass, plastic, paper, varnish, and dry wood. To calculate resistance, apply the formula:

$R = \dfrac{E}{I}$

where
R = resistance (in ohms)
E = voltage (in volts)
I = current (in amps)

Example: Calculating Resistance

An electrical circuit draws 8 A when connected to a 460 V supply. Calculate the resistance of the circuit. See Resistance Calculation.

$R = \dfrac{E}{I}$
$R = \dfrac{460}{8}$
$R = \mathbf{57.5\ \Omega}$

RESISTANCE CALCULATION

▪ Application — Using the Power Formula

Power Formula

The power formula is the relationship between the voltage, current, and power in an electrical circuit. The power formula states that the power in a circuit is equal to the voltage times the current. Any value in these relationships is found using the power formula, which is written $P = E \times I$, $E = \dfrac{P}{I}$, and $I = \dfrac{P}{E}$.

Power Calculation

Power (P) is the rate of doing work or using energy. Power may be expressed as true or apparent. *True power* is the actual power used in an electrical circuit. True power is measured using a wattmeter and is expressed in watts (W). *Apparent power* is the product of voltage and current in a circuit calculated without considering the phase shift that may be present between total voltage and current in the circuit. Apparent power is measured in volt-amperes (VA). A phase shift exists in most AC circuits that contain devices causing capacitance or inductance.

Capacitance is the property of an electric device that permits the storage of electrically separated charges when potential differences exist between the conductors. *Inductance* is the property of a circuit that causes it to oppose a change in current due to energy stored in a magnetic field. All coils (motor windings, transformers, solenoids, etc.) create inductance in an electrical circuit. True power equals apparent power in an electrical circuit containing only resistance. True power is less than apparent power in a circuit containing inductance or capacitance. To calculate apparent power, apply the formula:

$P_A = E \times I$

where
P_A = apparent power (in volt-amps)
E = measured voltage (in volts)
I = measured current (in amps)

To calculate true power, apply the formula:

$P_T = I^2 \times R$

where
P_T = true power (in watts)
I = total circuit current (in amps)
R = total resistive component of the circuit (in ohms)

Power factor (PF) is the ratio of true power used in an AC circuit to apparent power delivered to the circuit. Power factor is expressed as a percentage. True power equals apparent power when the power factor is 100%. When the power factor is less than 100%, the circuit is less efficient and has higher operating costs. To calculate power factor, apply the formula:

$PF = \dfrac{P_T}{P_A}$

where
PF = power factor (percentage)
P_T = true power (in watts)
P_A = apparent power (in volt-amps)

Example: Calculating Apparent Power

An electrical circuit has a measured current of 25 A and a measured voltage of 440 V. Calculate the apparent power of the circuit. See Apparent Power Calculation.

$P_A = E \times I$
$P_A = 440 \times 25$
$P_A = \mathbf{11{,}000 \ VA}$

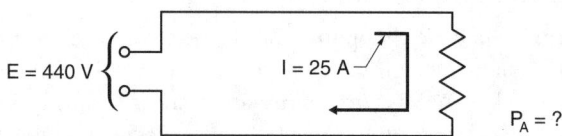

APPARENT POWER CALCULATION

Example: Calculating True Power

An electrical circuit has a current of 25 A and a total resistive component of 21 Ω. Calculate the true power of the circuit. See True Power Calculation.

$P_T = I^2 \times R$
$P_T = (25)^2 \times 21$
$P_T = 625 \times 21$
$P_T = \mathbf{13{,}125 \ W}$

TRUE POWER CALCULATION

Example: Calculating Power Factor

The apparent power of an electrical circuit is 11,000 VA and the true power is 13,125 W. Calculate the power factor of the circuit.

$PF = \dfrac{P_T}{P_A}$
$PF = \dfrac{11{,}000}{13{,}125} \times 100$
$PF = .838 \times 100$
$PF = \mathbf{83.80\%}$

Application — Temperature Conversion

Temperature is a measurement of the intensity of heat. *Ambient temperature* is the temperature of the air surrounding a device. Current flowing in a conductor produces heat. The heat produced in an electrical circuit may be by design, such as from a heating element, or unintentional, as from a bad conductor splice. Unintentional heat deteriorates insulation and lubrication, and can destroy electric devices.

Temperature rise is an increase in temperature above ambient temperature. All electric and electronic devices produce heat and are designed to function correctly within a given temperature rise. See Temperature Measurement. Temperature is usually measured in degrees Fahrenheit (°F) or degrees Celsius (°C). See Farenheit and Celsius Scales.

Converting Fahrenheit to Celsius

To convert a Fahrenheit temperature reading to Celsius, subtract 32 from the Fahrenheit reading and divide by 1.8. To convert Fahrenheit to Celsius, apply the formula:

$$°C = \frac{°F - 32}{1.8}$$

where
°C = degrees Celsius
°F = degrees Fahrenheit
32 = difference between bases
1.8 = ratio between bases

Example: Converting Fahrenheit to Celsius

A digital thermometer indicates a reading of 185°F. Convert the Fahrenheit temperature to Celsius.

$$°C = \frac{°F - 32}{1.8}$$
$$°C = \frac{185 - 32}{1.8}$$
$$°C = \frac{153}{1.8}$$
$$°C = \mathbf{85°C}$$

* WITH VARIOUS PROBES

Converting Celsius to Fahrenheit

To convert a Celsius temperature reading to Fahrenheit, multiply 1.8 by the Celsius reading and add 32. To convert Celsius to Fahrenheit, apply the formula:

$°F = (1.8 \times °C) + 32$

where
$°F$ = degrees Fahrenheit
1.8 = ratio between bases
$°C$ = degrees Celsius
32 = difference between bases

Example: Converting Celsius to Fahrenheit

A temperature monitor indicates 55°C. Convert the Celsius temperature to Fahrenheit.

$°F = (1.8 \times °C) + 32$
$°F = (1.8 \times 55) + 32$
$°F = 99 + 32$
$°F = \mathbf{131°F}$

Application — AC Sine Wave

An *AC sine wave* is a symmetrical waveform that contains 360 electrical degrees. The wave reaches its peak positive value at 90°, returns to 0 at 180°, increases to its peak negative value at 270°, and returns to 0 at 360°. See AC Sine Wave.

Cycle

A *cycle* is one complete wave of alternating voltage or current. An *alternation* is half of a cycle. A sine wave has one positive alternation and one negative alternation per cycle.

Period

Period (T) is the time required to produce one complete cycle of a waveform. Period is usually measured in seconds.

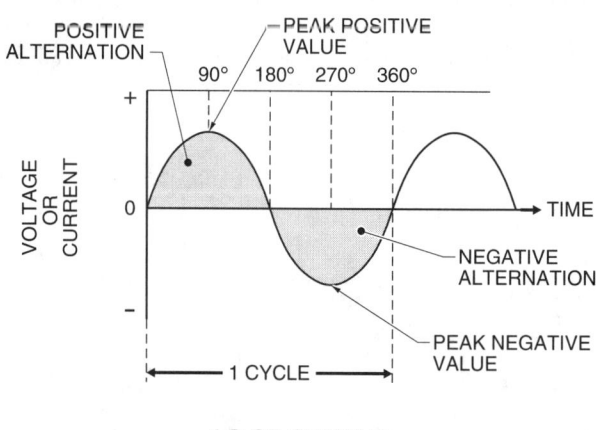

AC SINE WAVE

Frequency

Frequency (f) is the number of cycles per second (cps) in an AC sine wave. Frequency is measured in hertz (Hz). One hertz equals one cycle per second. See Frequency. Standard power is distributed at a frequency of 60 Hz. An audio amplifier amplifies all frequencies between 20 Hz and 20 kHz. To calculate frequency, apply the formula:

$$f = \frac{1}{T}$$

where
f = frequency (in hertz)
1 = constant
T = period (in seconds)

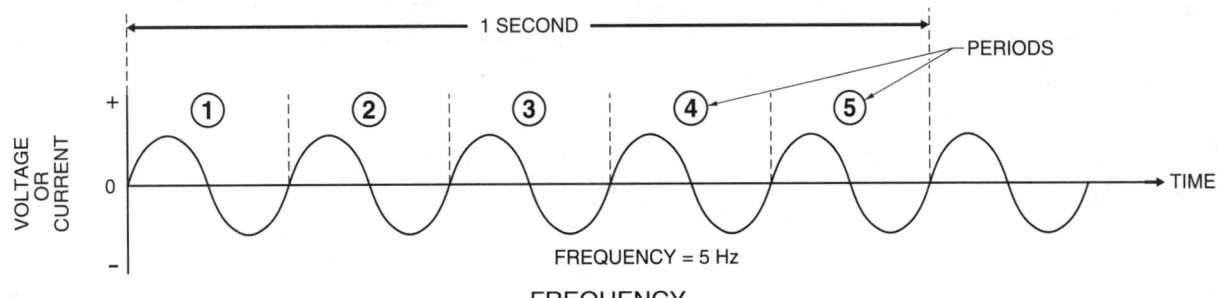

FREQUENCY

Example: Calculating Frequency

A waveform requires .02 seconds to complete one cycle. Calculate the frequency of the waveform. See Frequency Calculation.

$$f = \frac{1}{T}$$
$$f = \frac{1}{.02}$$
$$f = \mathbf{50\ Hz}$$

Period is calculated by applying the formula:
$$T = \frac{1}{f}$$
where
T = period (in seconds)
1 = constant
f = frequency (in hertz)

Example: Calculating Period

The frequency of an electrical signal is the standard 60 Hz power distribution frequency. Calculate the period required to complete one cycle. See Period Calculation.

$$T = \frac{1}{f}$$
$$T = \frac{1}{60}$$
$$T = \mathbf{.0167\ sec}$$

FREQUENCY CALCULATION

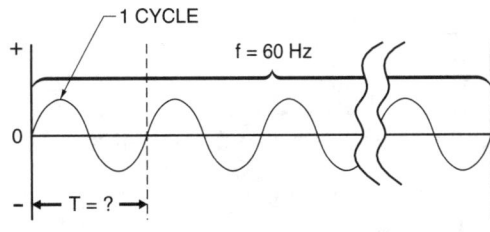

PERIOD CALCULATION

Amplitude

The *amplitude* of a sine wave is the height of the waveform. Voltmeters may specify a sine wave as peak, average, or root-mean-square.

Peak Value

The *peak value* (V_{max}) of a sign wave is the maximum value of either the positive or negative alternation. The positive and negative alternation are equal in a sign wave. See Sine Wave Peak Values.

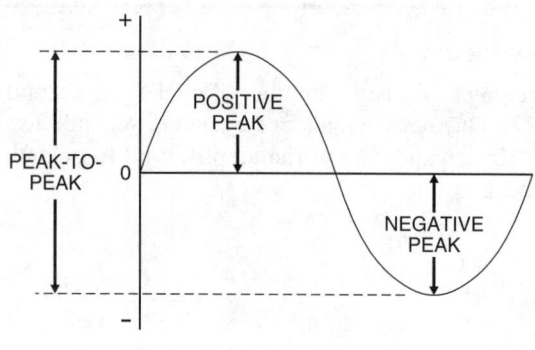

SINE WAVE PEAK VALUES

Peak-to-Peak

The *peak-to-peak value* ($V_{p\text{-}p}$) of a sine wave is the value measured from the maximum positive alternation to the maximum negative alternation. Peak-to-peak value is twice the peak value. To calculate peak-to-peak value, apply the formula:

$V_{p\text{-}p} = 2 \times V_{max}$
where
2 = constant
$V_{p\text{-}p}$ = peak-to-peak value
V_{max} = peak value

Example: Calculating Peak-to-Peak Voltage

A voltage source has a positive peak alternation of 165 V. Calculate the peak-to-peak voltage. See Peak-to-Peak Calculation.

$V_{p\text{-}p} = 2 \times V_{max}$
$V_{p\text{-}p} = 2 \times 165$
$V_{p\text{-}p} = \mathbf{330\ V}$

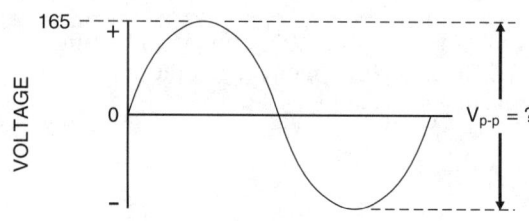

PEAK-TO-PEAK CALCULATION

Average Value

The *average value* (V_{avg}) of a sine wave is the mathematical mean of all instantaneous voltage values in the sine wave. Average value is equal to .637 of the peak value of a sine wave. See Sine Wave Average Values. To calculate average value, apply the formula:

$V_{avg} = V_{max} \times .637$
where
V_{avg} = average value (in volts)
V_{max} = peak value (in volts)
.637 = constant

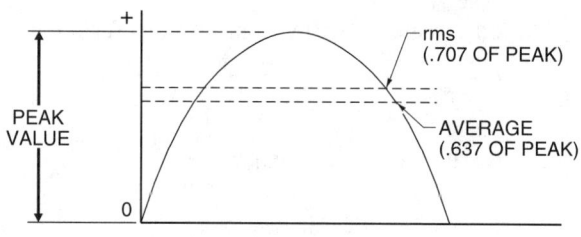

SINE WAVE AVERAGE VALUES

Example: Calculating Average Value

A voltage source has a positive peak alternation of 165 V. Calculate the average value. See Average Calculation.

$V_{avg} = V_{max} \times .673$
$V_{avg} = 165 \times .673$
$V_{avg} = \mathbf{111.045\ V}$

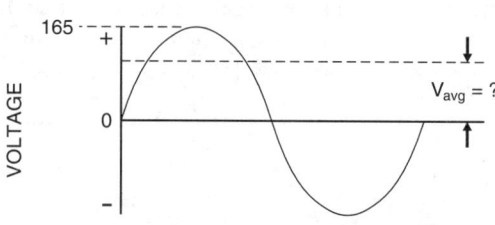

AVERAGE CALCULATION

Root-Mean-Square (Effective) Value

The *root-mean-square* (V_{rms}) value, or effective value, of a sine wave is the value that produces the same amount of heat in a pure resistive circuit as a DC of the same value. The rms value is obtained by squaring the instantaneous voltages along a waveform, averaging the squared values, and taking the square root of that number. The rms value is indicated by most AC voltmeters and ammeters. The rms value is equal to .707 of the peak in a sine wave. To calculate rms value, apply the formula:

$V_{rms} = V_{max} \times .707$
where
V_{rms} = rms value (in volts)
V_{max} = peak value (in volts)
.707 = constant

Example: Calculating Root-Mean-Square Value

A voltage source has a positive peak alternation of 165 V. Calculate the rms value. See rms Calculation.

$V_{rms} = V_{max} \times .707$
$V_{rms} = 165 \times .707$
$V_{rms} = \mathbf{116.655\ V}$

rms CALCULATION

VOLTAGE CONVERSIONS		
To Convert	**To**	**Multiply By**
rms	Average	.9
rms	Peak	1.414
Average	rms	1.111
Average	Peak	1.567
Peak	rms	.707
Peak	Average	.637
Peak	Peak-to-peak	2

Phase Shift

Phase shift is the state when voltage and current in a circuit do not reach their maximum amplitude and zero level simultaneously. Phase shift occurs in inductive or capacitive circuits.

Resistive Circuits

Alternating voltage and current are in-phase in resistive circuits. *In-phase* is the state when voltage and current reach their maximum amplitude and zero level simultaneously. A *resistive circuit* is a circuit that contains only resistive components, such as heating elements and incandescent lamps. See In-Phase AC Sine Wave.

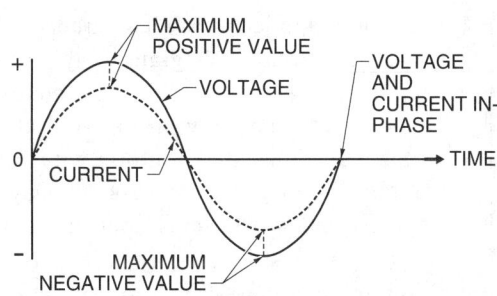

IN-PHASE AC SINE WAVE

True power is the actual power in a circuit. Apparent power is the product of voltage (in volts) and current (in amps) in a circuit. Apparent power equals true power when voltage and current are in-phase. Apparent power equals true power in a pure resistive circuit. True power is less than apparent power in circuits containing inductance or capacitance.

Inductive Circuits

A phase shift exists between alternating voltage and current in an inductive circuit. An *inductive circuit* is a circuit in which current lags voltage. The greater the inductance in a circuit, the larger the phase shift. True power in a circuit decreases as phase shift increases. Inductance reduces true power in a circuit. See Inductive Circuit AC Sine Wave.

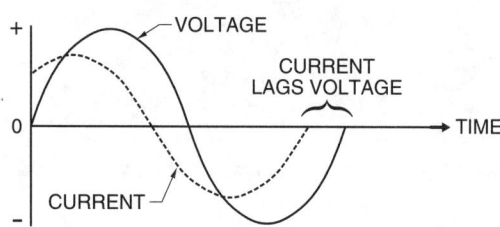

INDUCTIVE CIRCUIT AC SINE WAVE

Capacitive Circuits

A phase shift exists between voltage and current in a capacitive circuit. A *capacitive circuit* is a circuit in which current leads voltage. The greater the capacitance in a circuit, the larger the phase shift. Capacitance (like inductance) reduces the true power in a circuit. Capacitance is added to a circuit to increase true power when a circuit includes resistive and inductive loads. See Capacitive Circuit AC Sine Wave.

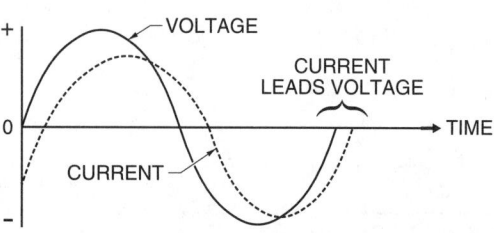

CAPACITIVE CIRCUIT AC SINE WAVE

Application — Drawings and Diagrams

Drawings and diagrams are used in the electrical/electronic field to convey facts, ideas, information and directions, and to provide information on the operation of devices and circuits. Standard practices for drawing electrical and electronic diagrams are detailed in Electrical and Electronic Diagrams, ANSI Y14.15, Reaffirmed 1988. This standard gives definitions, general information, and recommended drafting practices. See Drawing Standards.

General rules applicable to all diagrams include:
- Combined forms of diagrams may be used when the net result is helpful to the user.
- Line thickness and lettering shall conform with ANSI 14.2. Uppercase letters are generally used. Lowercase letters may be used to avoid ambiguity.
- Graphic symbols shall conform with ANSI Y32.2. Special symbols shall be explained by a note on the diagram. The terminal symbol (o) may be added as required to any basic symbol.
- Switch symbols and relay contacts shall be shown in the OFF position.
- Abbreviations shall conform with ANSI Z32.13.
- Diagrams should be drawn so that the main features are prominent. Crowding of parts should be avoided.
- Grouping of parts may be shown by a phantom line enclosing the parts.
- Wire colors may be indicated by abbreviations or numerical codes.

Basic Electrical Theory

DRAWING STANDARDS

WIRE COLORS		
Color	Abbreviation	Numerical Code
Black	BK	0
Brown	BR	1
Red	R	2
Orange	O	3
Yellow	Y	4
Green	G	5
Blue	BL	6
Violet	V	7
Gray	GY	8
White	W	9

DIAGRAM LINE CONVENTIONS		
Line Application	Thickness	Line
General use	Medium	————
Mechanical connection	Medium	- - - - - -
Bracket-connecting dash line	Medium	- - - - - -
Brackets, leaders, etc.*	Thin	————
Mechanical-grouping boundary line*	Thin	— - — - —
For emphasis*	Thick	▬▬▬▬

* use of line thicknesses is optional

Standard types of electrical diagrams include line (one-line), schematic (elementary), wiring (connection), interconnection, and terminal diagrams. Other drawings and diagrams include pictorial, sectional view, charts and graphs, application, operational, and block.

Line (One-Line) Diagrams

A *line (one-line) diagram* is a diagram which shows, with single lines and graphic symbols, the logic of an electrical circuit or system of circuits and components. Line diagrams show the connection of major equipment, protective relays, meters, and instruments. Line diagrams show the relationship between circuits and their components but do not show the actual location of the components. Line diagrams provide fast, easy understanding of the connections and use of each component. See Line Diagram.

36 TROUBLESHOOTING ELECTRICAL/ELECTRONIC SYSTEMS

LINE DIAGRAM

The arrangement of a line diagram should promote clarity. Graphic symbols, abbreviations, and device designations are drawn per standards. Connecting lines may be shown by the dot or no dot method. The main circuit is shown in the most direct path and logical sequence. Lines between symbols should be horizontal or vertical. Line crossing should be minimized.

Schematic (Elementary) Diagrams

A *schematic (elementary) diagram* is a diagram which shows with graphic symbols the electrical connections and functions of a specific circuit arrangement. A *function* is a mathematical, graphical, or tabular statement of the influence one device or component has on another device or component.

A schematic diagram does not show the physical relationship of the components in a circuit. A schematic diagram is used to trace a circuit and its functions without regard to the actual size, shape, or location of the component device or parts. See Schematic Diagram.

SCHEMATIC DIAGRAM

Wiring (Connection) Diagrams

A *wiring (connection) diagram* is a diagram that shows the connection of an installation or its component devices or parts. A wiring diagram shows, as closely as possible, the actual location of each component in a circuit, including the control circuit and the power circuit. Internal and external connections may be shown in sufficient detail to make or trace connections. See Wiring Diagram.

Wiring diagrams are used in troubleshooting because they show the actual component layout and connections. Wiring diagrams are limited in quickly showing circuit logic because the conductor connections are often hard to follow. When a wiring diagram is used with its line diagram, a clear understanding of the circuit wiring and operation is given.

Symbols used with wiring diagrams may be the symbols used with schematic diagrams or may be simple rectangles and circles. No attempt is made to show exact sizes of component devices or parts. Terminals should be shown as circles even when their physical appearances differ. Three types of wiring diagrams include the continuous-line type, interrupted-line type, and tabular type.

WIRING DIAGRAM

Interconnection Diagrams

An *interconnection diagram* is a diagram that shows the external connection between unit assemblies or equipment. Interconnection diagrams may be either the wiring type or cable type. The wiring type shows each wire. The cable type only shows cables. Combinations of the two types may also be used. Connection lines representing wires or cables shall be drawn horizontally or vertically and shall run parallel in a direct and logical path between terminals. For wiring, interconnection diagrams shall include wire color information, wire size, and wire type. For cables, interconnection diagrams shall include the part number, reference or designation, and cable type designation. See Interconnection Diagram.

INTERCONNECTION DIAGRAM

Terminal Diagrams

A *terminal diagram* is a diagram that relates the internal circuit of a piece of equipment to its terminal physical configuration. Terminal diagrams may be orthographic or pictorial. Reference letters and/or numbers are used to associate internal circuits with external connectors. See Terminal Diagram.

Pictorial Drawings

A *pictorial drawing* is a drawing that shows the length, height, and depth of an object in one view. True views of the surfaces are not obtained. Each surface is skewed allowing the three major surfaces to be seen. Pictorial drawings are used as working drawings to show the general location, function, and appearance of parts and/or assemblies. Pictorial drawings may be isometric, oblique, or perspective.

TERMINAL DIAGRAM

An *isometric drawing* is a pictorial drawing with a 120° included axis. Isometric drawings are used in promotional literature showing and describing products for sale. Isometrics are also used to show steps of assembly. Exploded isometrics are often used in service bulletins to show the relationship of individual components. The components are keyed to a parts list that provides statistical information. See Isometric Drawing.

ISOMETRIC DRAWING

Sectional View Drawings

A *sectional view drawing* is a drawing that shows the internal features of an object. An imaginary cutting plane is passed through the object perpendicular to the line of sight. The portion of the object between the cutting plane and the observer is removed, revealing the internal features. Sectional view drawings may be pictoral or orthographic.

Section lines are drawn on all surfaces cut by the cutting plane. General purpose section lines are commonly used. Whenever adjacent parts are shown in sectional views, the direction of the lines is changed. Specific section lines representing various materials, such as metals, soil, insulation, etc., may also be used. See Sectional Views.

SECTIONAL VIEWS

SELECTED SYMBOLS FOR SECTION LINING

Charts and Graphs

Charts and graphs are synonymous terms for graphical (pictorial) representations of data. Charts and graphs are used to show relationships between two or more variables. Charts and graphs include tabular, line, bar, flow, pie, and alignment. See Charts and Graphs.

A *tabular chart* is a chart that presents data in columnar form. Tabular charts are the least graphical of the various charts and graphs. Pictorial representations are often referenced to show the tabulated data.

A *line graph* is a graph that shows two related variables plotted on coordinate paper. The plotted data is joined by a continuous line. On a line graph, one known variable is plotted horizontally and another is plotted vertically. The relationship between the two variables is represented by a straight or curved line. The point at which either variable line intersects the straight or curved line represents the value of the unknown variable.

A *bar graph* is a graph that compares statistical data with vertical or horizontal bars. The bars may be color-coded or cross-hatched for comparative purposes. If so, a legend should accompany the bar graph.

A *flow chart* is a chart that shows a logical sequence of steps for a given set of conditions. Symbols and interconnecting lines are used to provide direction. Symbols used with flow charts include an ellipse, rectangle, diamond, and arrow. An ellipse indicates the beginning and end of a flow chart or section of a flow chart. A rectangle contains a set of instructions. A diamond contains a question stated so that a yes or no answer is achieved. The yes or no answer determines the direction to follow through the flow chart. An arrow indicates the direction to follow based on the answers.

A *pie chart* is a chart that compares parts of a whole in relation to their total use. A circle is divided into segments representing data on a percentage basis. Pie charts represent amounts clearly, as opposed to trends.

An *alignment chart* is a chart that shows the relationship among three variables. An alignment chart is used to solve three-variable equations. A straight line is drawn joining known values. The line is extended to intersect the scale of the third variable. The point where the line intersects the third scale is the value that solves the equation.

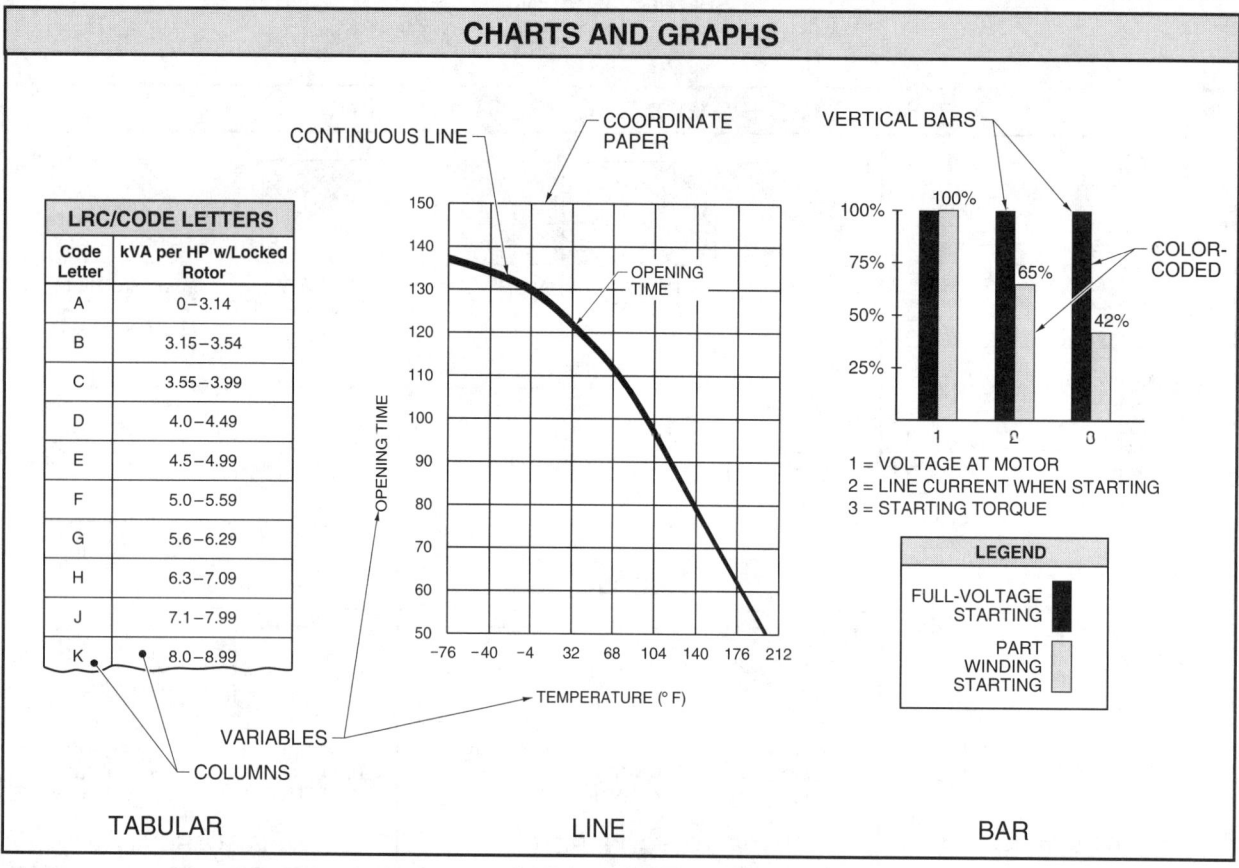

CHARTS AND GRAPHS

continued

CHARTS AND GRAPHS

FLOW
- ELLIPSE INDICATES BEGINNING OR END OF FLOW CHART
- ARROW INDICATES DIRECTION
- DIAMOND CONTAINS QUESTION
- RECTANGLE CONTAINS SET OF INSTRUCTIONS

PIE — ENERGY CONSUMPTION BASED ON MOTOR SIZE
- CIRCLE, SEGMENTS
- 3φ 21-125 HP — 36%
- 3φ 126-500 HP — 21%
- 3φ 500+ HP — 11%
- DC — 10%
- 1φ FRACTIONAL HP — 5%
- 3φ 1-20 HP — 17%

ALIGNMENT — HORSEPOWER TO TORQUE CONVERSION
- HORSEPOWER | TORQUE (IN LB-FT) | RPM
- TWO LINES ARE KNOWN VALUES OTHER LINE IS VARIABLE

Application Drawings

An *application drawing* is a drawing that shows the use of a particular piece of equipment or product in an application. Application drawings show product use and are not intended to show component connection or wiring. Application drawings present ideas on the use of a product in problem solving and are used by manufacturers to promote products. See Application Drawing.

Application drawings are used in troubleshooting to present ideas for the use of new or different components to replace older or troublesome components.

Operational Diagrams

An *operational diagram* is a diagram that shows the operation of individual components or circuits. Operational diagrams show the relationship between the input and output of a component or circuit. Operational diagrams provide an easily understood graphic representation of circuit operation. See Operational Diagram.

APPLICATION DRAWING

OPERATIONAL DIAGRAM

Block Diagrams

A *block diagram* is a diagram that shows the relationship between individual sections (blocks) of circuits or components. The circuit or component inside each block is not shown. Text describes the function of the block. Complicated circuits are not shown. See Block Diagram.

Block diagrams are used in troubleshooting to show the contents provided in a circuit or component under test. Block diagrams are also used to help isolate problems by checking individual blocks that may have failed.

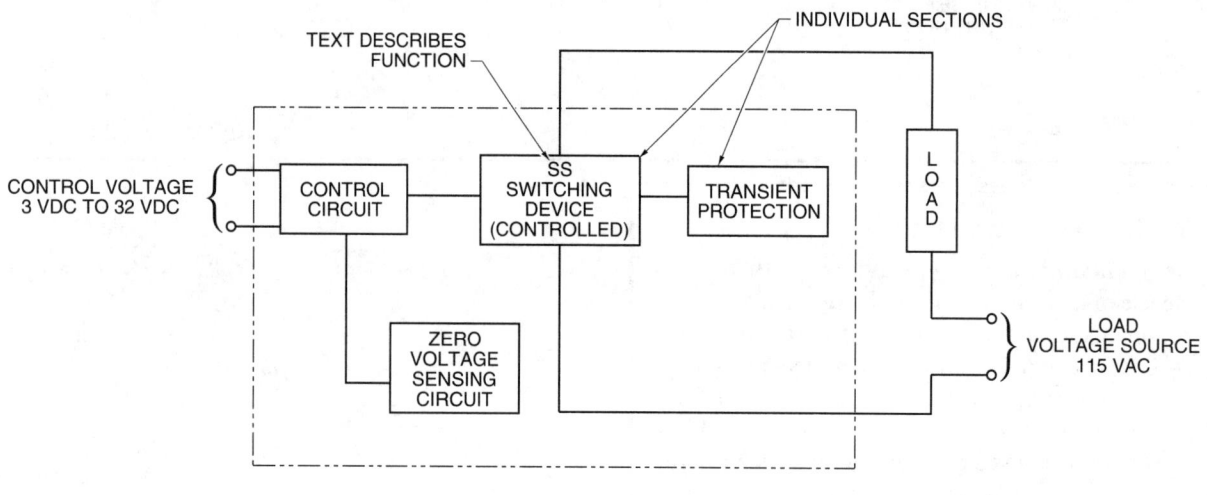

BLOCK DIAGRAM

BASIC ELECTRICAL THEORY

Name _____ Date _____

Activities

Activity 2-1. Electrical Prefixes

State the value for each meter reading in the units given.

Display: 1.25 V
1. _____ mV
2. _____ kV
3. _____ MV

Display: 1.1 mV
4. _____ V
5. _____ kV
6. _____ GV

Display: .003 mA
7. _____ A
8. _____ µA
9. _____ kA

Display: 1.05 µA
10. _____ A
11. _____ mA
12. _____ kA

Display: 101 Ω
13. _____ kΩ
14. _____ mΩ
15. _____ MΩ

Display: .03 MΩ
16. _____ Ω
17. _____ kΩ
18. _____ GΩ

Display: .01 V
19. _____ mV
20. _____ kV
21. _____ µV

Display: 103 V
22. _____ kV
23. _____ MV
24. _____ GV

43

44 TROUBLESHOOTING ELECTRICAL/ELECTRONIC SYSTEMS

Activity 2-2. Using Ohm's Law

1. A meter is normally set on the lowest range that produces a reading. Using the resistance and voltage values, determine the lowest meter range that produces a reading in each circuit.

_____ **A.** Range = _____   _____ **B.** Range = _____

_____ **C.** Range = _____   _____ **D.** Range = _____

_____ **E.** Range = _____   _____ **F.** Range = _____

2. Using the resistance values and ammeter readings, determine the voltage.

_____ **A.** _____ V
_____ **B.** _____ kV

_____ **C.** _____ V
_____ **D.** _____ mV

_____ **E.** _____ V
_____ **F.** _____ kV

_____ **G.** _____ V
_____ **H.** _____ kV

3. Using the ammeter and voltmeter readings, state the resistance values.

_____ **A.** R_1 = _____ Ω

_____ **B.** R_1 = _____ Ω
_____ **C.** R_1 = _____ kΩ

46 TROUBLESHOOTING ELECTRICAL/ELECTRONIC SYSTEMS

_____ D. R_1 = _____ Ω
_____ E. R_2 = _____ Ω

_____ F. R_1 = _____ Ω   _____ G. R_1 = _____ Ω

◯ Activity 2-3. Using the Power Formula

1. Determine the power that each heating unit produces when the control switch is closed.

_____ A. P = _____ W

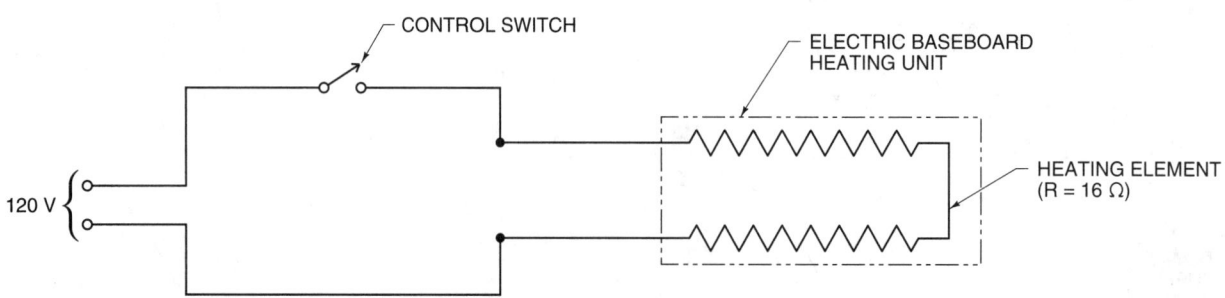

_____ B. P = _____ W

2. Determine the power that the coffeemaker produces when the thermostat is closed (brew) and open (warm). *Note*: When the thermostat is closed, the warm heating element is removed from the circuit.

_____ A. P = _____ W

_____ B. P = _____ W

48 TROUBLESHOOTING ELECTRICAL/ELECTRONIC SYSTEMS

3. A typical electric range includes two coils per heating element. The two coils are connected to a multiposition switch that connects the coils to the supply voltage. The switch applies 120 V or 230 V to each coil or any series or parallel combination of the coils. Ten different wattages are possible using two coils and two voltages. The wattage output of the coils varies from 160 W to 2880 W. Determine the power for each switching arrangement.

_____ A. P = _____ W   _____ B. P = _____ W

_____ C. P = _____ W   _____ D. P = _____ W

_____ E. P = _____ W   _____ F. P = _____ W

_____ **G.** P = _____ W

_____ **H.** P = _____ W

_____ **I.** P = _____ W

_____ **J.** P = _____ W

4. Determine the low, medium, and high heating position of the control switch, and also the power in each switch position.

_____ **A.** Position 1 = _____

_____ **B.** Power in circuit when switch is in Position 1 = _____ W

_____ **C.** Position 2 = _____

_____ **D.** Power in circuit when switch is in Position 2 = _____ W

_____ **E.** Position 3 = _____

_____ **F.** Power in circuit when switch is in Position 3 = _____ W

5. An automobile electrical system has a 12 VDC supply. The rear window defrost circuit has a resistance of 8 Ω. The starter motor draws between 20 A and 25 A. The fan motor draws between 2.5 A and 6 A, depending on the position of the speed-control switch. Each dashboard light is rated at 30 W. Each low beam and high beam headlight element is rated at 100 W. The dim/bright switch determines which elements are connected in the circuit. Using Automobile Electrical System, determine the maximum current and power in each circuit.

AUTOMOBILE ELECTRICAL SYSTEM

_____ A. Current in rear window defrost circuit = _____ A

_____ B. Power in rear window defrost circuit = _____ W

_____ C. Maximum current in starter motor circuit = _____ A

_____ D. Maximum power in starter motor circuit = _____ W

_____ E. Maximum current in fan motor circuit = _____ A

_____ F. Maximum power in fan motor circuit = _____ W

_____ G. Current in dashboard light circuit = _____ A

_____ H. Power in dashboard light circuit = _____ W

_____ I. Current in headlight circuit when headlight switch is in the dim position = _____ A

_____ J. Power in headlight circuit when headlight switch is in the dim position = _____ W

_____ K. Current in headlight circuit when headlight switch is in the bright position = _____ A

_____ L. Power in headlight circuit when headlight switch is in the bright position = _____ W

_____ M. Maximum total current in the circuit when all loads are energized = _____ A

_____ N. Maximum total power in the circuit when all loads are energized = _____ W

6. Using Standard Fuse Sizes and the calculated current values from Automobile Electrical System, determine the maximum standard fuse size for each circuit. *Note*: Round all fuse sizes to the next largest size.

STANDARD FUSE SIZES
¼, ½, ¾, 1, 1¼, 1½, 2, 2½, 3, 3½, 4, 4½, 5, 6, 7, 8, 9, 10, 12, 15, 20, 25, 30, 35, 40, 45, 50, 60, 70, 80, 90, 100, 110, 120, 130...

_____ A. The size of Fuse 1 using a safety factor of 115% is _____ A.

_____ B. The size of Fuse 2 using a safety factor of 125% is _____ A.

_____ C. The size of Fuse 3 using a safety factor of 135% is _____ A.

_____ D. The size of Fuse 4 using a safety factor of 125% is _____ A.

_____ E. The size of Fuse 5 using a safety factor of 125% is _____ A.

_____ F. The size of Fuse 6 using a safety factor of 125% is _____ A.

◐ Activity 2-4. Temperature Conversion

Calculate the equivalent temperature readings for the applications.

_____ 1. A temperature of 0°C is _____ °F.

_____ 2. The maximum operating temperature for a motor is _____ °C (140°F).

_____ 3. Ambient earth temperature is _____ °F (20°C).

_____ 4. Armored cable installed in thermal insulation shall have conductors rated at _____ °C (194°F).

_____ 5. Sealing compound shall provide a seal against passage of gas or vapors through the seal fitting, shall not have a melting point of less than _____ °F (93°C), and shall not be affected by the surrounding atmosphere or liquids.

_____ 6. Most motors have a temperature rise of _____ °F (40°C) listed on the motor nameplate.

_____ 7. When cleaning a water-soaked motor, the water temperature should not exceed _____ °C (200°F).

_____ 8. The melting point of most solder used in the electrical field is _____ °F (200°C).

◐ Activity 2-5. AC Sine Wave

1. Using the sine waves, determine the values.

_____ A. Number of cycles = _____

_____ B. Number of alternations = _____

_____ C. Period of one cycle = _____

_____ D. Frequency = _____ Hz

TIME = .25 SECONDS

52 TROUBLESHOOTING ELECTRICAL/ELECTRONIC SYSTEMS

_____ **E.** Peak-to-peak voltage = _____ V

_____ **F.** Average voltage = _____ V

_____ **G.** Root-mean-square voltage = _____ V

$V_P = 325\ V$

_____ **H.** Peak voltage = _____ V

_____ **I.** Average voltage = _____ V

_____ **J.** Root-mean-square voltage = _____ V

$V_{P-P} = 1358\ V$

◯ Activity 2-6. Phase Shift

Identify if the current is in-phase, leading, or lagging.

_____ **1.** Current is _____.

SPLIT-PHASE MOTOR — 115 VAC

_____ **2.** Current is _____.

RESISTANCE HEATING ELEMENT — 115 VAC

_____ **3.** Current is _____.

SOLENOID — 115 VAC

Activity 2-7. Drawings and Diagrams

1. Identify the drawings, diagrams, and types of symbols.

A. _____
B. _____
C. _____
D. _____
E. _____
F. _____
G. _____
H. _____
I. _____
J. _____
K. _____
L. _____

54 TROUBLESHOOTING ELECTRICAL/ELECTRONIC SYSTEMS

2. Identify the drawings, diagrams, and types of symbols.

_____ A.
_____ B.
_____ C.
_____ D.
_____ E.
_____ F.
_____ G.
_____ H.
_____ I.
_____ J.
_____ K.

SYMBOLS AND CIRCUITS — Applications

▍ Application — Electrical Symbols and Abbreviations

Electrical prints are used when troubleshooting, servicing, or repairing electrical systems or equipment. Electrical prints use standard symbols and abbreviations to show circuit operation and device use. See Selected Symbols and Selected Abbreviations. See Electrical Symbols and Abbreviations in Appendix.

SELECTED SYMBOLS

LIMIT SWITCHES		FULL-WAVE RECTIFIER	BATTERY	SILICON CONTROLLED RECTIFIER	CONTROL TRANSFORMER SINGLE VOLTAGE	FUSE POWER OR CONTROL
NO	NC					
HELD CLOSED	HELD OPEN					

SELECTED ABBREVIATIONS

AC	ALTERNATING CURRENT		NEG	NEGATIVE
CB	CIRCUIT BREAKER		NO	NORMALLY OPEN
CR	CONTROL RELAY		PB	PUSHBUTTON
DC	DIRECT CURRENT		POS	POSITIVE
DIO	DIODE		REC	RECTIFIER
DPDT	DOUBLE POLE, DOUBLE THROW		SCR	SILICON CONTROLLED RECTIFIER
DPST	DOUBLE POLE, SINGLE THROW		SOL	SOLENOID
LS	LIMIT SWITCH		SPDT	SINGLE POLE, DOUBLE THROW
FREQ	FREQUENCY		SPST	SINGLE POLE, SINGLE THROW
GRD	GROUND		TB	TERMINAL BOARD
NC	NORMALLY CLOSED		THS	THERMOSTAT SWITCH

▍ Application — Series, Parallel, and Series/Parallel Connections

Switches, loads, meters, and other components in electrical circuits are connected in series, parallel, or series/parallel combinations. A *series* connection has two or more components connected so that there is one path for current flow. A *parallel* connection has two or more components connected so that there is more than one path for current flow. A *series/parallel* connection is a combination of series- and parallel-connected components.

Switches and loads are connected in series or parallel. Ammeters are connected in series. Voltmeters are connected in parallel. See Series Connection and Parallel Connection.

56 TROUBLESHOOTING ELECTRICAL/ELECTRONIC SYSTEMS

SERIES CONNECTION   PARALLEL CONNECTION

> **Switch Connections**
> When connected in series, all switches in a circuit must be closed before current flows. Opening any one or more of the switches stops current flow. When connected in parallel, one or more switches must be closed before current flows. All closed switches must be opened to stop current flow. When connected in series/parallel, two or more switches must be closed before current flows. Any one or more series-connected switches, or all parallel-connected switches, must be opened to stop current flow. See Switch Connections.

SWITCH CONNECTIONS

Series Load Connections

The total voltage in a circuit containing series-connected loads equals the sum of the voltages across each load. Total voltage is divided across each load. The voltage across each load decreases if loads are added in series and increases if loads are removed.

The current in a circuit containing series-connected loads is the same throughout the circuit. The current in the circuit decreases if loads are added in series and increases if loads are removed. The total resistance in a circuit containing series-connected loads equals the sum of the resistances of all loads. The resistance in the circuit increases if loads are added in series and decreases if loads are removed. See Series Load Connections.

SERIES LOAD CONNECTIONS

Parallel Load Connections

The total voltage in a circuit containing parallel-connected loads is the same as the voltage across each load. The voltage across each load remains the same if parallel loads are added or removed. The total current in a circuit containing parallel-connected loads equals the sum of the current through all loads.

The total current increases if loads are added in parallel and decreases if loads are removed. The total resistance in a circuit containing parallel-connected loads is less than the smallest resistance value. The total resistance decreases if loads are added in parallel and increases if loads are removed. See Parallel Load Connections.

PARALLEL LOAD CONNECTIONS

Series/Parallel Load Connections

The total voltage in a circuit containing series/parallel-connected loads equals the sum of the voltage across all series-connected loads and the voltage through the parallel-connected load combinations.

The total current in a circuit containing series/parallel-connected loads equals the total current flowing through all series loads or the sum of the current flowing through all parallel combinations. The total resistance in a circuit containing series/parallel-connected loads equals the sum of the series loads and the equivalent resistance of the parallel combinations. See Series/Parallel Load Connections.

58 TROUBLESHOOTING ELECTRICAL/ELECTRONIC SYSTEMS

SERIES/PARALLEL LOAD CONNECTIONS

▪ Application — Circuit Polarity

All DC voltage sources have a positive and a negative terminal. The positive and negative terminals establish polarity in a circuit. All points in a DC circuit have polarity. *Polarity* is the positive (+) or negative (−) state of an object. The polarity of a component must be known before connecting it to a DC voltmeter or ammeter. See Circuit Polarity.

CIRCUIT POLARITY

▪ Application — Batteries and Photocells

A *battery* is a DC voltage source that converts chemical energy to electrical energy. Batteries are constructed by connecting individual cells in series, parallel, or series/parallel combinations.

A *cell* is a unit that produces electricity at a fixed voltage and current level. The voltage potential of a battery is increased by connecting cells in series. The current capacity of a battery is increased by connecting cells in parallel. Cells must have the same voltage and current rating when connected. See Battery Connections.

BATTERIES – SERIES CONNECTION

BATTERIES – PARALLEL CONNECTION

A *photocell* is a voltage source that converts light energy to electrical energy. Photocells are used in place of batteries in low-current devices, such as calculators, or are used to recharge battery-operated devices. Photocells are used individually or in combinations to produce high voltage and current.

Photocells are connected in series, parallel, or series/parallel. The voltage potential is increased when photocells are connected in series. The current capacity is increased when photocells are connected in parallel. See Photocell Connections.

PHOTOCELLS – SERIES CONNECTION

PHOTOCELLS – PARALLEL CONNECTION

▪ Application — Resistance In a Series Circuit

Series Circuit

A *series circuit* is a circuit that contains two or more loads and one path through which current flows. The total resistance of the loads (resistors) connected in series is equal to the sum of the individual resistances. To find the total resistance of a series circuit, apply the formula:

$$R_T = R_1 + R_2 + R_3 + ...$$

where
R_T = total resistance (in ohms)
R_1 = resistance 1 (in ohms)
R_2 = resistance 2 (in ohms)
R_3 = resistance 3 (in ohms)

Example: Finding Total Resistance — Series Circuit

A series circuit has resistances of 500 Ω, 700 Ω, and 100 Ω. Calculate the total resistance in the circuit. See Total Resistance—Series Circuit.

$$R_T = R_1 + R_2 + R_3$$
$$R_T = 500 + 700 + 100$$
$$R_T = \mathbf{1300 \ \Omega}$$

TOTAL RESISTANCE – SERIES CIRCUIT

Example: Finding Total Resistance — Parallel Circuit

A parallel circuit has resistances of 2 Ω and 4 Ω. Calculate the total resistance in the circuit. See Total Resistance—Parallel Circuit.

$$R_T = \frac{R_1 \times R_2}{R_1 + R_2}$$
$$R_T = \frac{2 \times 4}{2 + 4}$$
$$R_T = \mathbf{1.33 \ \Omega}$$

TOTAL RESISTANCE – PARALLEL CIRCUIT

▪ Application — Resistance In a Parallel Circuit

Parallel Circuit

A *parallel circuit* is a circuit that contains two or more loads and has more than one path through which current flows. To find the total resistance of a parallel circuit with two resistors, apply the formula:

$$R_T = \frac{R_1 \times R_2}{R_1 + R_2}$$

To find the total resistance of a parallel circuit with three or more resistors, apply the formula:

$$R_T = \frac{1}{\frac{1}{R_1} + \frac{1}{R_2} + \frac{1}{R_3} + ...}$$

Application — Resistance In a Series/Parallel Circuit

Series/Parallel Circuit

Most electrical circuits are combinations of series and parallel circuits. The total resistance of a series/parallel circuit is found by calculating the equivalent resistance of the parallel circuit(s) and adding the value to the resistance of the loads connected in series. To find the total resistance in a series/parallel circuit, apply the formula:

$$R_T = \frac{R_{P1} \times R_{P2}}{R_{P1} + R_{P2}} + R_{S1} + R_{S2} + \ldots$$

where
R_T = total resistance (in ohms)
R_{P1} = parallel resistance 1 (in ohms)
R_{P2} = parallel resistance 2 (in ohms)
R_{S1} = series resistance 1 (in ohms)
R_{S2} = series resistance 2 (in ohms)

Example: Finding Total Resistance — Series/Parallel Circuit

A series/parallel circuit has resistances of 150 Ω, 75 Ω, and 125 Ω connected in series and resistances of 60 Ω and 30 Ω connected in parallel. Calculate the total resistance. See Total Resistance—Series/Parallel Circuit.

$$R_T = \frac{R_{P1} \times R_{P2}}{R_{P1} + R_{P2}} + R_{S1} + R_{S2} + R_{S3}$$

$$R_T = \frac{60 \times 30}{60 + 30} + 150 + 75 + 125$$

$$R_T = \frac{1800}{90} + 350$$

$$R_T = 20 + 350$$

$$R_T = \mathbf{370\ \Omega}$$

TOTAL RESISTANCE – SERIES/PARALLEL CIRCUIT

Application — Capacitor Connection

Capacitors Connected in Series

Capacitors with different voltage ratings should not be connected. The applied voltage is divided over each capacitor for capacitors connected in series. The applied voltage is divided evenly between two capacitors of the same value. For example, two 110 V capacitors connected in series have 110 V across them when connected to a 220 V power source. See Capacitors Connected in Series.

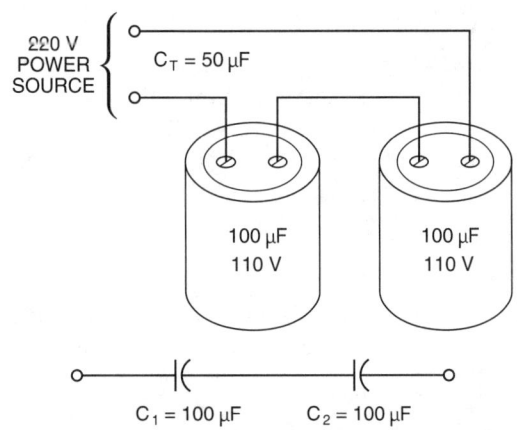

CAPACITORS CONNECTED IN SERIES

The total capacitance of two capacitors connected in series is less than the lowest-value capacitor. To find total capacitance of two capacitors connected in series, apply the formula:

$$C_T = \frac{C_1 \times C_2}{C_1 + C_2}$$

where
C_T = total capacitance (in µF)
C_1 = capacitance of capacitor 1 (in µF)
C_2 = capacitance of capacitor 2 (in µF)

To find the total capacitance of three or more capacitors connected in series, apply the formula:

$$\frac{1}{C_T} = \frac{1}{C_1} + \frac{1}{C_2} + \ldots$$

Example: Finding Total Capacitance — Three Capacitors Connected in Series

Capacitors of 10 µF, 50 µF, and 100 µF are connected in series. Find the total capacitance. See Three Capacitors Connected in Series.

$$\frac{1}{C_T} = \frac{1}{C_1} + \frac{1}{C_2} + \ldots$$

$$\frac{1}{C_T} = \frac{1}{10} + \frac{1}{50} + \frac{1}{100} \qquad \frac{1}{C_T} = .13$$

$$\frac{1}{C_T} = .1 + .02 + .01 \qquad C_T = \mathbf{7.69\ \mu F}$$

Capacitors Connected in Parallel

Capacitors connected in parallel have the same voltage across them. The voltage rating of each capacitor must be the same or greater than the supply voltage. For example, two 220 V capacitors connected in parallel have 220 V across them when connected to a 220 V power source. See Capacitors Connected in Parallel.

The equivalent capacitance of capacitors connected in parallel is equal to the sum of the individual capacitors. To find total capacitance of capacitors connected in parallel, apply the formula:

$$C_T = C_1 + C_2 + \ldots$$

Example: Finding Total Capacitance — Capacitors Connected in Series

Two 100 µF capacitors are connected in series. Find the total capacitance. See Two Capacitors Connected in Series.

$$C_T = \frac{C_1 \times C_2}{C_1 + C_2} \qquad C_T = \frac{10{,}000}{200}$$

$$C_T = \frac{100 \times 100}{100 + 100} \qquad C_T = \mathbf{50\ \mu F}$$

TWO CAPACITORS CONNECTED IN SERIES

THREE CAPACITORS CONNECTED IN SERIES

CAPACITORS CONNECTED IN PARALLEL

Example: Finding Total Capacitance — Capacitors Connected in Parallel

Two 50 µF capacitors are connected in parallel. Find the total capacitance. See Two Capacitors Connected in Parallel.

$C_T = C_1 + C_2 + ...$
$C_T = 50 + 50$
$C_T = \mathbf{100\ \mu F}$

TWO CAPACITORS CONNECTED IN PARALLEL

Capacitors Connected in Series/Parallel

To calculate the total capacitance of capacitors connected in series/parallel, apply the procedure:
1. Calculate the capacitance of each parallel branch.
2. Calculate the capacitance of each series branch.

Example: Finding Total Capacitance — Capacitors Connected in Series/Parallel

Two 30 µF capacitors are connected in parallel with a 50 µF capacitor connected in series. Find the total capacitance. See Capacitors Connected in Series/Parallel.

1. Calculate the capacitance of the parallel branch.
$C_T = C_1 + C_2$
$C_T = 30 + 30$
$C_T = \mathbf{60\ \mu F}$

2. Calculate the capacitance of the series combination.
$C_T = \dfrac{C_1 \times C_2}{C_1 + C_2}$
$C_T = \dfrac{60 \times 50}{60 + 50}$
$C_T = \dfrac{3000}{110}$
$C_T = \mathbf{27.27\ \mu F}$

CAPACITORS CONNECTED IN SERIES/PARALLEL

SYMBOLS AND CIRCUITS

Name _____ Date _____

Activities

Activity 3-1. Electrical Symbols and Abbreviations

Using the line and wiring diagrams, identify the symbols.

_____ 1. Fixed resistor
_____ 2. Control transformer
_____ 3. Disconnect switch
_____ 4. NC limit switch
_____ 5. NO limit switch
_____ 6. Three-phase motor
_____ 7. NC flow switch
_____ 8. NC pressure switch

_____ 9. Fuse
_____ 10. Pilot light
_____ 11. Horn
_____ 12. NO contact
_____ 13. NC temperature switch
_____ 14. NO, timed closed contact
_____ 15. NC pushbutton

63

Activity 3-2. Series, Parallel, and Series/Parallel Connections

1. Determine if the light is ON or OFF for each combination of switch positions.

SWITCH COMBINATIONS				
	Switch			
	1	2	3	4
A	X	X	X	X
B	O	O	O	O
C	X	O	O	X
D	X	O	X	X
E	O	X	X	O
F	X	X	X	O
G	X	X	O	X

X = switch closed
O = switch open

__ON__ A. Light is __ON__.

__OFF__ B. Light is __OFF__.

__OFF__ C. Light is __OFF__.

__ON__ D. Light is __ON__.

__OFF__ E. Light is __OFF__.

__ON__ F. Light is __ON__.

__OFF__ G. Light is __OFF__.

2. Determine if the light is ON or OFF for each combination of switch positions.

SWITCH COMBINATIONS								
	Switch							
	1	2	3	4	5	6	7	8
A	O	X	X	X	X	X	X	X
B	X	X	O	O	O	X	O	O
C	X	O	O	O	X	O	O	X
D	X	X	X	O	X	O	O	O
E	O	O	O	X	X	X	O	X
F	X	X	O	O	O	O	X	X
G	X	X	X	O	O	O	O	X
H	X	X	X	O	O	O	X	O
I	X	O	X	X	X	X	X	O

X = switch closed
O = switch open

__ON__ A. Light is __ON__.

__OFF__ B. Light is __OFF__.

__ON__ C. Light is __ON__.

__OFF__ D. Light is __OFF__.

__ON__ E. Light is __ON__.

__ON__ F. Light is __ON__.

__ON__ G. Light is __ON__.

__OFF__ H. Light is __OFF__.

__ON__ I. Light is __ON__.

3. Connect Meter 1 in parallel with the battery to measure the voltage at the battery. Connect Meter 2 in parallel with the collision stop lights to test the voltage to all three lights. Connect Meter 3 so that it reads the voltage at the left turn signal light. Connect Meter 4 so that it reads the voltage from the fuse that protects the running lights.

AUTO LIGHTING SYSTEM

66 TROUBLESHOOTING ELECTRICAL/ELECTRONIC SYSTEMS

4. Using Auto Lighting System, determine whether the devices are connected in series or parallel.

_____ A. The collision stop lights are connected in _____.

_____ B. The two brake lights are connected in _____.

_____ C. The fuses are connected in _____ with the lights in the circuit.

_____ D. The control switches are connected in _____ with the lights.

_____ E. The battery is connected in _____ with the lights.

Activity 3-3. Circuit Polarity

1. Using One-Station Power Window Circuit, complete the following.

A. Mark each pushbutton as up or down.

B. Connect the voltmeter to measure the voltage at the motor when the Up Pushbutton is pressed.

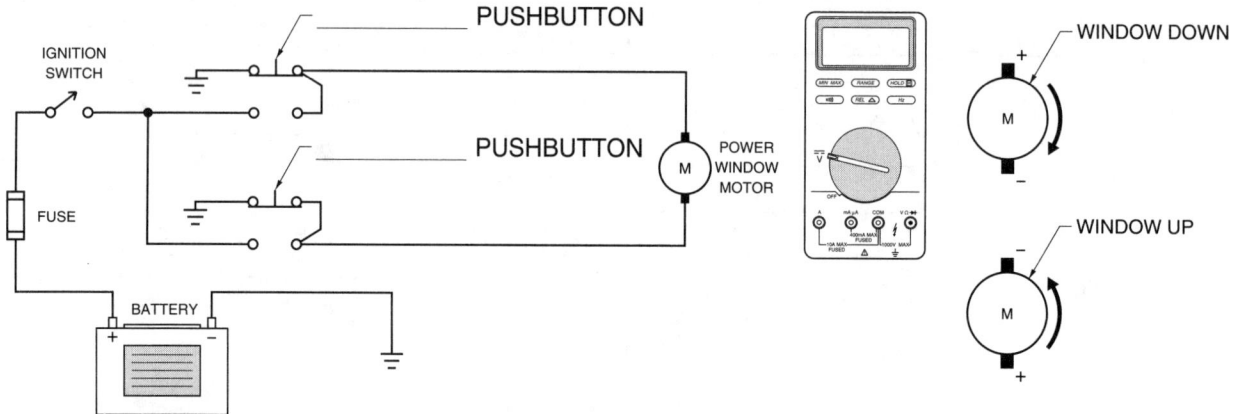

ONE-STATION POWER WINDOW CIRCUIT

2. Using Two-Station Power Window Circuit, complete the following.

A. Mark each pushbutton as up or down.

B. Connect the voltmeter to measure the voltage at the motor when the Down Pushbutton is pressed.

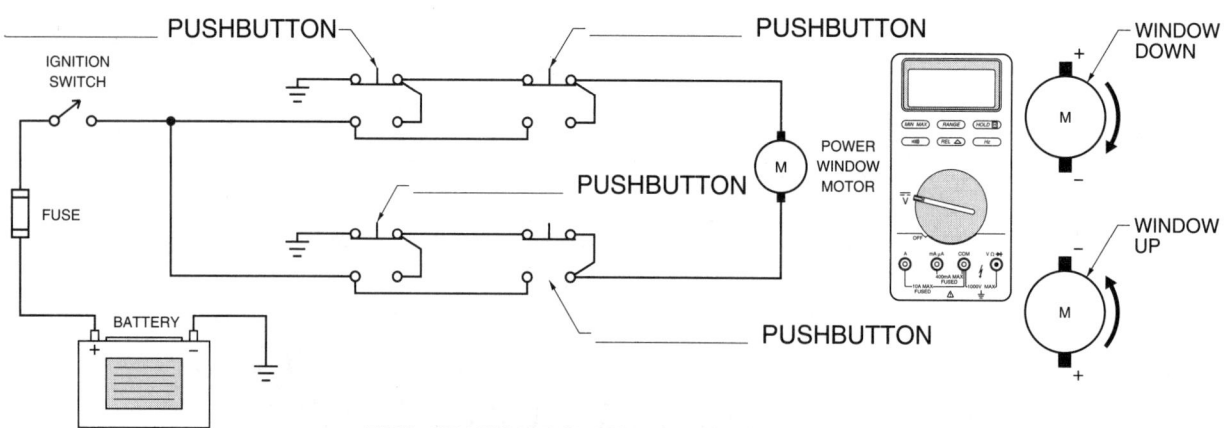

TWO-STATION POWER WINDOW CIRCUIT

Activity 3-4. Batteries and Photocells

1. Determine the output voltage and available current for each battery connection.

_____ A. E = _____ V   _____ C. E = _____ V
_____ B. I = _____ A   _____ D. I = _____ A

_____ E. E = _____ V
_____ F. I = _____ A

_____ G. E = _____ V   _____ I. E = _____ V
_____ H. I = _____ A   _____ J. I = _____ A

         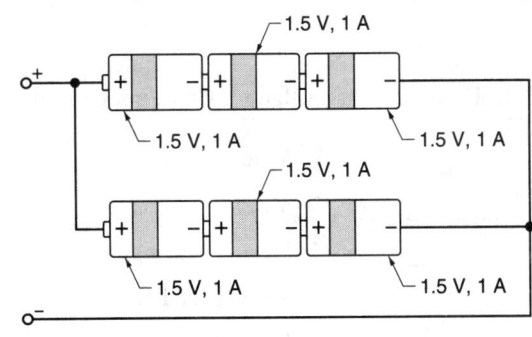

68 TROUBLESHOOTING ELECTRICAL/ELECTRONIC SYSTEMS

2. Using Photocell Battery Charging Circuit, mark the correct polarity (+) or (−) for each meter lead.

_____ A. _____   _____ D. _____
_____ B. _____   _____ E. _____
_____ C. _____   _____ F. _____

PHOTOCELL BATTERY CHARGING CIRCUIT

3. Using Photocell Battery Charging Circuit, list the meter readings for the light conditions.

_____ A. Meter 1 reading is _____ mA when light shines on photocells.
_____ B. Meter 2 reading is _____ V when light shines on photocells.
_____ C. Meter 3 reading is _____ V when light shines on photocells.
_____ D. Meter 1 reading is _____ mA when no light shines on photocells.
_____ E. Meter 2 reading is _____ V when no light shines on photocells.
_____ F. Meter 3 reading is _____ V when no light shines on photocells.

Activity 3-5. Resistance In a Series Circuit

Determine the total resistance and unknown electrical quantity.

_____ 1. $R_T =$ _____ Ω
_____ 2. $I_T =$ _____ mA

$E = 240$ V; $R_1 = 500\ \Omega$; $R_2 = 28.5$ kΩ; $R_3 = 1$ kΩ

_____ 3. $R_T =$ _____ Ω
_____ 4. $P_T =$ _____ W

$E = 20$ V; $R_1 = 1.5$ kΩ; $R_2 = .75$ kΩ; $R_3 = 1.5$ kΩ

_____ 5. $R_T =$ _____ kΩ
_____ 6. $P_T =$ _____ mW

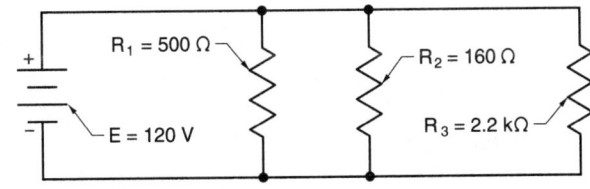
$E = 90$ V; $R_1 = 5.4$ kΩ; $R_2 = .6$ MΩ; $R_3 = 3.4$ kΩ

Activity 3-6. Resistance In a Parallel Circuit

Determine the total resistance and unknown electrical quantity.

_____ 1. $R_T =$ _____ Ω
_____ 2. $I_T =$ _____ mA

$E = 120$ V; $R_1 = 10$ kΩ; $R_2 = 10$ kΩ; $R_3 = 10$ kΩ

_____ 3. $R_T =$ _____ Ω
_____ 4. $I_T =$ _____ mA

$E = 24$ V; $R_1 = 250\ \Omega$; $R_2 = 80\ \Omega$

_____ 5. $R_T =$ _____ Ω
_____ 6. $P_T =$ _____ W

$E = 120$ V; $R_1 = 500\ \Omega$; $R_2 = 160\ \Omega$; $R_3 = 2.2$ kΩ

Activity 3-7. Resistance In a Series/Parallel Circuit

1. Determine the total resistance in the circuit by solving each resistance combination.

_____ A. $R_T =$ _____ Ω

_____ B. $R_T =$ _____ Ω

_____ C. $R_T =$ _____ Ω

_____ D. $R_T =$ _____ Ω

_____ **E.** $R_T =$ _____ Ω

2. Determine the total resistance in the circuit for each switch combination. Round the answer to the nearest whole number.

	SWITCH COMBINATIONS										
	Switch										
	1	2	3	4	5	6	7	8	9	10	11
A	X	O	O	X	O	O	O	X	O	O	O
B	X	X	X	O	O	O	X	O	O	O	X
C	X	O	O	X	X	O	O	O	O	X	O
D	O	O	O	X	O	O	O	O	O	O	X
E	X	X	X	O	O	X	X	X	X	O	O
F	X	X	O	O	O	X	O	X	O	O	O
G	X	O	O	O	X	O	O	O	O	X	X
H	O	X	X	X	X	X	X	O	X	O	O
I	X	O	X	X	O	X	X	O	O	O	X
J	X	X	O	O	O	X	X	O	X	O	O
K	X	X	X	O	O	O	X	X	O	O	O
L	X	X	O	O	X	X	O	O	O	O	X
M	X	X	X	X	O	O	X	O	X	O	O
N	X	X	X	O	X	O	X	O	O	X	O

X = switch closed
O = switch open

_____ **A.** $R_T =$ _____ Ω   _____ **H.** $R_T =$ _____ Ω
_____ **B.** $R_T =$ _____ Ω   _____ **I.** $R_T =$ _____ Ω
_____ **C.** $R_T =$ _____ Ω   _____ **J.** $R_T =$ _____ Ω
_____ **D.** $R_T =$ _____ Ω   _____ **K.** $R_T =$ _____ Ω
_____ **E.** $R_T =$ _____ Ω   _____ **L.** $R_T =$ _____ Ω
_____ **F.** $R_T =$ _____ Ω   _____ **M.** $R_T =$ _____ Ω
_____ **G.** $R_T =$ _____ Ω   _____ **N.** $R_T =$ _____ Ω

Activity 3-8. Capacitor Connection

Determine the total capacitance in each circuit.

_____ 1. $C_T =$ _____ µF

_____ 2. $C_T =$ _____ µF

_____ 3. $C_T =$ _____ µF

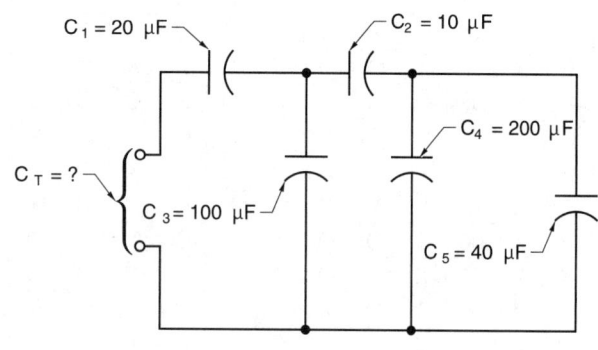

_____ 4. $C_T =$ _____ µF

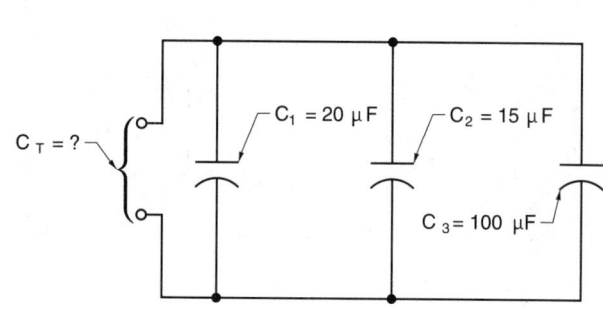

_____ 5. $C_T =$ _____ µF

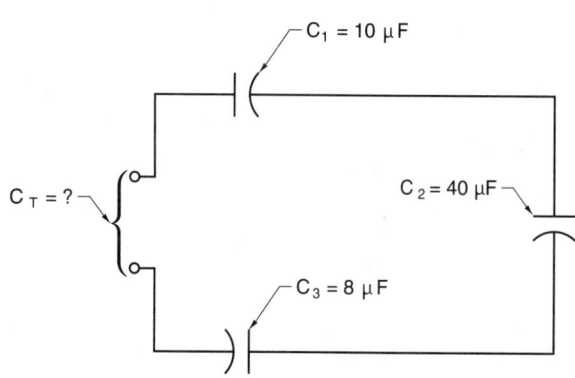

_____ 6. $C_T =$ _____ µF

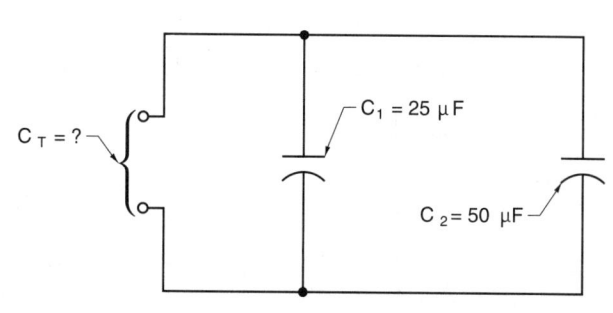

METER SYMBOLS AND TERMINOLOGY

Applications — 4

■ Application — Measurement Precautions

A variety of analog and digital meters measure electrical quantities such as voltage, current, resistance, power, speed, temperature, etc. Each meter includes an instruction manual. The instruction manual shows and explains individual meter specifications and features, proper operating procedure and safety, and specific warnings and applications.

Read the manufacturer's instruction manual before using any test equipment. Always refer to the instruction manual for information concerning specific use, safety, and meter capacity. Safety precautions are required when using meters because of the high voltage and currents in some circuits. Safety precautions for using meters include:

- Ensuring that the meter leads are connected to the correct meter jacks.
- Ensuring that the selector switch is set to the proper range and function before applying test leads to a circuit.
- Knowing meter accuracy limits. Any reading taken varies from the true reading by the accuracy limits of the meter.
- Averaging several readings when accuracy is critical. This reduces the effect of stray alternating fields on some meters.
- Starting with the highest range when measuring unknown values.
- Checking meter leads for frayed or broken insulation.
- Avoiding taking measurements under humid or damp conditions.
- Ensuring that hands, shoes, and work areas are dry.

■ Application — Meter Abbreviations and Symbols

Meter switches, scales, displays, and manuals use abbreviations and symbols to convey information. An *abbreviation* is a letter or combination of letters that represent a word. Abbreviations are dependent on a particular language. A *symbol* is a graphic element that represents a quantity or unit. Symbols are independent of language because a symbol can be recognized regardless of the language a person speaks. See Selected Meter Abbreviations and Selected Meter Symbols.

SELECTED METER ABBREVIATIONS					
AC	Alternating current or voltage	kΩ	Kilohms	RPM	Revolutions per minute
DC	Direct current or voltage	MΩ	Megohms	COM	Common
V	Volts	Hz	Hertz	OL	Overload
mV	Millivolts	kHz	Kilohertz	T	Time
kV	Kilovolts	μF	Microfarads	LSD	Least significant digit
A	Amperes	nF	Nanofarads	MAX	Maximum
mA	Milliamperes	°F	Degrees Farenheit	MIN	Minimum
μA	Microamperes	°C	Degrees Celsius	AVG	Average

74 TROUBLESHOOTING ELECTRICAL/ELECTRONIC SYSTEMS

SELECTED METER SYMBOLS

Symbol	Meaning	Symbol	Meaning	Symbol	Meaning
⚡	DANGEROUS OR HIGH VOLTAGE	🔧	SEE SERVICE MANUAL	○	SWITCH POSITION OFF (POWER)
∼	AC	⚠	SEE MANUAL FOR EXPLANATION			SWITCH POSITION ON (POWER)
⎓	DC	▢	DOUBLE INSULATION	■	MANUAL RANGE MODE
≈	AC OR DC	▭	FUSE	WARNING	HAZARD THAT COULD RESULT IN PERSONAL INJURY
+	POSITIVE	(battery cells)	BATTERY	CAUTION	HAZARD THAT COULD RESULT IN EQUIPMENT DAMAGE
−	NEGATIVE				
⏚	GROUND	H	HOLD	1000 V MAX	TERMINALS MUST NOT BE CONNECTED TO A CIRCUIT WITH HIGHER THAN LISTED VOLTAGE
±	PLUS OR MINUS	)))))	AUDIO-BEEPER		
Ω	OHMS–RESISTANCE	⊣⊢	CAPACITOR	△	RELATIVE MODE – DISPLAYED VALUE IS DIFFERENCE BETWEEN PRESENT MEASUREMENT AND PREVIOUS STORED MEASUREMENT
▶⊢	DIODE	%	PERCENT		
▶⊢)))))	DIODE TEST				

Abbreviations
All meters use standard abbreviations to represent a quantity or term. Abbreviations can be used individually (100 V) or in combinations (100 kV).

Symbols
Meters use symbols to display information. Most meters use standard international symbols that are designed to represent an electrical component (battery, diode, etc.), term (ground, positive, etc.), or message to the user (warning, see manual, etc.).

A component symbol used on a meter may differ from the same symbol used in wiring or schematic diagrams. A *component* is an individual device. For example, the symbol for a battery on a meter is not the same as the symbol for a battery in a schematic diagram. See Battery Symbols.

BATTERY SYMBOLS

Application — Meter Terminology

Meter Terminology

All meters use terms to describe displayed information. Understanding of meter terminology is required for proper meter use. For example, a meter may display an electrical quantity as an average, effective, peak, or rms value. The difference between each value must be understood to correctly interpret the measurements. See Meter Terminology.

METER TERMINOLOGY

TERM	SYMBOL	DEFINITION
AC	∿	Continually changing current that reverses direction at regular intervals. Standard U.S. voltage is 60 Hz.
AC/DC	≡∿	Indicates ability to read or operate on alternating and direct current.
ACCURACY ANALOG METER	(analog scale)	Largest allowable error (in percent of full scale) made under normal operating conditions. The reading of a meter set on the 250 V range with an accuracy rating of ± 2% could vary ± 5 V. Analog meters have greater accuracy when readings are taken on the upper half of the scale.
ACCURACY DIGITAL METER	100.0 AC V	Largest allowable error (in percent of reading) made under normal operating conditions. A reading of 100.0 V on a meter with an accuracy of ±2% is between 98.0 V and 102.0 V. Accuracy may also include a specified amount of digits (counts) that is added to the basic accuracy rating. For example, an accuracy of ± 2% (± 2 digits) means that a display reading of 100.0 V on the meter is between 97.8 V and 102.2 V.
ALLIGATOR CLIP	(clip symbol)	Long-jawed, spring-loaded clamp connected to the end of a test lead. Used to make temporary electrical connections.
AMBIENT TEMPERATURE		Temperature of air surrounding a meter or equipment to which the meter is connected.
AMMETER		Meter that measures electric current.
AMMETER SHUNT		Low-resistance conductor that is connected in parallel with the terminals of an ammeter to extend the range of current values measured by the ammeter.
AMPLITUDE	(sine wave)	Highest value reached by a quantity under test.
AUDIBLE	))))	Sound that can be heard.
AUTORANGING		Function that automatically selects a meter's range based on signals received.
AVERAGE VALUE	(sine wave with dashed line)	Value equal to .637 times the amplitude of a measured value.

continued

continued

METER TERMINOLOGY

TERM	SYMBOL	DEFINITION
BANANA PLUG		Long, thick terminal connection on one end of a test lead used to make a connection to a meter.
BANANA JACK		Meter jack that accepts a banana plug.
CELSIUS	°C	Temperature measured on a scale for which the freezing point of water is 0° and the boiling point is 100°.
CLOSED CIRCUIT		Circuit in which two or more points allow a predesigned current to flow.
COUNTS		Unit of measure of meter resolution. A 1999 count meter cannot display a measurement of $\frac{1}{10}$ of a volt when measuring 200 V or more. A 3200 count meter can display a measurement of $\frac{1}{10}$ of a volt up to 320 V.
DC		Current that constantly flows in one direction.
DIGITS		Indication of the resolution of a meter. A $3\frac{1}{2}$ digit meter can display three full digits and one half digit. The three full digits display a number from 0 to 9. The half digit displays a 1 or is left blank. A $3\frac{1}{2}$ digit meter displays readings up to 1999 counts of resolution. A $4\frac{1}{2}$ digit meter displays readings up to 19,999 counts of resolution.
DIODE		Semiconductor that allows current to flow in only one direction.
DISCHARGE		Removal of an electric charge.
EFFECTIVE VALUE		Value equal to .707 of the amplitude of a measured quantity.
FAHRENHEIT	°F	Temperature measured on a scale for which the freezing point of water is 32° and the boiling point is 212°.
FREQUENCY		Number of complete cycles occurring per unit of time.
FUNCTION SWITCH		Switch that selects the function (AC voltage, DC voltage, etc.) that a meter is to measure.
GROUND		Common connection to a point in a circuit whose potential is taken as zero.
HOLD BUTTON	HOLD H	Button that allows a meter to capture and hold a stable measurement.
MEASURING RANGE		Minimum and maximum quantity that a meter can safely and accurately measure.

continued

continued

METER TERMINOLOGY

TERM	SYMBOL	DEFINITION
OVERFLOW		Condition of a meter that occurs when a quantity to be measured is greater than the quantity the meter can display.
OVERLOAD	O.L.	Condition of a meter that occurs when a quantity to be measured is greater than the quantity the meter can safely handle for the meter range setting.
OPEN CIRCUIT		Circuit in which two (or more) points do not provide a path for current flow.
PEAK		Highest value reached when measuring.
POLARITY		Orientation of the positive (+) and negative (−) side of direct current or voltage.
PROBE		Pointed metal tip of a test lead used to make contact with the circuit under test.
RESOLUTION		Sensitivity of a meter. A meter may have a resolution of 1 V or 1 mV.
ROOT-MEAN-SQUARE		Value equal to .707 of the amplitude of a measured value.
SHORT CIRCUIT		Two or more points in a circuit that allow an unplanned current flow.
TERMINAL		Point to which meter test leads are connected.
TERMINAL VOLTAGE		Voltage level that meter terminals can safely handle.

▢ Application — Reading Analog Displays

Analog Displays

An *analog display* is an electromechanical device that indicates readings by the mechanical motion of a pointer. Analog displays use scales to display measured values. Analog scales may be linear or nonlinear. A *linear scale* is a scale that is divided into equally spaced segments. A *nonlinear scale* is a scale that is divided into unequally spaced segments. See Analog Displays.

ANALOG DISPLAYS

Analog scales are divided using primary divisions, secondary divisions, and subdivisions. A *primary division* is a division with a listed value. A *secondary division* is a division that divides primary divisions in halves, thirds, fourths, fifths, etc. A *subdivision* is a division that divides secondary divisions in halves, thirds, fourths, fifths, etc. Secondary divisions and subdivisions do not have listed numerical values.

When reading an analog scale, add the primary, secondary, and subdivision readings. See Reading Analog Scales.

	READING
① PRIMARY DIVISION	2
② SECONDARY DIVISION	.5
③ SUBDIVISION	.3
METER READING	2.8

READING ANALOG SCALES

To read an analog scale, apply the procedure:
1. Read the primary division.
2. Read the secondary division if the pointer moves past a secondary division. *Note*: This may not occur with very low readings.
3. Read the subdivision if the pointer is not directly on a primary or secondary division. Round the reading to the nearest subdivision if the pointer is not directly on a subdivision. Round the reading to the next highest subdivision if rounding to the nearest subdivision is unclear.

Add the primary division, secondary division, and subdivision readings to obtain the analog reading.

Application — Reading Digital Displays

Digital Displays

A *digital display* is an electronic device that displays readings as numerical values. Digital displays help eliminate human error when taking readings by displaying exact values measured. Errors occur when reading a digital display if the displayed prefixes, symbols, and decimal point are not properly applied.

Digital displays display values using either a light-emitting diode (LED) or a liquid crystal display (LCD). LED displays are easier to read and use more power than LCD displays. Most portable digital meters use an LCD.

The exact value on a digital display is determined from the numbers displayed and the position of the decimal point. A range switch determines the placement of the decimal point.

Typical voltage ranges on a digital display are 3 V, 30 V, and 300 V. The highest possible reading with the range switch on 3 V is 2.999 V. The highest possible reading with the range switch on 30 V is 29.99 V. The highest possible reading with the range switch on 300 V is 299.9 V. Accurate readings are obtained by using the range that gives the best resolution without overloading. See Digital Display.

DIGITAL DISPLAY

Bar Graph

Most digital displays include a bar graph to show changes and trends in a circuit. A *bar graph* is a graph composed of segments that function as an analog pointer. The displayed bar graph segments increase as the measured value increases and decrease as the measured value decreases. Reverse the polarity of test leads if a negative sign is displayed at the beginning of the bar graph. See Bar Graph Display.

BAR GRAPH DISPLAY

Wrap-Around Bar Graph

A *wrap-around bar graph* is a bar graph that displays a fraction of the full range on the graph. The pointer wraps around and starts over when the limit of the bar graph is reached. See Wrap-Around Bar Graph.

Using a Bar Graph

A bar graph reading is updated 30 times per second. A digital reading is updated 4 times per second. The bar graph is used when quickly changing signals cause the digital display to flash or when there is a change in the circuit that is too rapid for the digital display to detect.

For example, mechanical relay contacts may bounce open when exposed to vibration. Contact bounce causes intermittent problems in electrical equipment. Frequency and severity of contact bounce increase as a relay ages.

WRAP-AROUND BAR GRAPH

A contact's resistance changes momentarily from zero to infinity and back when a contact bounces open. A digital display cannot indicate contact bounce because most digital displays require more than 250 milliseconds (ms) to update their displays. The quick response of bar graphs enable detection of most contact bounce problems. The contact bounce is displayed by the movement of one or more segments the moment the contact opens.

Ghost Voltages

A meter set to measure voltage may display a reading before the meter is connected to a powered circuit. The displayed voltage is a ghost voltage that appears as changing numbers on a digital display or as a vibrating analog display. A *ghost voltage* is a voltage that appears on a meter that is not connected to a circuit.

Ghost voltages are produced by the magnetic fields generated by current-carrying conductors, fluorescent lighting, and operating electrical equipment. Ghost voltages enter a meter through the test leads because test leads not connected to a circuit act as antennae for stray voltages. See Ghost Voltages.

Ghost voltages do not damage a meter. Ghost voltages may be misread as circuit voltages when a meter is connected to a circuit that is believed to be powered. A circuit that is not powered can also act as an antenna for stray voltages. To ensure true circuit voltage readings, connect a meter to a circuit for a long enough time so that the meter displays a constant reading.

GHOST VOLTAGES

METER SYMBOLS AND TERMINOLOGY

Name _____   Date _____   **Activities**

◐ Activity 4-1. Measurement Precautions

Answer the questions using Model 01 Meter Specifications.

_____ 1. The meter has greater accuracy when measuring _____ voltage.

_____ 2. The meter has greater accuracy when measuring _____ current.

_____ 3. _____ VAC is the minimum value that can be displayed when the meter is set to autoranging.

_____ 4. _____ VAC is the maximum value that can be displayed when the meter is set to autoranging.

_____ 5. _____ mA is the minimum value of continuous AC current that can be displayed when the meter is set to autoranging.

_____ 6. _____ A is the maximum value of continuous AC current that can be displayed when the meter is set to autoranging.

_____ 7. _____ VDC is the minimum value that can be displayed when the meter is set to autoranging.

_____ 8. _____ VDC is the maximum value that can be displayed when the meter is set to autoranging.

_____ 9. _____ mA is the minimum value of continuous DC current that can be displayed when the meter is set to autoranging.

| MODEL 01 METER SPECIFICATIONS |||
Meter Function	Range	Accuracy
$\widetilde{V}$	400.0 mV	±.7%
	4.000 V	±.7%
	40.00 V	±.7%
	400.0 V	±.7%
	1000 V	±.7%
$\overline{\overline{V}}$	400.0 mV	±.1%
	4.000 V	±.1%
	40.00 V	±.1%
	400.0 V	±.1%
	1000 V	±.1%
Ω	400.0 Ω	±.2%
	4.000 kΩ	±.2%
	40.00 kΩ	±.2%
	400.0 kΩ	±.2%
	4.000 MΩ	±.2%
	40.00 MΩ	±1%
$\widetilde{mA}$	40.00 mA	±1%
	400.0 mA	±1%
	4000 mA	±1%
A∼	10.00 A*	±1%
$\overline{\overline{mA}}$	40.00 mA	±.2%
	400.0 mA	±.2%
	4000 mA	±.2%
A ⎓	10.00 A*	±.2%
Frequency	199.99 Hz	±.005%
	1999.9 Hz	±.005%
	19.999 kHz	±.005%
	199.99 kHz	±.005%

* 10 A continuous, 20 A for 30 seconds maximum

82 TROUBLESHOOTING ELECTRICAL/ELECTRONIC SYSTEMS

_____ 10. _____ A is the maximum value of continuous DC current that can be displayed when the meter is set to autoranging.

_____ 11. _____ Ω is the minimum resistance value that can be displayed when the meter is set to autoranging.

_____ 12. _____ MΩ is the maximum resistance value that can be displayed when the meter is set to autoranging.

_____ 13. _____ Hz is the minimum frequency value that can be displayed when the meter is set to autoranging.

_____ 14. _____ kHz is the maximum DC frequency value that can be displayed when the meter is set to autoranging.

_____ 15. The meter measures _____ most accurately when measuring different electrical quantities.

Activity 4-2. Meter Abbreviations and Symbols

Identify the abbreviations and symbols.

_____ 1. Direct current
_____ 2. Audio beeper
_____ 3. Millivolt
_____ 4. See manual
_____ 5. DC or AC
_____ 6. Diode
_____ 7. DC voltage setting
_____ 8. AC voltage setting
_____ 9. Millisecond
_____ 10. Maximum
_____ 11. Relative mode
_____ 12. Average
_____ 13. Positive or negative
_____ 14. Alternating current
_____ 15. Hold
_____ 16. Battery
_____ 17. Minimum
_____ 18. Capacitor
_____ 19. Danger
_____ 20. Hertz
_____ 21. Ohms

Activity 4-3. Meter Terminology

> Using the analog meter readings and specifications, determine the actual value range and accuracy of the readings.

1. A meter has a range setting of 50 V and an accuracy of ±2%.

   _____ A. The actual value is between _____ V and _____ V when a measured reading of 5 V is obtained.

   _____ B. The accuracy of the reading is _____%.

   _____ C. The actual value is between _____ V and _____ V when a measured reading of 12 V is obtained.

   _____ D. The accuracy of the reading is _____%.

   _____ E. The actual value is between _____ V and _____ V when a measured reading of 18 V is obtained.

   _____ F. The accuracy of the reading is _____%.

   _____ G. The actual value is between _____ V and _____ V when a measured reading of 25 V is obtained.

   _____ H. The accuracy of the reading is _____%.

2. A meter has a range setting of 500 mA and an accuracy of ±1.5%.

   _____ A. The actual value is between _____ mA and _____ mA when a measured reading of 16 mA is obtained.

   _____ B. The accuracy of the reading is _____%.

   _____ C. The actual value is between _____ mA and _____ mA when a measured reading of 48 mA is obtained.

   _____ D. The accuracy of the reading is _____%.

   _____ E. The actual value is between _____ mA and _____ mA when a measured reading of 90 mA is obtained.

   _____ F. The accuracy of the reading is _____%.

   _____ G. The actual value is between _____ mA and _____ mA when a measured reading of 260 mA is obtained.

   _____ H. The accuracy of the reading is _____%.

> Using the digital meter readings and specifications, determine the actual value range of the readings.

3. A meter has a range setting of 40.00 V and an accuracy of ±.5%.

   _____ A. The actual value is between _____ V and _____ V when a measured reading of 3.00 V is obtained.

   _____ B. The actual value is between _____ V and _____ V when a measured reading of 3.50 V is obtained.

   _____ C. The actual value is between _____ V and _____ V when a measured reading of 3.55 V is obtained.

D. The actual value is between _____ V and _____ V when a measured reading of 12.50 V is obtained.

E. The actual value is between _____ V and _____ V when a measured reading of 18.09 V is obtained.

4. A meter has a range setting of 400.0 V and an accuracy of ±.5%.

A. The actual value is between _____ V and _____ V when a measured reading of 30.0 V is obtained.

B. The actual value is between _____ V and _____ V when a measured reading of 35.0 V is obtained.

C. The actual value is between _____ V and _____ V when a measured reading of 35.5 V is obtained.

D. The actual value is between _____ V and _____ V when a measured reading of 125.0 V is obtained.

E. The actual value is between _____ V and _____ V when a measured reading of 180.9 V is obtained.

5. A meter has a range setting of 32.00 mA and an accuracy of ±.75%.

A. The actual value is between _____ mA and _____ mA when a measured reading of 00.32 mA is obtained.

B. The actual value is between _____ mA and _____ mA when a measured reading of 00.99 mA is obtained.

C. The actual value is between _____ mA and _____ mA when a measured reading of 01.01 mA is obtained.

D. The actual value is between _____ mA and _____ mA when a measured reading of 12.50 mA is obtained.

E. The actual value is between _____ mA and _____ mA when a measured reading of 18.66 mA is obtained.

6. A meter has a range setting of 320.0 Ω and an accuracy of ±.3%.

A. The actual value is between _____ Ω and _____ Ω when a measured reading of 001.2 Ω is obtained.

B. The actual value is between _____ Ω and _____ Ω when a measured reading of 015.0 Ω is obtained.

C. The actual value is between _____ Ω and _____ Ω when a measured reading of 025.6 Ω is obtained.

D. The actual value is between _____ Ω and _____ Ω when a measured reading of 100.1 Ω is obtained.

E. The actual value is between _____ Ω and _____ Ω when a measured reading of 180.9 Ω is obtained.

◯ **Activity 4-4. Reading Analog Displays**

List the correct reading for each range setting.

_____ 1. The reading equals _____ V with the range switch set on 1 V.

_____ 2. The reading equals _____ V with the range switch set on 10 V.

_____ 3. The reading equals _____ V with the range switch set on 100 V.

_____ 4. The reading equals _____ Ω with the range switch set on R × 1.

_____ 5. The reading equals _____ Ω with the range switch set on R × 100.

_____ 6. The reading equals _____ Ω with the range switch set on R × 1k.

_____ 7. The reading equals _____ V with the range switch set on 1 V.

_____ 8. The reading equals _____ V with the range switch set on 10 V.

_____ 9. The reading equals _____ V with the range switch set on 100 V.

_____ 10. The reading equals _____ V with the range switch set on 1 V.

_____ 11. The reading equals _____ V with the range switch set on 10 V.

_____ 12. The reading equals _____ V with the range switch set on 100 V.

_____ 13. The reading equals _____ Ω with the range switch set on R × 1.

_____ 14. The reading equals _____ Ω with the range switch set on R × 100.

_____ 15. The reading equals _____ Ω with the range switch set on R × 1 k.

86 TROUBLESHOOTING ELECTRICAL/ELECTRONIC SYSTEMS

Activity 4-5. Reading Digital Displays

1. Determine the resolution of the meter readings.

_____ A. The resolution equals _____ V.

[15 V]

_____ B. The resolution equals _____ V.

[1.555 V]

_____ C. The resolution equals _____ V.

[30.1 V]

_____ D. The resolution equals _____ V.

[12.25 V]

_____ E. The resolution equals _____ V.

[230.8 V]

_____ F. The resolution equals _____ V.

[115 V]

2. Determine the value of the individual segments on the bar graphs.

_____ A. Each segment equals _____ V.

RANGE SWITCH SET ON 30 V

_____ B. Each segment equals _____ V.

RANGE SWITCH SET ON 300 V

_____ C. Each segment equals _____ V.

RANGE SWITCH SET ON 3 V

_____ D. Each segment equals _____ V.

RANGE SWITCH SET ON 300 mV

_____ E. Each segment equals _____ V.

RANGE SWITCH SET ON 3000 V

_____ F. Each segment equals _____ V.

RANGE SWITCH SET ON 30 mV

METERS

5 Applications

▮ Application — Multimeters

A *multimeter* is a meter that is capable of measuring two or more electrical quantities. Multimeters may be analog or digital. The most common multimeter is a volt-ohm-milliammeter (VOM). A *VOM* is a meter that measures voltage, resistance, and current. The function switch is set on AC when measuring alternating voltage (VAC) and is set on DC when measuring direct voltage or current (DC).

If a multimeter does not have a separate setting for resistance, the positive DC setting is used. The function and range switches must be set at the correct quantity and the correct scale must be read when taking electrical measurements using a multimeter. See Multimeter.

MULTIMETER

Multimeter Settings

The normal setting for measuring DC voltage or current is +DC. This setting makes the red (+) lead positive. The alternative setting for measuring DC voltage or current is −DC. This setting makes the red (+) lead negative. This setting is used when it is preferable to have the test lead with the alligator clip (black) positive and the test lead with the pointer (red) negative.

Multimeter Use

Multimeters measure many different electrical quantities. Care must be taken to ensure that a multimeter is set on the correct settings, connected to a circuit correctly, and the scale is read accurately. Ensure that a multimeter is properly set before connecting it to a circuit. See Multimeter Use.

To use a multimeter, apply the procedure:
1. Determine the required function (voltage, current, resistance, etc.) and set the range and/or function switch to the electrical quantity and function required.

Caution: A multimeter is damaged if it is set to measure current but is connected as a voltmeter.

2. Set the meter to the highest quantity expected. Select the highest range for unknown readings.
3. Connect the meter to the circuit per the manufacturer's recommendations. Ensure that the polarity is correct when measuring DC or voltage. Connect one lead when connecting a meter to an unknown circuit. Slowly connect the other lead, observing the meter. Remove the lead immediately if the meter is overloaded.
4. Read the value on the meter.
5. Disconnect the meter from the circuit.

88 TROUBLESHOOTING ELECTRICAL/ELECTRONIC SYSTEMS

MULTIMETER USE

◻ Application — Test Lights

A *test light* is a light that is connected to two test leads and gives a visual indication when voltage is present in a circuit. The most common test light is a neon light. A *neon light* is a light that is filled with neon gas and uses two electrodes to ignite the gas.

The lamp glows if voltage is present in a circuit. If both sides of the bulb glow, the voltage in the circuit is AC. If one side glows, the voltage in the circuit is DC. The glowing side is the negative side of the DC circuit. The lamp glows brightly when high voltage is present. See Neon Test Light.

NEON TEST LIGHT

Two or more test lights are connected in series to measure high-voltage circuits. Connecting test lights in series divides the voltage in the circuit between the lamps, allowing for a larger total applied voltage. When testing a circuit, one side of the neon lamp is touched to the circuit. Voltage is present if the lamp glows. High voltage makes the neon lamp glow before it is fully connected to the circuit.

Test Light Use

Exercise caution when testing voltages over 24 V. See Test Light Use.

Warning: Ensure that no body part contacts any part of the live circuit, including the metal contact points at the tip of the test light.

TEST LIGHT USE

To use a test light, apply the procedure:
1. Ensure that the test light has a voltage rating higher than the highest potential voltage in the circuit. Connect two or more test lights in series if the voltage rating of the test light is less than the highest potential voltage in the circuit.
2. Connect one side of the test light to one side of the circuit or ground.
3. Connect the other side of the test light to the other side of the circuit. Voltage is present if the test light lights. Voltage is less than the rating of the test light if the test light lights dimly. Voltage is not present or is present at a very low level if the test light does not light.

Warning: If the test light does not light, a small voltage that can cause an electrical shock may be present.

Application — Voltage Testers

A *voltage tester* is a device that indicates approximate voltage level and type (AC or DC) by the movement and vibration of a pointer on a scale. Voltage testers contain a scale marked 120 VAC, 240 VAC, 480 VAC, 600 VAC, 120 VDC, 240 VDC, and 600 VDC.

Some voltage testers include a colored plunger to indicate the polarity of the test leads. If the red indicator is up, the red test lead is connected to the positive DC voltage. If the black indicator is up, the black test lead is connected to the positive DC voltage. See Voltage Tester.

VOLTAGE TESTER

Most voltage testers contain solenoids that vibrate when connected to AC. The solenoids do not vibrate when connected to DC. Most voltage testers include a neon light that determines if voltage is AC or DC. Both sides of the neon light glow when connected to AC. One side of the neon light glows when connected to DC. Voltage testers are designed for intermittent duty and should not be connected to a power supply for more than 15 seconds.

Voltage Tester Use

Exercise caution when testing AC voltages over 24 V. See Voltage Tester Use.

Warning: Ensure that no body part contacts any part of the live circuit, including the metal contact points at the tip of the tester.

To use a voltage tester, apply the procedure:
1. Connect one test probe to one side of the circuit.
2. Connect the other test probe to the other side of the circuit. The indicator shows a reading and vibrates if the current is AC. The indicator shows a reading and does not vibrate if the current is DC.

VOLTAGE TESTER USE

Application — Voltmeters

A *voltmeter* is a meter that measures voltage. Voltmeters produce a visual display (analog or digital) of the amount of voltage present in a circuit. A voltmeter is used when exact voltage values are required. A voltmeter is connected directly across the part of a circuit requiring measurement. Voltage measurements are taken while equipment is operating. See Voltmeter.

High-voltage test leads extend the range of a voltmeter. High-voltage test leads measure voltage in TV receivers, oscilloscopes, and ignition systems. Typical high-voltage test leads available with most standard voltmeters are not intended for measuring high-voltage power circuits, such as substations, broadcast transmitters, and X-ray equipment. High-voltage power circuit measurements must be performed only by well-trained personnel.

VOLTMETER

All electric devices are designed to operate at a certain voltage level or range. Most manufacturers list the acceptable voltage range in which their equipment operates properly. Problems occur if the supply voltage is not within the equipment's acceptable range. See Acceptable AC Voltage Ranges.

Low voltage occurs due to loose connections, corrosion at a connection, undersized feeder conductors, damaged conductors, or when additional loads are added to a circuit.

ACCEPTABLE AC VOLTAGE RANGES*		
Voltage Rating (60 Hz Power)	**Minimum Voltage**	**Maximum Voltage**
24	22	26
115	109	121
120	114	126
208	197	218
230	218	242
240	228	252
277	263	291
460	436	484
480	456	504

* in volts

Voltmeter — DC Voltage Measurement

Exercise caution when measuring DC voltages over 60 V. See Voltmeter — DC Voltage Measurement.

Warning: Ensure that no body part contacts any part of a live circuit, including the metal contact points at the tip of the test leads.

To measure DC voltages with a voltmeter, apply the procedure:
1. Set the selector switch to DC voltage. Select a setting high enough to measure the highest possible circuit voltage if the meter has more than one voltage position. Set the meter on the highest voltage setting if the circuit voltage is unknown.
2. Plug the black test lead into the common jack. The common jack may be marked com (common), – (negative), or lo (low).
3. Plug the red test lead into the voltage jack. The voltage jack may be marked V (voltage), + (positive), or hi (high).
4. Turn the power to the circuit or device being tested OFF. Discharge all capacitors if possible.
5. Connect the meter test leads in the circuit. Connect the black (–) test lead to circuit ground and the red (+) test lead to the point at which the voltage is under test. Reverse the black and red test leads if a negative sign appears in front of the reading on a digital meter. Reverse the black and red test leads or change the polarity switch (if applicable) to –DC if the needle moves in the reverse direction on an analog meter.
6. Turn the power to the circuit ON.
7. Read the voltage displayed on the meter.
8. Turn the power to the circuit OFF and remove the meter.

VOLTMETER – DC VOLTAGE MEASUREMENT

92 TROUBLESHOOTING ELECTRICAL/ELECTRONIC SYSTEMS

Voltmeter — AC Voltage Measurement

Exercise caution when measuring AC voltages over 24 V. See Voltmeter — AC Voltage Measurement.

Warning: Ensure that no body part contacts any part of the live circuit, including the metal contact points at the tip of the test leads.

To measure AC voltages with a voltmeter, apply the procedure:
1. Set the selector switch to AC voltage. Select the highest setting if the meter has more than one voltage position. Set the meter on the highest voltage setting if the voltage in the circuit is unknown.
2. Plug the black test lead into the common jack. The common jack may be marked com (common), – (negative), or lo (low).
3. Plug the red test lead into the voltage jack. The voltage jack may be marked V (voltage), + (positive), or hi (high).
4. Turn the power to the circuit or device OFF. Discharge all capacitors if possible.
5. Connect the test leads in the circuit. The position of the test leads is arbitrary. Common industrial practice is to connect the black test lead to the grounded (neutral) side of the AC voltage.
6. Turn the power to the circuit ON.
7. Read the voltage displayed on the meter.
8. Turn the power to the circuit OFF and remove the meter.

VOLTMETER – AC VOLTAGE MEASUREMENT

Application — In-Line Ammeters

An *in-line ammeter* is a meter that measures current in a circuit by inserting the meter in series with the component under test. In-line ammeters measure either DC or AC. The ammeter leads must be connected with the correct polarity at the test opening when measuring DC with an in-line ammeter. An in-line ammeter is limited to measuring AC current of 10 A or less.

When a motor is started, its current draw is 6 to 8 times the normal running current of the motor. An in-line ammeter is kept on a high scale to prevent damage to the meter if loads are switched ON when the ammeter is connected to the circuit.

In-Line Ammeter — DC Measurement

Care must be taken to protect the meter, circuit, and person using the meter when measuring DC with an in-line ammeter. Always apply the following rules when using an in-line ammeter:
- Check that the power to the test circuit is OFF before connecting and disconnecting test leads.
- Do not change the position of any switches or settings on the meter while the circuit under test is energized.
- Turn the power to the meter and circuit OFF before any meter settings are changed.
- Connect the ammeter in series with the components to be tested.
- Do not take current readings from any circuit in which the current may exceed the limit of the meter.

Many meters include a fuse in the low ampere range to prevent meter damage caused by excessive current. Before using a meter, check to see if the meter is fused on the current range being used. The meter is marked as fused or not fused at the test lead terminals. If the meter is not fused, an external fuse may be connected in series with the meter test leads. To protect the meter, the fuse rating should not exceed the current range of the meter. See In-Line Ammeter — DC Measurement.

IN-LINE AMMETER DC–MEASUREMENT

Warning: Ensure that no body parts contact any part of the live circuit, including the metal contact points at the tip of the test leads.

To measure DC using an in-line ammeter, apply the procedure:
1. Set the selector switch to DC. Select a setting high enough to measure the highest possible circuit current if the meter has more than one DC position.
2. Plug the black test lead into the common jack. The common jack may be marked com (common), – (negative), or lo (low).
3. Plug the red test lead into the current jack. The current jack may be marked + (positive), mA (milliamps), or hi (high).

4. Turn the power to the circuit or device under test OFF and discharge all capacitors if possible.
5. Open the circuit and connect the test leads to each side of the opening. The black (negative) test lead is connected to the negative side of the opening and the red (positive) test lead is connected to the positive side of the opening. Reverse the black and red test leads if a negative sign appears in front of the displayed reading.
6. Turn the power to the circuit under test ON.
7. Read the current displayed on the meter.
8. Turn the power OFF and remove the meter from the circuit.

The same procedure is used to measure AC with an in-line ammeter, except that the selector switch is set on AC current.

Application — Clamp-On Ammeters

A *clamp-on ammeter* is a meter that measures current in a circuit by measuring the strength of the magnetic field around a conductor. Clamp-on ammeters measure currents from .01 A or less to 1000 A or more. Most clamp-on ammeters measure AC, but DC clamp-on ammeters (or AC/DC) are also available. See Clamp-On Ammeter.

A clamp-on ammeter operates on the magnetic field present around a conductor as current flows through a circuit. Clamp-on ammeters cannot be connected incorrectly because there is no direct connection to the live circuit.

CLAMP-ON AMMETER

Clamp-on ammeters take current readings without opening a circuit. The jaws of a clamp-on ammeter are opened and encircled around the conductor under test. Readings are taken on bare or insulated conductors. When the jaws close, a current reading is indicated on the meter's scale. The reading indicates the amount of current drawn by loads connected to the conductor.

Clamp-On Ammeter — AC Measurement

Clamp-on ammeters measure the current in a circuit by measuring the strength of the magnetic field around a conductor. Care must be taken to ensure that the meter does not pick up stray magnetic fields. Whenever possible, separate the conductors under test from other surrounding conductors by a few inches. See Clamp-On Ammeter — AC Measurement.

To measure AC using a clamp-on ammeter, apply the procedure:
1. Set the selector switch to AC current. Select a setting high enough (10 A or 100 A) to measure the highest possible circuit current if the meter has more than one current position. Set the clamp-on ammeter on the highest current setting if the circuit current is unknown.
2. Remove the test leads from the meter to prevent the leads from accidentally becoming part of the circuit.
3. Open the meter's jaws by pressing against the trigger.
4. Enclose one conductor in the jaws and release the trigger. Ensure that the jaws are completely closed before taking readings.
5. Read the current displayed on the meter.
6. Remove the meter from the circuit.

CLAMP-ON AMMETER AC–MEASUREMENT

The same procedure is used to measure DC with a clamp-on ammeter, except that the selector switch is set to the DC position.

The jaws are opened and closed immediately when taking a DC measurement to reduce the possibility of error. To further reduce the possibility of error, a reading is taken, the jaws are reversed, a second reading is taken, and the two readings are averaged.

Application — Ohmmeters

An *ohmmeter* is an instrument that measures resistance. The higher the resistance reading, the lower the current flow. The lower the resistance reading, the higher the current flow.

Ohmmeter leads are connected across the device under test with all power removed from the circuit. An ohmmeter supplies power through batteries inside the meter. A standard ohmmeter cannot measure high resistance values, such as several million ohms. A high applied voltage is required to measure high resistance values.

Ohmmeter Resistance Measurement

Ensure that no voltage is present in the circuit or component under test before taking resistance measurements. Low voltage applied to a meter set to measure resistance causes inaccurate readings. High voltage applied to a meter set to measure resistance causes meter damage. Check for voltage using a voltmeter. See Ohmmeter Resistance Measurement.

To measure resistance using an ohmmeter, apply the procedure:
1. Check that all power is OFF in the circuit or component under test.
2. Set the selector switch to the resistance position, which is marked R × 1, R × 100, R × 10,000, etc., on analog meters. The resistance position is marked Ω on digital meters.
3. Plug the black test lead into the common jack. The common jack may be marked com (common), − (negative), or lo (low).
4. Plug the red test lead into the Ω jack.

5. Ensure that the meter batteries are in good condition and then zero the meter. On analog meters, the condition of the batteries is determined by the movement of the pointer. An analog meter is zeroed by touching the two test leads together and setting the zero adjustment to place the pointer on the "0" resistance mark. On digital meters, the battery symbol is displayed when the batteries are low. Digital meters are zeroed by the manufacturer.
6. Connect the meter test leads across the component under test. Ensure that contact between the test leads and the circuit is good. Dirt, solder flux, oil, and other foreign substances greatly affect resistance readings.
7. Read the resistance displayed on the meter. Check the circuit schematic for parallel paths. Paths in parallel with the resistance under test cause reading errors. Remove one end of the component under test from the circuit if parallel paths exist.
8. Turn the meter OFF after measurements are taken to save battery life.

OHMMETER RESISTANCE MEASUREMENT

Application — Continuity Checkers

A *continuity checker* is an instrument that indicates an open or closed circuit in a circuit in which all power is OFF. A continuity checker is an economical tester that is used to test switches, fuses, grounds, and is also used for identifying individual conductors in a multiwire cable.

A bulb on the continuity checker lights when there is a path for current to flow. Expensive models also give an audible indication when there is a complete path. Most multimeters have a built in continuity checker. See Continuity Test.

CONTINUITY TEST

Short or Open Conductor Testing

Multiconductor cable is used in communication, entertainment, automotive, and security applications. A short or open may develop in a conductor if the cable is abused. A continuity checker tests for a shorted or open conductor. See Short Conductor Testing and Open Conductor Testing.

To test a multiconductor cable for a short between two conductors, apply the procedure:
1. Set selector switch to continuity setting.
2. Connect one test lead to one conductor.
3. Connect the other test lead to each of the remaining conductors. The meter sounds if any two conductors are shorted. Connect the first test lead to a different conductor and the other test lead to each of the remaining conductors until all are checked.

To test a multiconductor cable for an open conductor, apply the procedure:
1. Set selector switch to continuity setting.
2. Connect one pair of conductors.
3. Connect the meter test leads to the other ends of the conductors. The meter sounds if a conductor is continuous. The meter does not sound if a conductor is open. Connect each conductor to a different conductor in the cable to determine which conductor is open. Connect two different conductors until all are checked.

SHORT CONDUCTOR TESTING

OPEN CONDUCTOR TESTING

Application — Wattmeters

A *wattmeter* is an instrument that measures true power in a circuit. A wattmeter accounts for the phase difference between current and voltage. See Wattmeter. Wattmeters are designed for measuring power in single-phase and three-phase circuits. Apparent power is calculated after voltage and current are measured separately.

Wattmeter Power Measurement

Most wattmeters have more than one power range. Always start with the highest range if a wattmeter has more than one power range. See Wattmeter Power Measurement.

WATTMETER

WATTMETER POWER MEASUREMENT

To measure power with a wattmeter, apply the procedure:
1. Ensure that the voltage in the circuit does not exceed the voltage limit of the meter. Check the circuit voltage using a voltmeter. A voltmeter is used on high-voltage circuits because it has a higher voltage range than a wattmeter.
2. Ensure that the current in the circuit does not exceed the current limit of the meter. Check the current using a clamp-on ammeter. A clamp-on ammeter is used on high-current circuits because it has a higher current range than a wattmeter.
3. Set the wattmeter at the highest wattage range.
4. Connect the circuit under test to the wattmeter with the power OFF. Check that the power is OFF by using a voltmeter.
5. Turn the power to the test circuit ON after the meter is properly connected.
6. Read the voltage, current, and power displayed on the meter. *Note:* The displayed power measurement is true power.
7. Turn the power OFF and remove the meter from the circuit.
   Multiply the voltage reading by the current reading to determine the apparent power in the circuit. Divide apparent power by true power and multiply by 100 to determine the efficiency of the circuit.

METERS

Name _____ Date _____ **Activities**

◯ Activity 5-1. Multimeters

Set the selector switches to measure the electrical quantities required. Draw the position on each meter. Select the setting that places the pointer in the center of the scale if more than one setting is possible.

100 TROUBLESHOOTING ELECTRICAL/ELECTRONIC SYSTEMS

◐ Activity 5-2. Test Lights

Determine the problem. Select the number of the problem from the list of possible problems.

Number	Possible Problems
1	Open fuse
2	Loose or corroded connection
3	Load has an internal open circuit
4	Broken wire
5	Faulty control switch
6	Open ground wire

1. The rear window defroster does not melt ice.

    _____ **A.** The problem is _____.

    _____ **B.** Potential problem(s) are _____.

2. The high beams do not work on both headlights.

    _____ **A.** The problem is _____.

    _____ **B.** Potential problem(s) are _____.

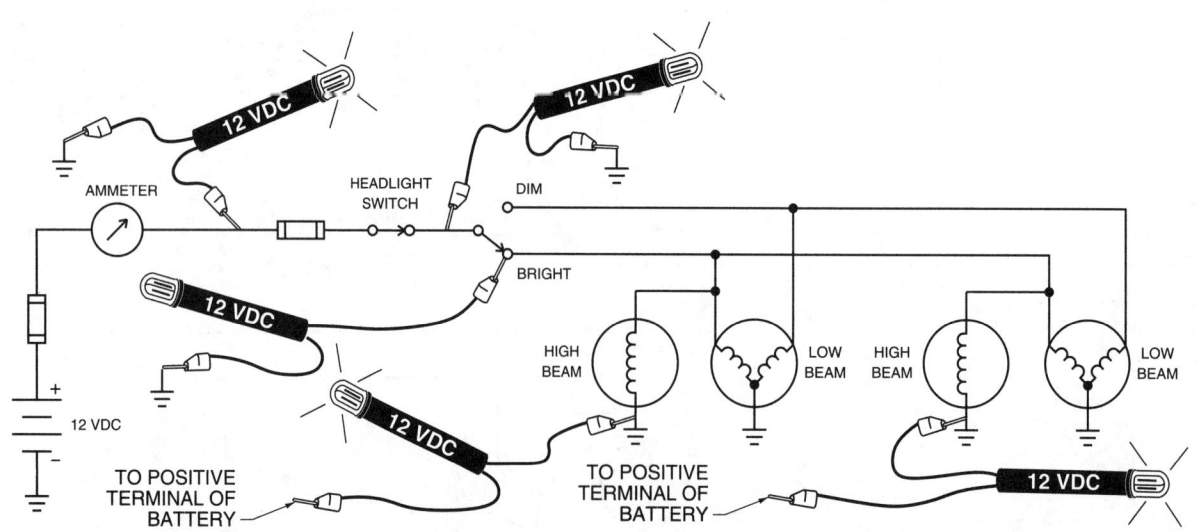

3. The fan motor operates only in high speed.

   _____ A. The problem is _____.

   _____ B. Potential problem(s) are _____.

4. One dashboard light does not work even after a new bulb is installed.

   _____ A. The problem is _____.

   _____ B. Potential problem(s) are _____.

102 TROUBLESHOOTING ELECTRICAL/ELECTRONIC SYSTEMS

Activity 5-3. Voltage Testers

Use the Voltage Test tables to determine the problems.

1. A voltage tester is used to test a split-wired receptacle circuit. In the split-wired circuit, the top half of the receptacle is controlled by the switch. The bottom half of the receptacle has power at all times.

_____ **A.** The problem is _____. _____ **B.** The problem is _____.

VOLTAGE TEST 1		
Voltage Tester	Test Position	Voltage Tester Reading
1	1	115 V (at all times)
1	2	0 V (switch in ON positon)
1	2	0 V (switch in OFF position)
2	1	0 V (switch in ON position)
2	1	0 V (switch in OFF position)
2	2	115 V (switch in ON position)
2	2	115 V (switch in OFF position)

VOLTAGE TEST 2		
Voltage Tester	Test Position	Voltage Tester Reading
1	1	115 V (at all times)
1	2	0 V (switch in ON position)
1	2	0 V (switch in OFF position)
2	1	115 V (switch in ON position)
2	1	0 V (switch in OFF position)
2	2	115 V (switch in ON position)
2	2	115 V (switch in OFF position)

2. A voltage tester is used to test a lamp that is controlled by two 3-way switches.

_____ **A.** The problem is _____.   _____ **B.** The problem is _____.

VOLTAGE TEST 3		
Voltage Tester	Test Position	Voltage Tester Reading
1	1	115 V (at all times)
1	2	0 V (switches in any positon)
2	1	115 V (switch 1 in up position)
2	1	0 V (switch 1 in down position)
2	2	0 V (switch 1 in down position)
3	1	0 V (at all times)
3	2	0 V (switch 2 in up position)
3	2	0 V (switch 2 in down position)

VOLTAGE TEST 4		
Voltage Tester	Test Position	Voltage Tester Reading
1	1	115 V (at all times)
1	2	0 V (switches in any positon)
2	1	115 V (switch 1 in up position)
2	1	115 V (switch 1 in down position)
2	2	115 V (at all times)
3	1	115 V (switch 2 in up position)
3	1	115 V (switch 2 in down position)

104 TROUBLESHOOTING ELECTRICAL/ELECTRONIC SYSTEMS

Activity 5-4. Voltmeters

1. Set the selector switch(es) and connect each voltmeter to measure the circuit voltage. Draw the position on each meter. Connect Voltmeter 1 to measure the voltage out of the transformer feeding the motor power circuit. Connect Voltmeter 2 to measure the voltage out of the fuse in the motor power circuit. Connect Voltmeter 3 to measure the voltage into the magnetic motor starter. Connect Voltmeter 4 to measure the voltage at the motor's armature when the motor is operating in the forward direction.

Activity 5-5. In-Line Ammeters

1. Complete the Power Pack Charging Circuit adding the in-line ammeters. Set the selector switches to measure the circuit current. Draw the position on each meter. Connect Ammeter 1 to measure current through the 120 VAC circuit. Connect Ammeter 2 to measure current in the stepped down AC circuit. Connect Ammeter 3 to measure current out of the positive side of the rectifier circuit.

Activity 5-6. Clamp-On Ammeters

Current in a circuit is indirectly measured using a clamp-on ammeter by wrapping a current-carrying conductor around the meter jaws. The indirect method is used when the current in the circuit is lower than the lowest limit of the meter. Wrapping the conductor around the meter jaws increases the strength of the magnetic field. The actual current value in the circuit is calculated by dividing the current reading on the meter by the number of times the conductor is wrapped around the jaws. Determine the actual value of current flowing in the circuit. List the current value in the units given.

_____ 1. The reading equals _____ mA.

_____ 2. The reading equals _____ mA.

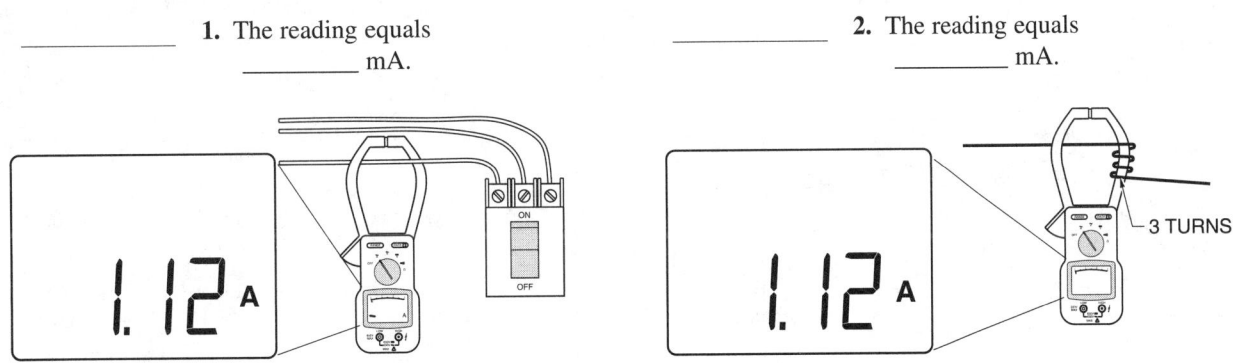

106 TROUBLESHOOTING ELECTRICAL/ELECTRONIC SYSTEMS

_____ 3. The reading equals _____ A.

_____ 4. The reading equals _____ μA.

◐ Activity 5-7. Ohmmeters

List the correct reading for the function/range settings.

_____ 1. The reading equals _____ kΩ.

_____ 2. The reading equals _____ Ω.

_____ 3. The reading equals _____ Ω.

_____ 4. The reading equals _____ Ω.

_____ 5. The reading equals
            _____ Ω.

_____ 6. The reading equals
            _____ MΩ.

◯ Activity 5-8. Continuity Checkers

Using the meter connections, determine whether the device under test is good or bad.

_____ 1. The SPST switch is _____.

_____ 2. The three-way switch is _____.

108 TROUBLESHOOTING ELECTRICAL/ELECTRONIC SYSTEMS

_____ 3. The four-way switch is _____.

_____ 4. Fuse 1 is _____.   _____ 5. Fuse 2 is _____.

◯ Activity 5-9. Wattmeters

Determine the circuit apparent power, true power, and efficiency.

_____ 1. Apparent power = _____ W   _____ 4. Apparent power = _____ W
_____ 2. True power = _____ W   _____ 5. True power = _____ W
_____ 3. Efficiency = _____ %   _____ 6. Efficiency = _____ %

SPECIAL METERS

Applications

Application — Contact Thermometers

A *contact thermometer* is an instrument that measures temperature at a single point. Temperature is measured when troubleshooting because the resistance of most materials changes as the temperature of the materials change. An increase in temperature decreases the performance of electrical equipment and destroys insulation.

Temperature is measured with a contact thermometer by placing a temperature probe in direct contact with the area to be measured. See Contact Thermometer. Contact thermometers measure the temperature of various solids, liquids, and gases, depending on the probe used. Most contact thermometer temperature probes are interchangeable.

CONTACT THERMOMETER

Contact Thermometer Temperature Measurement

Measuring hot devices presents a safety hazard. Ensure that proper equipment is used and that safety precautions are taken. Use a temperature probe that is rated higher than the highest possible temperature and always wear safety glasses. See Contact Thermometer Temperature Measurement.

Warning: Avoid contact with any material that can cause burns.

CONTACT THERMOMETER TEMPERATURE MEASUREMENT

To measure temperature using a contact thermometer, apply the procedure:
1. Select a temperature probe. A probe must have a higher temperature rating than the highest temperature it may contact and should have a shape that allows good contact with the device under test.
2. Connect the temperature probe to the meter.
3. Set the meter to the correct temperature range. Select the highest range if the temperature is unknown.
4. Place the temperature probe tip on the object or in the area to be measured.
5. Read the temperature displayed on the meter.
6. Remove the temperature probe from the object or area under test.

Application — Infrared Meters

An *infrared meter* is a meter that measures heat energy by measuring the infrared energy that a material emits. All materials emit infrared energy in proportion to their temperature.

Infrared meters detect small amounts of heat in electrical distribution systems, motors, and switching circuits. A problem is usually caused by a poor electrical connection. The poorer the connection, the higher the heat level in the system.

An infrared meter displays an image on a screen that shows different temperatures indicated by different colors or gives a direct digital temperature readout. See Infrared Meter.

INFRARED METER

Loose, corroded, or dirty electrical connections generate unwanted resistance and heat. The temperature rise at a connection depends on the current flowing through the connection and the resistance of the connection. A temperature rise of 20°F above ambient temperature indicates a fault that requires routine maintenance. Routine maintenance is performed so that the problem is fixed before it causes damage. A 100°F increase requires immediate action. Immediate action involves immediate shutting down of a system and repairing the fault.

Infrared Meter Temperature Measurement

Resistance in a normal circuit occurs at the load. Resistance in a circuit that has poor connections, corrosion, or other high resistance paths occurs at points other than the load. Infrared meter temperature measurements prevent problems by locating unwanted heat. See Infrared Meter Temperature Measurement.

To measure temperature using an infrared meter, apply the procedure:
1. Aim meter at area to be measured. Focus meter based on distance the object is from meter.
2. Take ambient temperature reading for reference.
3. Take temperature reading of any areas suspected to have temperatures above ambient temperature.

To determine temperature readings in such areas, subtract the ambient temperature reading from the reading obtained.

EQUIPMENT CONDITION	
Temperature Difference*	Indication
45	Light load on circuit
60	Heavy load on circuit
85	Possible problem. Schedule routine maintenance
100	Dangerous problem. Take immediate corrective action

* in °F

Special Meters 111

FINDING TEMPERATURE READING

What is the temperature reading if the temperature of an area is 145°F and the ambient temperature is 60°F?

1. T = Temperature Reading − Ambient Reading
   T = 145 − 60

T = **85°F (Schedule Routine Maintenance)**

INFRARED METER TEMPERATURE MEASUREMENT

Application — Contact Tachometers

A *tachometer* is a device that measures the speed of a moving object. Types of tachometers include contact, photo, and strobe tachometers. See Contact, Photo, and Strobe Tachometers. A *contact tachometer* is a device that measures the rotational speed of an object through direct contact of the tachometer tip with the object to be measured. A contact tachometer is used to measure the speed of rotating objects, such as motors, gears, belts, pulleys, and linear speeds, such as moving conveyors, webs, and printing presses. A contact tachometer measures speeds from .1 rpm to 25,000 rpm.

CONTACT TACHOMETER   PHOTO TACHOMETER   STROBE TACHOMETER

Electricity drives motors that produce a rotating mechanical force. The amount of force a motor produces depends on the size and speed of the motor. Speed may be rotational or linear.

Rotational speed is distance traveled per unit of time in a circular direction. Rotational speed is measured in revolutions per minute (rpm). *Linear speed* is distance traveled per unit of time in a straight line. Linear speed is measured in miles per hour (mph), feet per second (fps), or inches per second (ips). See Rotational Speed and Linear Speed.

Contact Tachometer Speed Measurement

Exercise caution when working around moving objects. Use a photo or strobe tachometer if there is danger of contact with a moving object. See Contact Tachometer Speed Measurement.

To measure speed with a contact tachometer, apply the procedure:
1. Place the tip of the tachometer in direct contact with the moving object.
2. Read the speed displayed on the meter.

Rotating speeds are displayed in rpm, and linear speeds are displayed in feet per minute (fpm) or meters per minute (mpm).

ROTATIONAL SPEED AND LINEAR SPEED

CONTACT TACHOMETER SPEED MEASUREMENT

▢ Application — Photo Tachometers

A *photo tachometer* is a device that measures the speed of an object without direct contact with the object. A photo tachometer measures speed by focusing a light beam on a reflective area or a section of reflective tape.

A photo tachometer is used in applications in which the rotating object cannot be reached or touched. A photo tachometer cannot measure linear speeds. Photo tachometers measure speeds from 1 rpm to 99,999 rpm.

Photo Tachometer Speed Measurement

Exercise caution when working around moving objects. Turn OFF, lockout, and tag disconnect switches of motors and rotating machinery as required. Use a photo or strobe tachometer if there is danger of contact with a moving object. See Photo Tachometer Speed Measurement.

PHOTO TACHOMETER SPEED MEASUREMENT

To measure speed with a photo tachometer, apply the procedure:
1. Place reflective tape on the object to be measured.
2. Point the light beam of the photo tachometer at the reflective tape.
3. Read the speed displayed on the meter.

Rotating speeds are displayed in rpm.

Application — Strobe Tachometers

A *strobe tachometer* is a device that uses a flashing light to measure the speed of a moving object. A strobe tachometer measures speed by synchronizing its light's flash rate with the speed of the moving object.

A strobe tachometer allows speed measurements through glass. This eliminates the possibility of direct exposure with hazardous areas. Strobe tachometers are also used for analysis of motion and vibration. Strobe tachometers measure speeds from 20 rpm to 100,000 rpm.

Strobe Tachometer Speed Measurement

Exercise caution when working around moving objects. Use a photo or strobe tachometer if there is danger of contact with a moving object. See Strobe Tachometer Speed Measurement.

STROBE TACHOMETER SPEED MEASUREMENT

To measure speed with a strobe tachometer, apply the procedure:
1. Set the meter for the best speed range for the application.
2. Turn the tachometer ON and align the visible light beam with the object to be measured.
3. Read the speed displayed on the meter.

Application — Megohmmeters

A *megohmmeter* is a device that detects insulation deterioration by measuring high resistance values under high test voltage conditions. Megohmmeter test voltages range from 50 V to 5000 V. A megohmmeter detects insulation failure or potential failure of insulation caused by excessive moisture, dirt, heat, cold, corrosive vapors or solids, vibration, and aging. See Megohmmeter.

MEGOHMMETER

Insulation separates conductors from each other and from earth ground. Insulation must have a high resistance in order to prevent current from passing through it. All insulation has a resistance value less than infinity and allows some leakage current. *Leakage current* is current that flows through insulation. Under normal operating conditions, the amount of leakage current is so small (microamperes) that it has no effect on the operation or safety of a circuit.

The total leakage current of insulation is the combination of conductive leakage current, capacitive leakage current, and surface leakage current. All three contribute to the total leakage current in a circuit.

Conductive Leakage Current

Conductive leakage current is the small amount of current that normally flows through the insulation of a conductor. Conductive leakage current flows from conductor to conductor or from a hot conductor to ground. See Conductive Leakage Current. Ohm's law is used to determine conductive leakage current. To calculate conductive leakage current, apply the formula:

$$I_L = \frac{E_A}{R_I}$$

where
I_L = leakage current (in microamperes)
E_A = applied voltage (in volts)
R_I = insulation resistance (in megohms)

CONDUCTIVE LEAKAGE CURRENT

Example: Calculating Conductive Leakage Current

Calculate the conductive leakage current when the applied voltage is 240 V and the resistance of the insulation is 10 MΩ. See Leakage Current Calculation.

$$I_L = \frac{E_A}{R_I}$$

$$I_L = \frac{240}{10,000,000}$$

$$I_L = .000024 \text{ or } 24 \text{ μA}$$

LEAKAGE CURRENT CALCULATION

The resistance of insulation decreases as the insulation ages and is exposed to damaging elements. Conductive leakage current increases as the resistance of the conductor insulation decreases. Additional insulation deterioration results as conductive leakage current increases. Conductive leakage current is minimized by keeping insulation clean and dry.

Capacitive Leakage Current

Capacitive leakage current is leakage current that flows through conductor insulation due to a capacitive effect. A *capacitor* is an electronic device used to store an electric charge. A capacitor is made by separating two plates with a dielectric. Two conductors that are run together act as a low-level capacitor. The insulation between the conductors is the dielectric, and the conductors are the plates. See Capacitive Leakage Current.

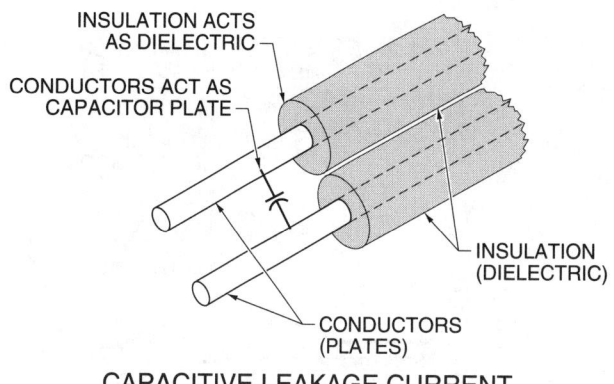

CAPACITIVE LEAKAGE CURRENT

DC voltages produce little capacitive leakage current. Conductors are charged as DC voltage is applied. The capacitive leakage current lasts for a few seconds and then stops. AC voltages produce continuous capacitive leakage current. Conductors are continuously charged and discharged as AC voltage is applied. This is due to the alternating characteristic of AC voltage. Capacitive leakage current flows continuously as AC voltage alternates. Capacitive leakage current is minimized by separating or twisting the conductors.

Surface Leakage Current

Surface leakage current is current that flows from areas on conductors where insulation is removed to allow electrical connections. Conductors are terminated to wire nuts, splices, spade lugs, terminal posts, and other fastening devices at different points along an electrical circuit. The point at which the insulation is removed provides a low resistance path for surface leakage current. Dirt and moisture allow additional surface leakage current to flow. See Surface Leakage Current. Surface leakage current results in increased heat at a connection. Increased heat contributes to insulation deterioration and makes a conductor brittle. Surface leakage current is minimized by making good, tight, clean connections.

SURFACE LEAKAGE CURRENT

Measuring Resistance with a Megohmmeter

Megohmmeters deliver a high voltage to the circuit under test. See Megohmmeter Resistance Measurement.

Warning: Ensure that no voltage is present in a circuit or component under test before taking any resistance measurements. Ensure that no body part contacts the high voltage.

116 TROUBLESHOOTING ELECTRICAL/ELECTRONIC SYSTEMS

MEGOHMMETER RESISTANCE MEASUREMENT

To use a megohmmeter, apply the procedure:
1. Ensure that all power is OFF in the circuit or component under test. Test for voltage using a voltmeter if uncertain.
2. Set the selector switch to the voltage at which the circuit is to be tested. The test voltage should be as high or higher than the highest voltage to which the circuit under test is exposed.
3. Plug the black test lead into the negative (earth) jack.
4. Plug the red test lead into the positive (line) jack.
5. Ensure that the batteries are in good condition. The meter contains no batteries if the meter includes a crank. The meter contains no batteries or crank if the meter plugs into a standard outlet.
6. Connect the line test lead to the conductor under test.
7. Connect the earth test lead to a second conductor in the circuit or earth ground.
8. Press the test button or turn the crank and read the resistance displayed on the meter. Change the meter resistance or voltage range if required.
9. Consult the equipment manufacturer or meter manufacturer for the minimum recommended resistance values. The insulation is good if the meter reading is equal to or higher than the minimum value.

Application — Digital Logic Probes

Digital logic circuits make decisions in control circuits. A digital signal is a signal represented by one of two states. The signal is either high (1) or low (0). A high signal is normally 5 V, but can be from 2.4 V to 5 V. A low signal is normally 0 V, but can be from 0 V to .8 V.

Digital logic gates are used to control electrical circuits. The AND, OR, and NOT logic gates are the three basic logic functions that make up most digital circuit logic. The NOT gate is used to invert the incoming signal to the gate. The NOR gate is a NOT, OR, or inverted OR gate. The NAND gate is a NOT, AND, or inverted AND gate. AND, OR, NOT, NOR, and NAND logic has the same meaning for digital logic, hard-wired electrical logic, and relay logic. See Basic Logic Functions.

BASIC LOGIC FUNCTIONS

FUNCTION	DIGITAL SYMBOL	DESCRIPTION
AND	(AND gate circuit with two inputs, 5 VDC)	**ENERGIZED** Output energized if all inputs are activated **DE-ENERGIZED** Output de-energized if any one input is deactivated
OR	(OR gate circuit with two inputs, 5 VDC)	**ENERGIZED** Output energized if one or more inputs are activated **DE-ENERGIZED** Output de-energized if any input is deactivated
NOT	(NOT gate circuit with one input, 5 VDC)	**ENERGIZED** Output energized if input is not activated **DE-ENERGIZED** Output de-energized if input is activated
NOR	(NOR gate circuit with two inputs, 5 VDC)	**ENERGIZED** Output energized if no inputs are activated **DE-ENERGIZED** Output de-energized if one or more inputs are activated
NAND	(NAND gate circuit with two inputs, 5 VDC)	**ENERGIZED** Output energized unless all inputs are activated **DE-ENERGIZED** Output de-energized if all inputs are activated

Digital Logic Probe

A *digital logic probe* is a special DC voltmeter that detects the presence or absence of a signal. Displays on a digital logic probe include logic high, logic low, pulse light, memory, and TTL/CMOS. See Digital Logic Probe.

The high light-emitting diode (LED) lights when the logic probe detects a high logic level (1). The low LED lights when the logic probe detects a low logic level (0).

The pulse LED flashes relatively slowly when the probe detects logic activity present in a circuit. Logic activity indicates that the circuit is changing between logic levels. The pulse light displays the changes between logic levels because the changes are usually too fast for the high and low LEDs to display.

DIGITAL LOGIC PROBE

The memory switch sets the logic probe to detect short pulses, usually lasting a few nanoseconds. Any change from the original logic level causes the memory LED to light and remain ON. The memory LED is the pulse LED switch in the memory position. The memory switch is manually moved to the pulse position and back to the memory position to reset the logic probe.

The TTL/CMOS switch selects the logic family of integrated circuits (ICs) to be tested. *Transistor-transistor logic (TTL) ICs* are a broad family of ICs that employ a two-transistor arrangement. The supply voltage for TTL ICs is 5.0 VDC, ±.25 V.

Complementary metal-oxide semiconductor (CMOS) ICs are a group of ICs that employ MOS transistors. CMOS ICs are designed to operate on a supply voltage ranging from 3 VDC to 18 VDC. Check circuit schematics for CMOS circuit voltages. The supply voltage for CMOS ICs should be greater than −5% of the rated voltage. CMOS ICs are noted for their exceptionally low power consumption.

Pull-Up Resistor

A *floating input* is a digital input that is not high or low at all times. A floating input should not occur in a digital circuit. A pull-up resistor prevents a floating input. A *pull-up resistor* is a resistor that has one side connected to the power supply at all times and the other side connected to one side of the input switch. The connection produces a high signal on one side of the input switch. The other side of the input switch is connected to ground. The high signal becomes a low signal when the input switch is closed. See Pull-Up Resistor Use.

PULL-UP RESISTOR USE

Digital Logic Probe Use

Digital circuits fail because the signal is lost somewhere between the circuit input and output stages. The problem is usually solved by finding the point where the signal is missing and repairing that area. Repairing normally involves replacing a component, section, or the entire PC board.

The power supply voltage is checked with a voltmeter if the digital circuit or logic probe has intermittent problems. A logic probe may indicate a high signal, but the supply voltage may be too low for proper circuit operation. See Digital Logic Probe Use.

To use a digital logic probe, apply the procedure:
1. Connect the positive (red) power lead to the positive side of the circuit's power supply. The positive power supply is +5 VDC for TTL circuits.
2. Connect the negative (black) power lead to the ground side of the circuit's power supply.
3. Set the selector switch to the logic family (TTL or CMOS) under test.

4. Touch the probe tip to the point being tested. Start at the input side of the circuit and move to the output of the circuit.
5. Note the condition of the LEDs on the logic probe.

Single-shot pulses are stored indefinitely by placing the switch in the memory position.

DIGITAL LOGIC PROBE USE

LOGIC PROBE FUNCTIONS	
LED Condition	Meaning
No LEDs lit	Open circuit
High LED lit	Logic 1
Low LED lit	Logic 0
Pulse LED lit	Circuit activity

PC BOARD REPAIR

Circuit Board Breaks

Broken traces (conducting paths) result when a PC board is subjected to mechanical stress. The break is repaired by soldering a bridge over the break. A bridge is made by soldering a short piece of wire on both sides of the break. See PC Board Repair.

Application — Oscilloscopes/Scopemeters

A graphical display of circuit voltage has more meaning than a numerical value in some troubleshooting situations. When circuits include rapidly fluctuating signals, stray signals, and phase shifts, problems are easily detected on a graphical display.

A *scope* is a device that gives a visual display of voltages. A scope shows the shape of a circuit's voltage and allows the voltage level, frequency, and phase to be measured. The two basic types of scopes used in troubleshooting are oscilloscopes and scopemeters. See Oscilloscope and Scopemeter.

OSCILLOSCOPE

SCOPEMETER

Oscilloscopes

An *oscilloscope* is an instrument that displays an instantaneous voltage. An oscilloscope is used to display the shape of a voltage waveform when bench testing electronic circuits. *Bench testing* is testing performed when equipment under test is brought to a designated service area. Oscilloscopes are used to troubleshoot digital circuits, communication circuits, TVs, VCRs, and computers. Oscilloscopes are available in basic and specialized types that can display different waveforms simultaneously.

Scopemeters

A *scopemeter* is a combination oscilloscope and digital multimeter. A scopemeter is used to display the shape of a voltage waveform when troubleshooting circuits in the field. A scopemeter is portable and can be used as a multimeter and a scope. A scopemeter does not have all the features of a specialized oscilloscope.

Scope Displays

A scope displays the voltage under test on the scope screen. The scope screen contains scribed horizontal and vertical axes. The horizontal (x) axis represents time. The vertical (y) axis represents the amplitude of the voltage waveform.

The scribed lines divide the screen into equal divisions. The divisions help to measure the voltage level and frequency of the displayed waveforms. See Scope Screen.

Scope Trace

A trace is established on the screen before a circuit under test is connected. A *trace* is a reference point/line that is visually displayed on the face of the scope screen. The trace is normally positioned over the horizontal center line on the screen.

The starting point of the trace is located near the left side of the screen. *Sweep* is the movement of the displayed trace across the scope screen. The sweep of the scope trace is from left to right. See Scope Trace.

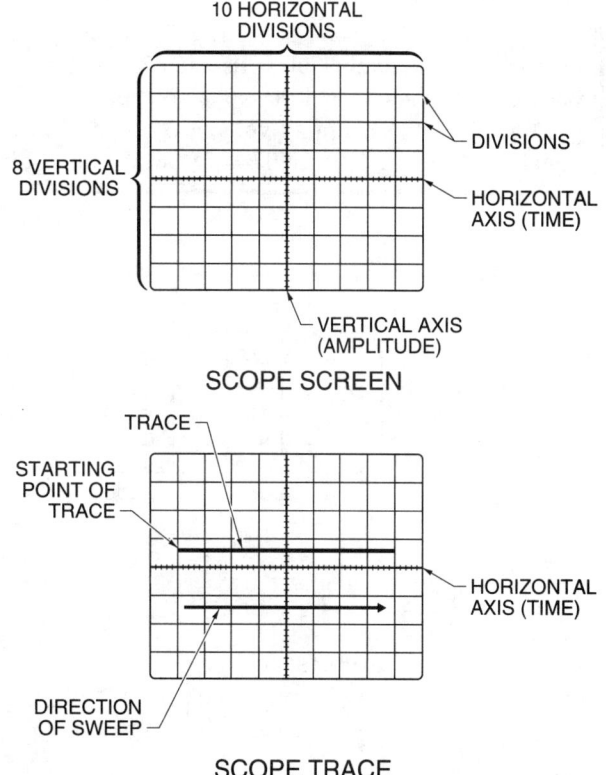

Manually Operated Controls

Manually operated controls are adjusted to view a waveform. Typical scope adjustment controls include intensity, focus, horizontal positioning, vertical positioning, volts/division, and time/division.

Intensity is the level of brightness. The intensity control sets the level of brightness on the displayed voltage trace. The intensity level is kept as low as possible to keep the trace in focus. The focus control adjusts the sharpness of the displayed voltage trace.

The horizontal control adjusts the left and right positions of the displayed voltage trace. The horizontal control sets the starting point of the trace. The vertical control adjusts the up and down positions of the displayed voltage trace. See Horizontal Positioning and Vertical Positioning.

HORIZONTAL POSITIONING

VERTICAL POSITIONING

The *volts/division (volts per division) control* selects the height of the displayed waveform. The setting determines the number of volts each horizontal screen division represents. For example, if a waveform occupies 4 divisions and the volts/division control is set on 20, the peak-to-peak voltage ($V_{p\text{-}p}$) equals 80 V (4 × 20 = 80 V). Eighty volts peak-to-peak equals 40 V peak (V_{max}) (80 ÷ 2 = 40 V). Forty volts peak equals 28.28 V_{rms} (40 × .707 = 28.28 V). See Volts Per Division Control.

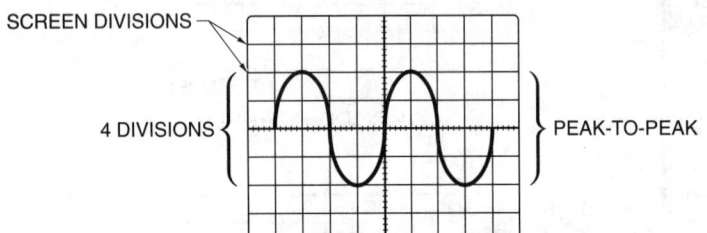

FINDING PEAK-TO-PEAK VALUE

What is the peak-to-peak value when a waveform occupies 4 divisions and the volts/division control is set on 20?

1. $V_{p\text{-}p}$ = Divisions x Volts/Division
   $V_{p\text{-}p}$ = 4 x 20

   $V_{p\text{-}p}$ = **80 V**

VOLTS PER DIVISION CONTROL

The *time/division (time per division) control* selects the width of the displayed waveform. The setting determines the length of time each cycle takes to move across the screen. For example, if the time/division control is set on 10, each vertical screen division equals 10 milliseconds (ms). If one cycle of a waveform equals 4 divisions, the displayed time equals 40 ms (4 × 10 = 40 ms). See Time Per Division Control.

To determine the frequency of the displayed waveform, apply the formula:

$$f = \frac{1}{T}$$

where
f = frequency (in hertz)
1 = constant
T = time period (in seconds)

Example: Calculating Frequency

Find the frequency of a waveform if one cycle of the waveform occupies 6 divisions and the time/division control is set on 5 ms (5 ms = .005 s).

1. Calculate time period.
   $T = 6 \times 5$
   $T = 6 \times .005$
   $T = \mathbf{.03\ s}$

2. Calculate frequency.
   $f = \dfrac{1}{T}$
   $f = \dfrac{1}{.03}$
   $f = \mathbf{33.3\ Hz}$

TIME PER DIVISION CONTROL

FINDING FREQUENCY

What is the frequency of a waveform if one cycle occupies 4 divisions and the time/division control is set on 10 ms? (10 ms = .01 s)

1. $T = Divisions \times Time/Divisions$
   $T = 4 \times .01$
   $T = .04$ s
2. $f = \frac{1}{T}$
   $f = \frac{1}{.04}$
   $f = 25$ Hz

Scope Measurement — AC Voltage

A scope is connected in parallel with a circuit or component under test. The scope is connected by a probe on the end of a test lead. A 1X probe (1 to 1) is used to connect the input of the scope to the circuit under test when the test voltage is lower than the voltage limit of the scope.

A 10X probe (10 to 1) is used to divide the input voltage by 10. The scope voltage limit equals 10 times the normal rated voltage when a 10X probe is used. The amount of measured voltage displayed on the scope screen must be multiplied by 10 to obtain the actual circuit voltage when using a 10X probe. For example, if the measured scope voltage is 25 V when using a 10X probe, the actual circuit voltage is 250 V ($25 \times 10 = 250$ V). See Scope Measurement — AC Voltage.

To use a scope to measure AC voltage, apply the procedure:
1. Turn the power switch ON and adjust the trace brightness on the screen.
2. Set the AC/DC control switch to AC.
3. Set the volts/division control to display the voltage level under test. Set the control to the highest value if the voltage level is unknown.
4. Connect the scope probe to the AC voltage under test.
5. Adjust the volts/division control to display the full waveform of the voltage under test.
6. Set the time/division control to display several cycles of the voltage under test.
7. Use the vertical control to set the lower edge of the waveform on one of the lower lines.
8. Measure the vertical amplitude of the waveform by counting the number of divisions displayed (V_{p-p}).

To calculate V_{rms}, first calculate V_{p-p}. V_{p-p} is calculated by multiplying the number of divisions by the volts/division setting. For example, if a waveform occupies 4 divisions and the volts/division setting is 10, V_{p-p} equals 40 V ($4 \times 10 = 40$ V). V_{max} equals 20 V ($40 \div 2 = 20$ V).

V_{max} is multiplied by .707 to find V_{rms}. For example, if V_{max} is 20 V, V_{rms} equals 14.14 V ($20 \times .707 = 14.14$ V). V_{rms} is the value of the voltage under test as measured on a voltmeter.

FINDING rms VALUE

What is the rms value of a waveform if the vertical amplitude is 4 divisions and the volts/division control is set on 10?

1. $V_{P\text{-}P}$ = Divisions x Volts/Division
   $V_{P\text{-}P}$ = 4 x 10

   $V_{P\text{-}P}$ = **40 V**

   V_{max} = 20 V (40 ÷ 2 = 20 V)

2. V_{rms} = V_{max} x .707
   V_{rms} = 20 x .707

   V_{rms} = **14.14 V**

SCOPE MEASUREMENT– AC VOLTAGE

Scope Measurement — Frequency

In AC applications such as variable frequency motor drives, it may be necessary to measure the frequency in the circuit. A frequency meter should be used in applications where high accuracy is required. A scopemeter gives an adequate reading in most frequency measurement applications. A scopemeter also shows any distortion present in the circuit under test. To measure frequency, the scope probes are connected in parallel with the circuit or component under test. See Scope Measurement — Frequency.

To use a scope to measure frequency, apply the procedure:
1. Turn the power switch ON and adjust the trace brightness.
2. Set the AC/DC control switch to AC.
3. Set the volts/division control to display the voltage level under test. Set the control to the highest value if the voltage level is unknown.
4. Connect the scope probe to the AC voltage under test.
5. Adjust the volts/division control to display the vertical amplitude of the waveform under test.
6. Set the time/division control to display approximately two cycles of the waveform under test.
7. Set the vertical control so that the center of the waveform is on the center line of the scope screen.
8. Set the horizontal control so that the start of one cycle of the waveform begins at the vertical center line on the scope screen.
9. Measure the number of divisions between the start point and end point of one cycle.

To determine frequency, multiply the number of divisions by the time/division setting. This value is the time period for one cycle. To determine the frequency of the waveform, divide the time period by 1.

SCOPE MEASUREMENT– FREQUENCY

Example: Calculating Frequency

Calculate the frequency if one cycle of a waveform occupies 10 divisions and the time/division setting is 2 μs (microseconds).

1. Calculate time period.
   $T = 10 \times .000002$
   $T = .00002$ s

2. Calculate frequency.
   $f = \dfrac{1}{T}$
   $f = \dfrac{1}{.00002}$
   $f = 50,000$ Hz

Scope Measurement — DC Voltage

A test probe is connected to the point in the circuit where the DC voltage is to be measured. The ground lead of the scope is connected to the ground of the circuit. The voltage is positive if the trace moves above the center line. The voltage is negative if the trace moves below the center line. See Scope Measurement — DC Voltage.

SCOPE MEASUREMENT – DC VOLTAGE

To use a scope to measure DC voltage, apply the procedure:
1. Turn the power switch ON and adjust the trace brightness.
2. Set the AC/DC control switch to DC.
3. Set the volts/division control to display the voltage level under test. Set the control to the highest value if the voltage level is unknown.
4. Connect the scope probe to the ground point of the circuit under test.
5. Set the vertical control so that the displayed line is in the center of the screen. The displayed line represents 0 VDC.
6. Remove the scope probe from ground point and connect it to the DC voltage under test. The displayed voltage moves above or below the scope centerline depending on the polarity of the DC voltage under test.
7. Measure the vertical amplitude of the voltage from the centerline by counting the number of divisions from the centerline.

Multiply the number of displayed divisions by the volts/division setting to determine the DC voltage. For example, if a waveform is 3 divisions above the centerline and the volts/division control is set at 5 V, the voltage equals 15 VDC ($3 \times 5 = 15$ V).

SPECIAL METERS

Name _____ Date _____

Activities

◯ Activity 6-1. Contact Thermometers

Relative humidity may be measured using a contact thermometer by measuring the dry bulb and wet bulb temperatures of the air. *Dry bulb temperature* is the temperature of the air taken with a dry temperature probe. *Wet bulb temperature* is the temperature of the air taken with a wet sock covering the temperature probe. Dry bulb and wet bulb temperatures are plotted on a psychrometric chart to determine relative humidity. A *psychrometric chart* is a graph that is used to find various properties of air. See Psychrometric Charts.

The point where the dry bulb and wet bulb temperatures intersect is the relative humidity of the air. For example, if the dry bulb temperature is 75.1°F and the wet bulb temperature is 65.2°F, the relative humidity is 60%.

Using the test results and the Psychrometric Chart, determine the relative humidity.

PSYCHROMETRIC CHARTS

128 TROUBLESHOOTING ELECTRICAL/ELECTRONIC SYSTEMS

1. Relative humidity = _____ %
   DRY BULB TEMPERATURE: 70.2°F
   WET BULB TEMPERATURE: 65.1°F

2. Relative humidity = _____ %
   DRY BULB TEMPERATURE: 90.1°F
   WET BULB TEMPERATURE: 65.0°F

3. Relative humidity = _____ %
   DRY BULB TEMPERATURE: 49.9°F
   WET BULB TEMPERATURE: 39.8°F

4. Relative humidity = _____ %
   DRY BULB TEMPERATURE: 94.3°F
   WET BULB TEMPERATURE: 59.5°F

◯ Activity 6-2. Infrared Meters

An infrared meter may be used to check motor bearing temperature. When motor bearing temperature exceeds 100°C, the bearing hardness decreases, bearing life is reduced, and the allowable load is reduced. The load on the bearings should be lessened to reduce the temperature of the bearings. For example, the load on a motor with a bearing temperature of 210°C should be reduced to 70%.

Using the Bearing Temperature Derating chart, determine the percent that the load should be reduced to lower bearing temperature to an acceptable level.

BEARING TEMPERATURE DERATING

1. The load should be reduced to _____ %.

2. The load should be reduced to _____ %.

Activity 6-3. Contact Tachometers

Slip is the difference between the synchronous speed and actual speed of a motor. Normal rated slip is the difference between the synchronous speed and nameplate speed. Normal rated slip is approximately 4%. For example, if the synchronous speed of a motor is 1800 rpm and the nameplate speed is 1725 rpm, the normal rated slip is 75 rpm (1800 − 1725 = 75 rpm). Measured slip is the difference between the synchronous speed and measured speed.

Percent normal rated slip is found by dividing the normal rated slip by the synchronous speed and multiplying by 100. For example, if the normal rated slip equals 75 rpm and synchronous speed equals 1800 rpm, the percent normal rated slip equals 4% ($\frac{75}{1800} \times 100 = 4\%$). Higher slip indicates a problem.

Slip should not exceed 20% more than the normal rated slip in general applications. For example, the normal rated slip of a motor is 75 rpm (4% of synchronous speed). The actual (measured) motor speed should not drop an additional 15 rpm (75 rpm × 20% = 15 rpm). Determine the normal rated slip and measured slip.

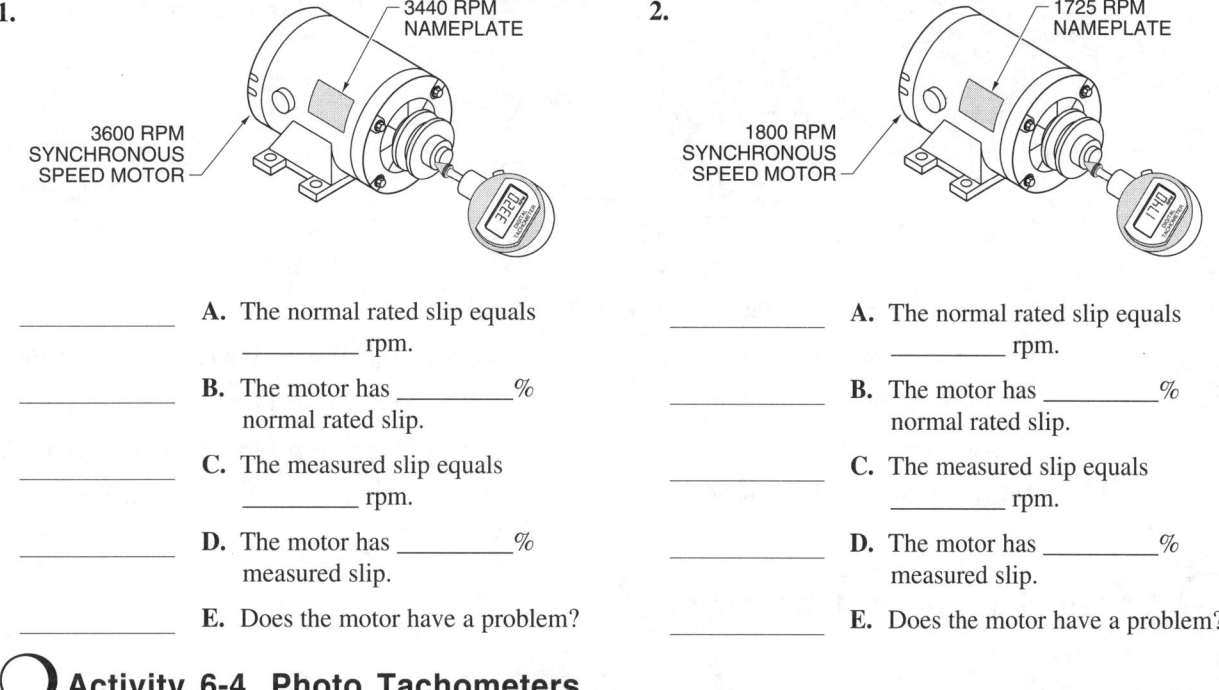

1.
 A. The normal rated slip equals _____ rpm.
 B. The motor has _____% normal rated slip.
 C. The measured slip equals _____ rpm.
 D. The motor has _____% measured slip.
 E. Does the motor have a problem?

2.
 A. The normal rated slip equals _____ rpm.
 B. The motor has _____% normal rated slip.
 C. The measured slip equals _____ rpm.
 D. The motor has _____% measured slip.
 E. Does the motor have a problem?

Activity 6-4. Photo Tachometers

A photo tachometer may be used to measure the speed of a pump. Pumps are rated by their maximum operating pressure and output in gallons per minute (gpm) at a given speed. The output of a pump decreases as the pump wears. The speed of a pump and the pump's output are measured when troubleshooting a pump. A pump's output is proportional to its driven speed. Most manufacturers provide a table or graph showing a pump's output at different pressures and speeds.

The pump's rated output may be used to calculate the pump flow rate. The more the pump is worn, the less the output is at higher speeds. To calculate the flow rate for a given drive speed, apply the formula:

$$Q = \frac{N \times V_d}{231}$$

where
Q = flow rate (in gpm)
N = pump drive speed (in rpm)
V_d = pump displacement (in cu in./rev)
231 = constant

1. Calculate the output of Pump 1.
 A. The output of Pump 1 is _____ gpm when the photo tachometer displays 1800 rpm.
 B. The output of Pump 1 is _____ gpm when the photo tachometer displays 1250 rpm.

_____ C. The output of Pump 1 is _____ gpm when the photo tachometer displays 900 rpm.

_____ D. The output of Pump 1 is _____ gpm when the photo tachometer displays 1640 rpm.

PUMP 1

The speed of a pump may be set to deliver the proper flow rate when a given pump output is required. To calculate the required drive speed for a given flow rate, apply the formula:

$$N = \frac{Q \times 231}{V_d}$$

where
N = pump drive speed (in rpm)
Q = pump flow rate (in gpm)
V_d = rated displacement of pump (in cu in./rev)
231 = constant

PUMP 2

2. Calculate the required pump speed of Pump 2.

_____ A. The tachometer should display _____ rpm when a 20 gpm output from Pump 2 is required.
_____ B. The tachometer should display _____ rpm when a 35 gpm output from Pump 2 is required.
_____ C. The tachometer should display _____ rpm when a 28 gpm output from Pump 2 is required.
_____ D. The tachometer should display _____ rpm when a 45 gpm output from Pump 2 is required.

Activity 6-5. Strobe Tachometers

Proximity switches must be able to activate at a speed faster than the speed of the objects moving down the conveyor to properly detect the objects. A strobe tachometer may be used to measure the speed of moving objects to determine if a given switch may be used. Tachometer readings must be converted from feet per minute (fpm) to operations per second (ops). To convert fpm to ops, apply the formula:

$$ops = \frac{npf \times V}{60}$$

where
ops = operations per second
npf = objects per foot
V = object speed (in fpm)
60 = constant

PROXIMITY SWITCH RATINGS					
Number	Type	Voltage	Sensing Distance (in mm)	Operations Per Second (max)	Cost ($)
M01	Mechanical	DC	100	5	25
M02	Mechanical	AC	100	4	25
S01	Solid-state	AC	75	15	20
S02	Solid-state	DC	50	150	25
S03	Solid-state	DC	50	300	27
S04	Solid-state	DC	25	500	27
S05	Solid-state	DC	25	700	29
S06	Solid-state	DC	15	1000	32
S07	Solid-state	DC	10	1500	35
S08	Solid-state	DC	5	2000	38
S09	Solid-state	DC	2	3000	40

Using the Proximity Switch Ratings, determine the least expensive proximity switch that may be used to measure the speed of the products moving down the conveyor.

_____ 1. Proximity switch number equals _____.

_____ 2. Proximity switch number equals _____.

Activity 6-6. Megohmmeters

Answer the questions using the megohmmeter readings.

_____ 1. Leakage current equals _____ mA.

_____ 2. Is the insulation resistance within the manufacturer's recommendations?

132 TROUBLESHOOTING ELECTRICAL/ELECTRONIC SYSTEMS

3. Is the leakage current enough to produce a shock that can be felt?

EFFECT OF ELECTRIC CURRENT

⏻ Activity 6-7. Digital Logic Probes

1. Two 115 V loads are controlled by two pushbuttons in a digital circuit. Using Test Circuit 1, complete the Truth Table for the switch positions.

TRUTH TABLE			
A	**B**	**Load 1**	**Load 2**
O	O	ON	OFF
O	X		
X	O		
X	X		

X = switch closed
O = switch open

TEST CIRCUIT 1

2. Digital logic probes and voltmeters are added to Test Circuit 1. Using the meter indications, determine the problem. Select the answer from the following list.

1 = No 115 VAC supply voltage in system
2 = Problem with Load 1
3 = Problem with Load 2
4 = Problem with Solid-state Relay 1
5 = Problem with Solid-state Relay 2
6 = Problem with Inverter Gate
7 = Problem with OR Gate
8 = Problem with Switch A
9 = Problem with Switch B

TEST CIRCUIT 1

	Switch Positions		Meter Indication								Load	
			Logic Probe					Voltmeter				
	A	B	1	2	3	4	5	1	2	3	1	2
A.	Open	Open	High	High	High	High	High	115 V	115 V	115 V	ON	ON
B.	Closed	Closed	Low	Low	High	Low	Low	115 V	0 V	115 V	OFF	OFF
C.	Closed	Closed	Low	High	Low	High	Low	115 V	115 V	0 V	ON	OFF
D.	Open	Closed	High	High	Low	High	High	115 V	115 V	0 V	ON	OFF
E.	Open	Open	High	High	Low	High	High	115 V	0 V	0 V	OFF	OFF
F.	Closed	Closed	Low	Low	High	Low	Low	0 V	0 V	0 V	OFF	OFF
G.	Closed	Open	Low	Low	High	Low	High	115 V	0 V	0 V	OFF	OFF
H.	Open	Open	High	High	Low	High	High	115 V	115V	0 V	OFF	OFF

_____ **A.** The problem equals __6__.   _____ **E.** The problem equals __4__.

_____ **B.** The problem equals __3__.   _____ **F.** The problem equals __1__.

_____ **C.** The problem equals __7__.   _____ **G.** The problem equals __7__.

_____ **D.** The problem equals __9__.   _____ **H.** The problem equals __2__.

134 TROUBLESHOOTING ELECTRICAL/ELECTRONIC SYSTEMS

Activity 6-8. Oscilloscopes/Scopemeters

1. Determine the peak-to-peak voltage, rms voltage, and frequency.

   _____ A. The peak-to-peak voltage equals _____ V.

   _____ B. The rms voltage equals _____ V.

   _____ C. The frequency equals _____ Hz.

2. Determine the voltage.

   _____ A. The voltage equals _____ V.

   _____ B. The voltage equals _____ V.

3. Determine the minimum and maximum peak voltage.

   _____ A. The minimum peak voltage equals _____ V.

   _____ B. The maximum peak voltage equals _____ V.

   _____ C. The minimum peak voltage equals _____ V.

   _____ D. The maximum peak voltage equals _____ V.

BASIC CIRCUIT MEASUREMENTS — Applications

Application — Branch Circuit Voltage Drop Measurement

Branch circuit conductors must be properly sized for proper and safe equipment operation. Conductors are sized large enough to prevent no more than a 3% voltage drop at the furthest outlet. Voltage is measured under no-load and full-load conditions to test for excessive voltage drop.

To measure voltage drop in a branch circuit, apply the procedure:
1. Turn all loads connected to the branch circuit OFF.
2. Measure the voltage at the furthest outlet on the branch circuit to obtain the circuit's no-load voltage.
3. Turn all loads connected to the branch circuit ON.
4. Measure the voltage at the furthest outlet on the branch circuit to obtain the circuit's full-load voltage.

The percentage of voltage drop in the branch circuit is calculated by applying the formula:

$$\%V_D = \frac{V_{NL} - V_{FL}}{V_{FL}} \times 100$$

where
$\%V_D$ = percent voltage drop (in volts)
V_{NL} = no-load voltage drop (in volts)
V_{FL} = full-load voltage drop (in volts)
100 = constant

Example: Calculating Percent Voltage Drop — Branch Circuit

A branch circuit has a measured full-load voltage of 117 V and a measured no-load voltage of 120 V. Calculate the percent voltage drop. See Branch Circuit.

$$\%V_D = \frac{V_{NL} - V_{FL}}{V_{FL}} \times 100$$

$$\%V_D = \frac{120 - 117}{120} \times 100$$

$$\%V_D = \frac{3}{120} \times 100$$

$$\%V_D = .025 \times 100$$

$$\%V_D = \mathbf{2.5\%}$$

BRANCH CIRCUIT

Application — Recording Minimum and Maximum Readings

Intermittent problems are found by recording a circuit's minimum and maximum readings. A meter must have a minimum/maximum (min/max) function to record readings over time. Most meters that have a min/max function hold any change that lasts more than one second. See Recording Minimum and Maximum Readings.

The meter test leads are connected to a circuit such as a voltmeter when measuring voltage or an ammeter when measuring current. The min/max button is pressed after the meter is connected to the circuit. The minimum or maximum value is read by pressing the min/max button anytime after the meter is connected to the circuit.

Min or max is displayed indicating which reading is being displayed. Most meters have a beeper that sounds each time a new minimum or maximum value is recorded. A continuous sound usually indicates the meter is overloaded.

Low voltage problems caused by large motor start-ups high voltage surges, or temporary voltage losses are detected when the meter is connected to a circuit such as a voltmeter and the min/max button is pressed.

The maximum current used in a circuit is detected when the meter is connected to a circuit such as an ammeter and the min/max button is pressed. This information is used to determine if a circuit is overloaded or if there is room for additional loads.

RECORDING MINIMUM AND MAXIMUM READINGS

Application — Meter Loading Effect

A meter connected to an electrical circuit causes the current and voltage values in the circuit to change. This occurs because the meter becomes an additional load in the circuit. *Meter loading effect* is the additional resistance a meter adds to the total resistance in a circuit. The meter is equivalent to adding a high resistance load in parallel with the load under test. A meter should load a circuit under test as little as possible for accurate readings. The better the meter, the less loading effect produced.

Loading effect is a problem when measuring voltages in circuits that include high resistances. For example, in a circuit that has two 200 kΩ resistors connected in series with a 120 V power supply, the voltage drop across each resistor is 60 V. The current through each resistor is .0003 A (300 μA). See Voltmeter Loading Effect.

VOLTMETER LOADING EFFECT

A meter with a 100 kΩ resistance significantly loads the circuit if used to measure the voltage drop across a resistor. The meter loads the circuit to the extent that the voltage drop across the measured resistor changes to 30 V. This change causes a 50% meter error. The current in the circuit also increases. A meter with at least 10 times the input resistance should be used in a high resistance circuit to prevent a large meter error.

The loading effect a meter has on a circuit depends on the input resistance of the meter. The higher the input resistance, the less the loading effect. Analog meters are rated at the amount of ohms per volt (Ω/V) resistance they have. For example, a meter may be rated at 20,000 Ω/V. This means that for every volt of a given range the voltmeter has 20 kΩ of resistance. The total meter loading effect is 200,000 Ω if a 20,000 Ω/V voltmeter is set on the 10 V range.

AC analog meters generally have a higher loading effect than DC meters because AC meters have less input resistance than DC meters. A typical analog meter has 5000 Ω/V for AC and 20,000 Ω/V for DC.

Analog meters generally have a greater loading effect than digital meters because digital meters have a high input resistance (typically 10 MΩ). A 10 MΩ input resistance usually does not cause a significant loading effect. Some analog voltmeters use a high resistance amplifier circuit to develop a high input resistance. This input circuit uses field effect transistors (FETs) for high resistance. Voltmeters that use FETs typically have a 10 MΩ input resistance regardless of the range on which the meter is set.

A meter with a resistance at least ten times greater than the resistance of the component under test does not significantly load the circuit. The higher the resistance of the meter, the more accurate the reading is due to less loading effect.

Application — Voltmeter Circuit Connections

Voltmeter Load Connections

Voltmeters are normally connected across (in parallel) loads such as lights, solenoids, motors, heating elements, and other devices. A voltmeter is connected across a load to determine if voltage is present and at the correct level.

When a voltmeter is connected across a load, the meter reads 0 V when the control switch in the circuit is open. The voltmeter reads 0 V if the load is good or has a fault because no current flows to the load. See Voltmeter Load Connection.

VOLTMETER LOAD CONNECTION

When the control switch is closed, the voltmeter reads the full power supply voltage (120 V), provided the supply voltage is present in the circuit. The full supply voltage is read if the load is operating or not. The meter reading indicates that the supply voltage is delivered to the load.

Voltmeter Switch Connections

A voltmeter may be connected across a control switch that controls a load. When a voltmeter is connected across an open control switch, the meter reads a voltage level provided the load does not have an open circuit. The meter is equivalent to adding a high resistance load in series with the load under test. See Voltmeter Switch Connection.

VOLTMETER SWITCH CONNECTION

The voltmeter reads a voltage level because the meter completes the path from the power lines. The voltage level the meter reads depends on the resistance of the load and the resistance of the meter.

A meter normally has a large resistance in relation to the resistance of the load. A meter normally reads almost the full supply voltage. The higher the meter resistance in relation to the load resistance, the higher the meter voltage reading. If a load has an open (incomplete circuit) the voltmeter reads 0 V when the switch is open.

When a voltmeter is connected across a closed control switch, the meter reads 0 V if the load is good, faulty, or has an open. The voltmeter reads 0 V because the closed control switch shorts the meter leads. No voltage reading is indicated on the meter.

Application — Harmonic Distortion

Harmonic distortion is the voltage and/or current in a power line that is a multiple of the fundamental line frequency. For example, the fundamental line frequency in the U.S. is 60 Hz. The 60 Hz fundamental line frequency may have harmonic distortion comprised of 120 Hz (second harmonic) and 180 Hz (third harmonic) frequencies. See Fundamental and Harmonic Waveforms.

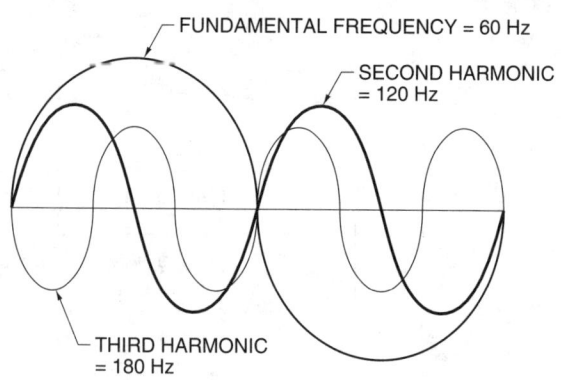

Note: Harmonic frequency does not have to be in-phase with the fundamental frequency

FUNDAMENTAL AND HARMONIC WAVEFORMS

Harmonic distortion is created by electronic circuits such as variable speed motor drives, electronic ballasts (used in lighting circuits), personal computers, printers, and medical test equipment that draw current from power lines in short pulses. Harmonic distortion is created on power lines as a result of the pulsed power draw. Equipment efficiency is improved when electronic circuits draw current in short pulses.

Harmonic distortion is created when diodes are used to rectify AC power and charge capacitors with the rectified AC power. The result is that the capacitors draw current only during the peak of the AC sine wave. A normal power line voltage appears on a scope as a near perfect sine wave. Harmonic distortion distorts the waveform. See Distorted Waveforms.

PURE SINE WAVE

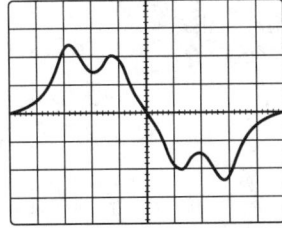
DISTORTED CURRENT WAVEFORM CAUSED BY HARMONIC DISTORTION

DISTORTED VOLTAGE WAVEFORM CAUSED BY HARMONIC DISTORTION

DISTORTED WAVEFORMS

Harmonic distortion overheats transformers and neutral conductors, burns out motors, and trips circuit breakers. Harmonic distortion in a pure 60 Hz AC power line, causes overheating by preventing the normal cancellation between conductors. Overheating may occur when voltage and current readings appear normal.

Harmonic Distortion Measurement

A harmonic analyzer measures harmonic distortion. A *harmonic analyzer* is a special meter that detects and measures harmonic distortion. An oscilloscope or scopemeter may be used to give a visual display of distortion on AC power lines. Two ammeters may be used to measure harmonic distortion when an analyzer or scope is not available. To measure harmonics with two ammeters, a meter that measures average current and one that measures true (rms) current are required. See Harmonic Distortion Measurement.

HARMONIC DISTORTION MEASUREMENT

To measure harmonic distortion using two meters, apply the procedure:
1. Measure the current in the power line with the average ammeter.
2. Measure the current in the power line with the true ammeter.

Divide the average meter reading by the true meter reading. A ratio of 1 indicates little or no harmonic distortion. A ratio less than 1 indicates harmonic distortion. A ratio of .5 indicates major harmonic distortion.

Example: Determining Harmonic Distortion

The average current measured in a circuit is 14.4 A. The rms current measured in a circuit is 15 A. Determine the amount of harmonic distortion. See Current Levels.

$$D_H = \frac{I_{avg}}{I_{rms}}$$

$$D_H = \frac{14.4}{15}$$

$$D_H = .96$$

The circuit has little harmonic distortion.

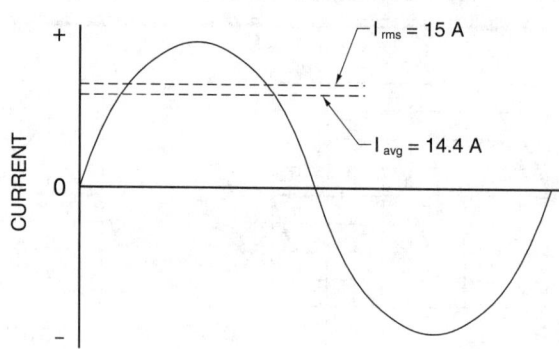

CURRENT LEVELS

The following steps may be taken to help eliminate the effect of harmonic distortion.
- Add harmonic filters to the circuit. The filters are added at the loads.
- Balance the loads in the circuit to help even the current in the lines.
- Oversize neutral conductors that are shared by hot conductors or add a neutral conductor for each hot power line.
- Derate transformers feeding harmonic loads. The amount of derating depends on the amount of harmonic distortion. The greater the harmonic distortion, the more the transformers should be derated.

Application — Troubleshooting Fuses

A *fuse* is an overcurrent protection device with a fusible link that melts and opens the circuit on an overcurrent condition. Fuses are connected in series with a circuit to protect a circuit from overcurrents or shorts. Fuses may be one-time or renewable. *One-time fuses* are fuses that cannot be reused after they have opened. One-time fuses are the most common. *Renewable fuses* are designed so that the fusible link can be replaced. A multimeter or voltmeter is used to test fuses. See Troubleshooting Fuses.

To troubleshoot fuses, apply the procedure:
1. Turn the handle of the safety switch or combination starter OFF.
2. Open the door of the safety switch or combination starter. The operating handle must be capable of opening the switch. If it is not, replace the switch.
3. Check the enclosure and interior parts for deformation, displacement of parts, and burning. Such damage may indicate a short, fire, or lightning strike. Deformation requires replacement of the part or complete device. Any indication of arcing damage or overheating, such as discoloration or melting of insulation, requires replacement of the damaged part(s).
4. Check the voltage between each pair of power leads. Incoming voltage should be within 10% of the voltage rating of the motor. A secondary problem exists if voltage is not within 10%. This secondary problem may be the reason the fuses have blown.

5. Test the enclosure for grounding if voltage is present and at the correct level. To test for grounding, connect one side of a voltmeter to an unpainted metal part of the enclosure and touch the other side to each of the incoming power leads. A voltage difference is indicated if the enclosure is grounded. The line-to-ground voltage probably does not equal the line-to-line voltage reading taken in Step 4.
6. Turn the handle of the safety switch or combination starter ON to test the fuses. One side of a voltmeter is connected to one side of an incoming power line at the top of one fuse. The other side of the voltmeter is connected to the bottom of each of the remaining fuses. A voltage reading indicates the fuse is good. If no voltage reading is obtained, the fuse is open and no voltage passes through. The fuse must be replaced (not at this time). Repeat this procedure for each fuse. When testing the last fuse, the voltmeter is moved to a second incoming power line.
7. Turn the handle of the safety switch or combination starter OFF to replace the fuses. Use a fuse puller to remove bad fuses. Replace all bad fuses with the correct type and size replacement. Close the door on the safety switch or combination starter and turn the circuit ON.

TROUBLESHOOTING FUSES

SINGLE ELEMENT FUSE OPERATION

Application — Troubleshooting Circuit Breakers

A *circuit breaker* (CB) is a reusable overcurrent protective device that opens a circuit automatically at a predetermined overcurrent. CBs are connected in series with the circuit. CBs protect a circuit from overcurrents or short circuits. CBs are thermally or magnetically operated and are reset after an overload. A multimeter is used to test CBs. Circuit breakers perform the same function as fuses and are tested the same way. See Troubleshooting Circuit Breakers.

142 TROUBLESHOOTING ELECTRICAL/ELECTRONIC SYSTEMS

TROUBLESHOOTING CIRCUIT BREAKERS

To troubleshoot CBs, apply the procedure:
1. Turn the handle of the safety switch or combination starter OFF.
2. Open the door of the safety switch or combination starter. The operating handle must be capable of opening the switch. Replace the operating handle if it does not open the switch.
3. Check the enclosure and interior parts for deformation, displacement of parts, and burning.
4. Check the voltage between each pair of power leads. Incoming voltage should be within 10% of the voltage rating of the motor.
5. Test the enclosure for grounding if voltage is present and at the correct level.
6. Examine the CB. It will be in one of three positions, ON, TRIPPED, or OFF.
7. If no evidence of damage is present, reset the CB by moving the handle to OFF and then to ON. CBs must be cooled before they are reset. CBs are designed so they cannot be held in the ON position if an overload or short is present. Check the voltage of the reset CB if resetting the CB does not restore power. Refer to Troubleshooting Fuses, Step 6. Replace all faulty CBs. Never try to service a faulty CB.

BASIC CIRCUIT MEASUREMENTS

Name _____   Date _____

Activities

⬤ Activity 7-1. Branch Circuit Voltage Drop Measurement

Determine the percentage of voltage drop.

_____ **1.** The voltage drop equals _____%.

_____ **2.** The voltage drop equals _____%.

_____ **3.** The voltage drop equals _____%.

4. The voltage drop equals _____%.

◯ Activity 7-2. Recording Minimum and Maximum Readings

Minimum and maximum readings are recorded when troubleshooting thermostat differential problems. Thermostat differential problems occur when there is too large or too small of a temperature difference between the turn ON and turn OFF points of a furnace or air conditioner.

Room temperature varies excessively and may be uncomfortable when the temperature differential is too large. The furnace or air conditioner cycles repeatedly when the temperature differential is too small. Overcycling causes motor burnout. The differential should remain between 3°F and 4°F for comfort and equipment reliability. For example, a thermostat set at 72°F turns ON the furnace when room temperature falls to 72°F, and turns OFF the furnace when the room temperature rises to 75°F. The thermostat differential is equal to 3°F.

Determine the differential temperature and the number of degrees higher or lower the thermostat differential adjustment should be set to maintain a 3°F differential.

1. The differential temperature equals _____ °F.
2. The required adjustment equals _____ °F.
3. The differential temperature equals _____ °F.
4. The required adjustment equals _____ °F.

● Activity 7-3. Meter Loading Effect

Determine the meter error caused by the meter loading effect.

1. The meter reading of Model A equals _____ V.
2. The meter error equals _____ %.

3. The meter reading of Model B equals _____ V.
4. The meter error equals _____ %.

Basic Circuit Measurements 145

146 TROUBLESHOOTING ELECTRICAL/ELECTRONIC SYSTEMS

_____ 5. The meter reading of Model C equals _____ V.
_____ 6. The meter error equals _____ %.

_____ 7. The meter reading of Model D equals _____ V.
_____ 8. The meter error equals _____ %.

◯ Activity 7-4. Voltmeter Circuit Connections

Determine the voltmeter reading(s).

_____ 1. The reading of the voltmeter equals _____ V.

_____ 2. The reading of Voltmeter 1 equals _____ V.

_____ 3. The reading of Voltmeter 2 equals _____ V.

_____ 4. The reading of Voltmeter 3 equals _____ V.

_____ 5. The reading of Voltmeter 4 equals _____ V.

_____ 6. The reading of Voltmeter 5 equals _____ V.

_____ 7. The reading of Voltmeter 6 equals _____ V.

◯ Activity 7-5. Harmonic Distortion

Using Circuits 1-4, answer the questions.

_____ 1. Circuit _____ has the greatest harmonic distortion.

_____ 2. Circuit _____ has the least harmonic distortion.

_____ 3. Circuit _____ has no harmonic distortion.

148 TROUBLESHOOTING ELECTRICAL/ELECTRONIC SYSTEMS

Activity 7-6. Troubleshooting Fuses

Using the meter readings, determine whether the fuse is good, bad, or questionable. A fuse is questionable when the meter readings do not provide enough information to determine whether the fuse is definitely good or bad.

	1. Fuse 1 is _____.		4. Fuse 1 is _____.
	2. Fuse 2 is _____.		5. Fuse 2 is _____.
	3. Fuse 3 is _____.		6. Fuse 3 is _____.

Activity 7-7. Troubleshooting Circuit Breakers

Using the meter readings, determine whether the circuit breaker is good, bad, or questionable. The circuit breaker is questionable when the meter readings do not provide enough information to determine whether the breaker is definitely good or bad.

_____ 1. The circuit breaker is _____.   _____ 2. The circuit breaker is _____.

RELAYS AND MOTOR STARTERS

Applications — Chapter 8

Application — Relays

A *relay* is an interface that controls one electrical circuit by opening and closing contacts in another circuit. The two types of relays are electromechanical and solid-state relays.

Electromechanical Relays

An *electromechanical relay* is a device that allows the connection of two different components, voltage levels, voltage types (AC/DC), or systems. An electromechanical relay is normally used to increase the number of contacts and allow a low current to switch a high current. By increasing the number of contacts, the current across an individual contact is reduced and allows the control of additional loads. See Electromechanical Relay.

ELECTROMECHANICAL RELAY

An electromechanical relay increases the number of contacts by using a coil to magnetically move several different contacts simultaneously. One coil normally controls three or more contacts.

Contact arrangements are described by their poles, throws, and breaks. *Poles* are the number of completely isolated circuits that can pass through the switch at one time. *Throws* are the number of different closed contact positions per pole that are available on the switch. *Breaks* are the number of separate contacts the switch uses to open or close each individual circuit. Abbreviations are used to simplify relay contact identification. See Relay Contact Abbreviations.

An electromechanical relay usually has multipole and multithrow contact arrangements that allow the relay to simultaneously switch several different circuits. The contacts switch to either AC or DC. See Relay Contact Arrangements.

An electromechanical relay has a low contact voltage drop; therefore, a heat sink is not required to dissipate heat produced by the switching of the contacts. A *heat sink* is a device that conducts and dissipates heat away from a component.

RELAY CONTACT ABBREVIATIONS	
Abbreviation	Meaning
SP	Single pole
DP	Double pole
ST	Single throw
DT	Double throw
NO	Normally open
NC	Normally closed
SB	Single break
DB	Double break

An electromechanical relay is relatively inexpensive, resistant to voltage transients, and does not have off-state leakage current through open contacts. *Off-state leakage current* is the amount of current that leaks through a solid-state switch when the switch is turned OFF.

Electromechanical relay contacts have a limited life which is shortened when used for rapid switching applications. An electromechanical relay has poor performance when switching high inrush currents which generate electromagnetic noise and interference on power lines.

Electromechanical Relay Life

Electromechanical relay life expectancy is rated in contact life and mechanical life. *Contact life* is the number of times a relay's contacts switch the load controlled by the relay before malfunctioning. Typical contact life ratings are 100,000 to 500,000 operations. *Mechanical life* is the number of times a relay's mechanical parts operate before malfunctioning. Typical mechanical life ratings are 1,000,000 to 10,000,000 operations.

Relay contact life expectancy is lower than mechanical life expectancy because the life of a contact depends on the application. The contact rating of a relay is based on the contact's full-rated power. Contact life is increased when contacts switch loads less than their full-rated power. Contact life is reduced when contacts switch loads that develop destructive arcs. *Arcing* is the discharge of an electric current across a gap such as when an electric switch is opened. Arcing causes contact burning and temperature rise. See Arcing.

Arcing is minimized by an arc suppressor and by using the correct contact material for the application. An *arc suppressor* is a device that dissipates the energy present across opening contacts. Arc suppression is used in applications that switch arc producing loads such as solenoids, coils, motors, and other inductive loads.

Arc suppression is also accomplished by using a contact protection circuit. A *contact protection circuit* is a circuit that protects contacts by providing a nondestructive path for generated voltage as a switch is opened. A contact protection circuit may contain a diode, a resistance/capacitance (RC) circuit, or a varistor.

A diode is used as contact protection in DC circuits. The diode does not conduct electricity when the load is energized. The diode conducts electricity and shorts the generated voltage when the switch is opened. By shorting the generated voltage, the voltage is dissipated across the diode and not the relay contacts. See Contact Protection Circuits.

An RC circuit and varistor are used as contact protection in AC circuits. The capacitor in an RC circuit is a high impedance to the 60 Hz line power and a short circuit to generated high frequencies. The short circuit dissipates generated voltage. A *varistor* is a resistor whose resistance is inversely proportional to the voltage applied to it. The varistor becomes a low impedance circuit when its rated voltage is exceeded. The low impedance circuit dissipates generated voltage when a switch is opened.

CONTACT PROTECTION CIRCUITS

Contact Material

Relay contacts are available in fine silver, silver-cadmium, gold-flashed silver, and tungsten. Fine silver has the highest electrical and thermal properties of all metals. Fine silver sticks, welds, and is subject to sulfidation when used for many applications. *Sulfidation* is the formation of film on the contact surface. Sulfidation increases the resistance of the contacts. Silver is alloyed with other metals to reduce sulfidation.

Silver is alloyed with cadmium to produce a silver-cadmium contact. Silver-cadmium contacts have good electrical characteristics and low resistance which helps the contact resist arcing but not sulfidation. Silver or silver alloy contacts are used in circuits that switch several amperes at more than 12 V, which burns off the sulfidation.

Sulfidation can damage silver contacts when used in intermittent applications. Gold-flashed silver contacts are used in intermittent applications to minimize sulfidation and provide a good electrical connection. Gold contacts are not used in high-current applications because the gold burns off quickly. Gold-flashed silver contacts are good for switching loads of one ampere or less.

Tungsten contacts are used in high-voltage applications because tungsten has a high melting temperature and is less affected by arcing. Tungsten contacts are used when high repetitive switching is required.

Contact Failure

In most applications a relay fails due to contact failure. In some low-current applications, the relay contacts may look clean but may have a thin film of sulfidation, oxidation, or contaminates on the contact surface. This film increases the resistance to the flow of current through the contact. Normal contact wiping or arcing usually removes the film. In low-power circuits this action may not take place. In most applications, contacts are oversized for maximum life. Low-power circuit contacts are not oversized to the extent that they switch just a small fraction of their rated value.

Contacts are often subject to high-current surges. High-current surges reduce contact life by accelerating sulfidation and contact burning. For example, a 100 W incandescent lamp has a current rating of about 1 A. The life of the contacts is reduced if a relay with 5 A contacts is used to switch the lamp because the lamp's filament has a low resistance when cold. When first turned ON, the lamp draws 12 or more amperes. The 5 A relay switches the lamp, but does not switch it for the rated life of the relay. Contacts are oversized in applications that have high-current surges.

Solid-State Relays

A *solid-state relay* is an electronic switching device that has no moving parts. A solid-state relay is used for applications requiring highly repetitive switching, such as a flashing light.

A solid-state relay provides faster switching, longer life, smaller size, and the ability to handle more complex functions than an electromechanical relay. A solid-state relay differs from an electromechanical relay in its operating characteristics, sensitivity, and ability to operate correctly in different environments. See Solid-State Relay.

SOLID-STATE RELAY

A solid-state relay has a long life when properly applied. A solid-state relay does not have contacts and does not produce arcing which generates electromagnetic interference. A solid-state relay is resistant to shock and vibration and has fast-switching capability. A solid-state relay has logic compatible to programmable controllers, digital circuits, and computers.

A solid-state relay normally has only one contact per relay. The voltage drop across the contact requires a heat sink. A solid-state relay can switch only AC or DC depending on the type used and has off-state leakage current when the contact is open. A solid-state relay must be the correct size and type for an application.

Solid-State Relay Temperature Problems

Temperature rise is the largest problem in applications using a solid-state relay. As temperature increases, the failure rate of a solid-state relay increases. The operation of a solid-state relay is directly affected by its operating temperature. The higher the heat in a solid-state relay, the more problems occur. See Temperature Effect On Solid-State Relays.

TEMPERATURE EFFECT ON SOLID-STATE RELAYS

The failure rate of most solid-state relays doubles for every 10°C temperature rise above an ambient temperature of 40°C. An ambient temperature of 40°C is considered standard by most manufacturers.

Solid-state relay manufacturers specify the maximum relay temperature permitted. The relay must be properly cooled to ensure that the temperature does not exceed the specified maximum safe value. Proper cooling is accomplished by installing the solid-state relay to the correct heat sink. A heat sink is chosen based on the maximum amount of load current controlled.

Heat Sinks

The performance of a solid-state relay is affected by ambient temperature. The ambient temperature of a relay is a combination of the temperature of the relay location and the type of enclosure used. The temperature inside an enclosure may be much higher than the ambient temperature of an enclosure that does not allow good air flow.

The temperature inside an enclosure increases if the enclosure is located next to a heat source or in the sun. The electronic circuit and solid-state relay also produce heat. Forced cooling is required in some applications.

Selecting Heat Sinks

A low resistance to heat flow is required to remove the heat produced by a solid-state relay. The opposition to heat flow is thermal resistance. *Thermal resistance (R_{TH}) is the ability of a device to impede the flow of heat.* Thermal resistance is a function of the surface area of a heat sink and the conduction coefficient of the heat sink material. Thermal resistance is expressed in degrees Celsius per Watt (°C/W). See Heat Sink Selections.

Heat sink manufacturers list the thermal resistance of heat sinks. The lower the thermal resistance number, the easier the heat sink dissipates heat. The larger the thermal resistance number, the less effectively the heat sink dissipates heat. The thermal resistance value of a heat sink is used with a solid-state relays Load Current/Ambient Temperature chart to determine the size of the heat sink required.

A relay can control a large amount of current when a heat sink with a low thermal resistance number is used. A relay can control the least amount of current when no heat sink (free air mounting) is used. See 40 A Relay Load Current/Ambient Temperatures.

To maximize heat conduction through a relay and into a heat sink:
- Use heat sinks made of a material that has a high thermal conductivity. Silver has the highest thermal conductivity rating. Copper has the highest practical thermal conductivity rating. Aluminum has a good thermal conductivity rating and is the most cost effective and widely used heat sink.
- Keep the thermal path as short as possible.

HEAT SINK SELECTIONS

Type	H × W × L (mm)	R_{TH} (°C/W)
01	15 × 79 × 100	2.5
02	15 × 100 × 100	2.0
03	25 × 97 × 100	1.5
04	37 × 120 × 100	.9
05	40 × 60 × 150	.5
06	40 × 200 × 150	.4

40 A RELAY LOAD CURRENT/AMBIENT TEMPERATURES

- Use the largest cross-sectional surface area in the smallest space.
- Always use thermal grease or pads between the relay housing and the heat sink to eliminate air gaps and aid in thermal conductivity.

Mounting Heat Sinks

A heat sink must be correctly mounted to assure proper heat transfer.

To properly mount a heat sink:
- Choose a smooth mounting surface. The surfaces between a heat sink and a solid-state device should be as flat and smooth as possible. Ensure mounting bolts and screws are securely tightened.
- Locate heat producing devices so that the temperature is spread over a large area. This helps prevent increased temperature areas.
- Use heat sinks with fins to maintain as large a surface area as possible.
- Ensure that the heat from one heat sink does not add to the other.

Relay Current Problems

The overcurrent passing through a solid-state relay must be kept below the maximum load current rating of the solid-state relay. An overload protection fuse is used to prevent overcurrents from damaging a solid-state relay.

An overload protection fuse opens the circuit when the current is increased to a higher value than the nominal-load current. The fuse should be an ultra-fast fuse used for the protection of semiconductors. See Overcurrent Protection.

OVERCURRENT PROTECTION

Relay Voltage Problems

Most AC power lines contain voltage spikes superimposed on the voltage sine wave. Voltage spikes are produced by switching motors, solenoids, transformers, motor starters, contactors, and other inductive loads. Large spikes are also produced by lightning striking the power distribution system.

The output element of a relay can exceed its breakdown voltage and turn ON for part of a half period if overvoltage protection is not provided. This short turn ON can cause problems in the circuit.

Varistors are added to the relay output terminals to prevent an overvoltage problem. A varistor should be rated for 10% higher than the line voltage of the output circuit. The varistor bypasses the transient current. See Overvoltage Protection.

OVERVOLTAGE PROTECTION

Voltage Drop

A voltage drop in the switching component is unavoidable in a solid-state relay. The voltage drop produces heat. The larger the current passing through the relay, the greater amount of heat produced. The generated heat affects relay operation and can destroy the relay if not removed. See Relay Voltage Drop.

The voltage drop in a solid-state relay is usually 1 V to 1.6 V, depending on the load current. For small loads (less than 1 A), the heat produced is safely dissipated through the relay's case. High-current loads require a heat sink to dissipate the extra heat. See Solid-State Relay Voltage Drop.

SOLID-STATE RELAY VOLTAGE DROP		
Load Current (in A)	Voltage Drop (in V)	Power at Switch (in W)
1	2	2
2	2	4
5	2	10
10	2	20
20	2	40
50	2	100

For example, if the load current in a circuit is 1 A and a solid-state relay switching device has a 2 V drop, the power generated in the device is 2 W. The 2 W of power generates heat that can be dissipated through the relay's case.

If the load current in a circuit is 20 A and the solid-state relay switching device has a 2 V power drop, the power generated in the device is 40 W. The 40 W of power generates heat that requires a heat sink to safely dissipate the heat.

Application — Contactors and Motor Starters

MOTOR STARTER

Contactors and motor starters are used to switch high-load currents. A *contactor* is a control device that uses a small control current to energize or de-energize the load connected to it. A *motor starter* is a contactor with overload protection added. A motor starter includes an overload relay that detects excessive current passing through a motor.

Contactors are used to switch non-motor loads such as lights, heating elements, and transformers. Motor starters are used to switch all types and sizes of motors. Contactors and motor starters control the loads connected to them either by using a magnetic coil and contacts, or by solid-state switching. Contactors and motor starters are available in sizes that can switch loads of a few amperes to several hundred amperes. See Motor Starter.

Overload Relays

An *overload relay* in a motor starter is a time-delay device that allows temporary overloads without disconnecting the load. The overload relay trips and disconnects the motor from the circuit when an overload occurs and lasts longer than the preset time.

Overload relays are electromagnetic or thermal. An *electromagnetic overload relay* is an electromechanical relay operated by the current flow in a circuit. When the level of current in the circuit reaches a preset value, the increased magnetic field opens a set of contacts.

A *thermal overload relay* is an electromechanical relay that operates by heat developed in the relay. When the level of current in a circuit reaches a preset value, the increased temperature opens a set of contacts. The increased temperature opens the contacts through a bimetallic strip or by melting an alloy that activates a mechanism that opens the contacts.

Overload relays are connected in series with a motor. The same amount of current passes through the overload relay and the motor. Overload relays are designed to operate with their contacts normally closed (NC). The contacts are connected in series with the starting coil. When the contacts open, the starting coil is de-energized and the motor is disconnected from power.

Overload Tripping

Overloads are installed in motor starters to protect the motor when running. Overloads protect a motor by disconnecting the motor from the power supply when the heat generated in the motor windings approaches a damaging level.

Motors draw 6 to 8 times their normal running current when starting. Time delay is required to permit a motor to start and allow overloads of short duration. Overloads trip when an overload exists for more than a short time. The time it takes for an overload to trip depends on the length of time the overload exists and the ambient temperature in which the overloads are located. See Motor Current Draw.

Overload relays on manual starters are reset after tripping by pressing the stop button and then the start button or by pressing the reset button. The overloads on magnetic starters set for manual reset are reset by pressing the reset button next to the overload contact. The overloads on magnetic starters set for automatic reset are reset automatically after the unit has cooled.

Trip Time

The time in which it takes an overload to trip depends on the length of time the overload current exists. A Heater Trip Characteristics chart shows the relationship between the time an overload takes to trip and the current flowing in the circuit. A Heater Trip Characteristics chart is based on the standard 40°C ambient temperature installation. The larger the overload (horizontal axis), the shorter the time required to trip the overload (vertical axis). See Heater Trip Characteristics.

Standard ambient temperature is 40°C unless otherwise stated. Any change from ambient temperature affects the tripping time of an overload. For temperatures higher than 40°C, the overloads trip at a current rating less than the value of the overload. For example, at 50°C the overloads trip at 90% of their rated value. For temperatures lower than 40°C, the overloads trip at a current rating greater than the rated value of the overload. See Heater Ambient Temperature Correction.

MOTOR CURRENT DRAW

HEATER TRIP CHARACTERISTICS

HEATER AMBIENT TEMPERATURE CORRECTION

Selecting Overloads

The ambient temperature in which a starter and motor is located must be considered when selecting overloads because a high ambient temperature reduces overload trip time. Reduced overload trip time can lead to nuisance tripping if a motor is located in a cooler ambient temperature than the starter. Reduced overload trip time can lead to motor burnout when the motor is located in a hotter ambient temperature than the starter.

An overload heater with a full-load current closest to the full-load current value listed on the motor nameplate is selected when the ambient temperature is the same at the motor and the starter. This provides the motor with overload protection between 110% and 120% (115% is standard for most manufacturers) of the motor's full-load current.

A higher overload heater value is selected when the ambient temperature at the starter is higher than the temperature at the motor. A lower overload heater value is selected when the ambient temperature at the starter is lower than the temperature at the motor.

RELAYS AND MOTOR STARTERS

8

Name _____ Date _____

Activities

◐ Activity 8-1. Relays

1. An electromechanical relay is used to control a solenoid-operated valve. The valve opens when the solenoid is energized and closes when the solenoid is de-energized. A new relay must be correctly wired and the contact pin numbers identified when replacing a faulty relay with a new one. Identify each set of relay contact pin numbers.

The temperature of a relay coil increases when the relay coil is energized, the relay is placed in a high ambient temperature, or the relay is turned ON and OFF repeatedly. The power draw of the coil increases as the temperature of the relay coil increases. Manufacturers rate coil power according to the coil's temperature. Using the coil power graph, determine the amount of coil power the wattmeter reads if the coil is good.

_____ 2. The wattmeter reads _____ W.

159

3. A 40 A solid-state relay is used in a control circuit. Using Heat Sink Selections and 40 A Relay Load Current/Ambient Temperatures, determine the minimum heat sink type (01, 02, etc.) required to dissipate the heat from the relay.

HEAT SINK SELECTIONS

Type	H × W × L (mm)	R_{TH} (°C/W)
01	15 × 97 × 100	2.5
02	15 × 100 × 100	2.0
03	25 × 97 × 100	1.5
04	37 × 120 × 100	.9
05	40 × 160 × 150	.5
06	40 × 200 × 150	.4

40 A RELAY LOAD CURRENT/AMBIENT TEMPERATURES

_____ A. A type _____ heat sink is required to switch a 15 A load.

_____ B. A type _____ heat sink is required to switch a 30 A load.

_____ C. A type _____ heat sink is required to switch a 27 A load.

_____ D. A type _____ heat sink is required to switch a 20 A load.

_____ **E.** A type _____ heat sink is required to switch a 15 A load.

_____ **F.** A type _____ heat sink is required to switch a 10 A load.

4. Determine the power produced at each solid-state switch.

_____ **A.** _____ W of power is produced at the solid-state switch.

_____ **B.** _____ W of power is produced at the solid-state switch.

162 TROUBLESHOOTING ELECTRICAL/ELECTRONIC SYSTEMS

_____ **C.** _____ W of power is produced at the solid-state switch.

_____ **D.** _____ W of power is produced at the solid-state switch.

_____ **E.** _____ W of power is produced at the solid-state switch.

Activity 8-2. Contactors and Motor Starters

1. Using Heater Trip Characteristics, determine the approximate time (in seconds) that it takes the overloads to trip.

_____ **A.** Overload trip time is _____ seconds.

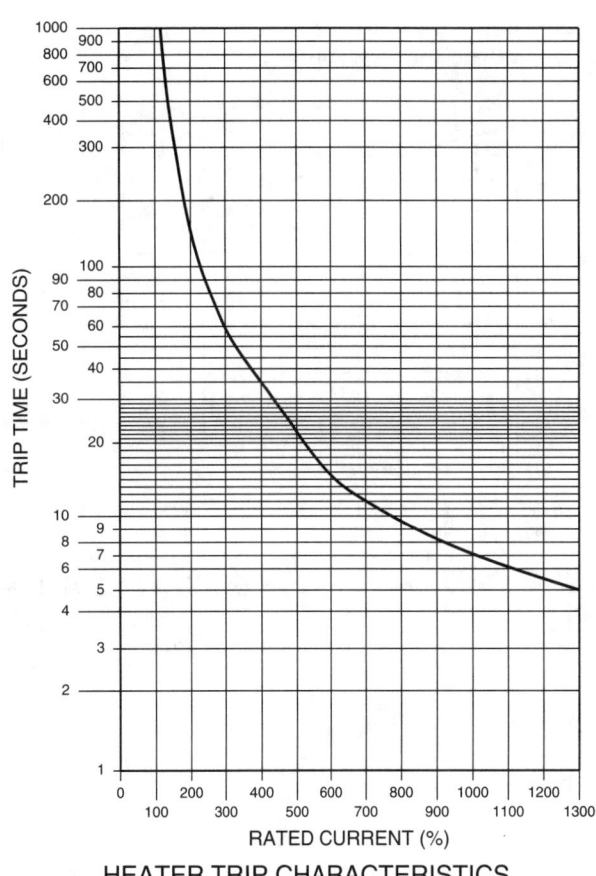

HEATER TRIP CHARACTERISTICS

_____ **B.** Overload trip time is _____ seconds.

_____ **C.** Overload trip time is _____ seconds.

164 TROUBLESHOOTING ELECTRICAL/ELECTRONIC SYSTEMS

2. Using Heater Ambient Temperature Correction, determine the corrected current value used to select an overload for each application.

HEATER AMBIENT TEMPERATURE CORRECTION

_____ **A.** The corrected current value is _____ A.

_____ **B.** The corrected current value is _____ A.

_____ **C.** The corrected current value is _____ A.

_____ **D.** The corrected current value is _____ A.

TROUBLESHOOTING RELAYS AND STARTERS

Applications

Application — Troubleshooting Electromechanical Relays

When troubleshooting an electromechanical relay, the input and output of the relay are checked to determine if the circuit on the input side of the relay is the problem, if the circuit on the output side of the relay is the problem, or if the relay itself is the problem. The relay coil and contacts are checked to determine if the relay is the problem. The correct voltage must be applied to the relay's coil before it energizes. The relay contacts are checked by energizing and de-energizing the coil. The contacts should have little to no voltage drop across them when closed. The contacts should have nearly full voltage across them when open.

Check for contact sticking or binding if the relay is not functioning properly. Tighten any loose parts. Replace any broken, bent, or badly worn parts. Check all contacts for signs of excessive wear and dirt buildup. Contacts are not harmed by discoloration or slight pitting. Vacuum or wipe contacts with a soft cloth to remove dirt. Never use a contact cleaner on relay contacts. Contacts require replacement when the silver surface has become badly worn. Replace all contacts when severe contact wear is evident on any contact. Replacing all contacts prevents uneven and unequal contact closing. Never file a contact.

Relay coils should be free of cracks and burn marks. Replace the coil if there is any evidence of overheating, cracking, melting, or burning. Check the coil terminals for the correct voltage level. Overvoltage or undervoltage conditions of more than 10% should be corrected. Use only replacement parts recommended by the manufacturer when replacing parts of a relay. Using non-approved parts can void any manufacturer warranty and may transfer product liability from the manufacturer. Relays are tested by manual operation and by using a multimeter.

Manual Relay Operation

Most relays can be manually operated. Manually operating a relay determines whether the circuit that the relay is controlling (output side) is working correctly. A relay is manually operated by pressing down at a designated area on the relay. This closes the relay contacts. Electromechanical relays may include a push-to-test button. See Manual Relay Operation.

When manually operating relay contacts, the circuit controlling the coil is bypassed. Troubleshoot from the relay through the control circuit when the load controlled by the relay operates manually. Troubleshoot the circuit that the relay is controlling if the load controlled by the relay does not operate when the relay is manually operated.

MANUAL RELAY OPERATION

Multimeter Test

A multimeter is also used to test an electromechanical relay. A multimeter is connected across the input and output side of a relay. Troubleshoot from the input of the relay through the control circuit when no voltage is present at the input side of the relay. The relay is the problem if the relay is not delivering the correct voltage.

Troubleshoot from the output of the relay through the power circuit when the relay is delivering the correct voltage. The supply voltage measured across an open contact indicates that the multimeter is completing the circuit across the contact and to the load. The contacts are not closing and the relay is defective if the voltage measured across the contact remains zero (or very low) when the coil is energized and de-energized. The contacts are welded closed and the relay is defective if the voltage measured across the contacts remains at full voltage when the coil is energized and de-energized. See Troubleshooting Electromechanical Relays.

TROUBLESHOOTING ELECTROMECHANICAL RELAYS

To troubleshoot an electromechanical relay, apply the procedure:
1. Measure the voltage in the circuit containing the control relay coil. The voltage should be within 10% of the voltage rating of the coil. The relay coil cannot energize if the voltage is not present. The coil may not energize properly if the voltage is not at the correct level. Troubleshoot the power supply when the voltage level is incorrect.
2. Measure the voltage across the control relay coil. The voltage across the coil should be within 10% of the coil's rating. Troubleshoot the switch controlling power to the coil when the voltage level is incorrect.
3. Measure the voltage in the circuit containing the control relay contacts. The voltage should be within 10% of the rating of the load. Troubleshoot the power supply if the voltage level is incorrect.
4. Measure the voltage across the control relay contacts. The voltage across the contacts should be less than 1 V when the contacts are closed and nearly equal to the supply voltage when open. The contacts have too much resistance and are in need of service if the voltage is more than 1 V when the contacts are closed. Troubleshoot the load when the voltage is correct at the contacts and the circuit does not work.

Application — Troubleshooting Solid-State Relays

Solid-state relays require periodic inspection. Dirt, burning, or cracking should not be present on a solid-state relay. Printed circuit (PC) boards should be properly seated. Ensure that the board locking tabs are in place if used. Consider adding locking tabs if a PC board without locking tabs loosens. Check to ensure that any cooling provisions are working and free of obstructions.

Troubleshooting a solid-state relay is accomplished by either the exact replacement method or the circuit analysis method. The *exact replacement method* replaces a bad relay with a relay of the same type and size. The exact replacement method involves making a quick check of the relay's input and output voltages. The relay is assumed to be the problem and is replaced when there is only an input voltage (no output voltage) being switched.

The *circuit analysis method* uses a logical sequence to determine the reason for the failure. Steps are taken to prevent the problem from recurring once the reason for a failure is known. The circuit analysis method of troubleshooting is based on three improper relay operations, which are:
- The relay fails to turn OFF the load
- The relay fails to turn ON the load
- Erratic relay operation

Relay Fails to Turn OFF Load

A relay may not turn OFF the load it is connected to when a relay fails. This condition occurs either when the load is drawing more current than the relay can withstand, the relay's heat sink is too small, or transient voltages are causing a breakover of the relay's output.

Transient voltages are temporary, unwanted voltages in an electrical circuit. Overcurrent permanently shorts the relay's switching device if the load draws more current than the rating of the relay. High temperature causes thermal runaway of the relay's switching device if the heat sink does not remove the heat. Replace the relay with one of a higher voltage rating and/or add a transient suppression device to the circuit if the power lines are likely to have transients (usually from inductive loads connected on the same line). See Relay Fails to Turn OFF Load.

RELAY FAILS TO TURN OFF LOAD

To troubleshoot a solid-state relay that fails to turn OFF a load, apply the procedure:
1. Disconnect the input leads from the solid-state relay. See Step 3 if the relay load turns OFF. The relay is the problem if the load remains ON and the relay is normally open.
2. Measure the voltage of the circuit that the relay is controlling. The line voltage should not be higher than the rated voltage of the relay. Replace the relay with a relay that has a higher voltage rating if the line voltage is higher than the relay's rating. Check to ensure that the relay is rated for the type of line voltage (AC or DC) being used.
3. Measure the current drawn by the load. The current draw must not exceed the relay's rating. For most applications, the current draw should not be more than 75% of the relay's maximum rating.
4. Reconnect the input lead(s) and measure the input voltage to the relay at the time when the control circuit should turn the relay OFF. The control circuit is the problem and needs to be checked if the control voltage is present. The relay is the problem if the control voltage is removed and the load remains ON. Before changing the relay, ensure that the control voltage is not higher than the relay's rated limit when the control circuit delivers the supply voltage. Ensure that the control voltage is not higher than the relay's rated drop-out voltage when the control circuit removes the supply voltage. This condition may occur in some control circuits using solid-state switching.

Relay Fails to Turn ON Load

When a relay fails, the relay may fail to turn ON the load connected to it. This condition occurs when the relay's switching device receives a very high voltage spike or the relay's input is connected to a higher-than-rated voltage. A high voltage spike blows open the relay's switching device, preventing the load from turning ON. Excessive voltage on the relay's input side destroys the relay's electronic circuit.

Replace the relay with one that has a higher voltage and current rating and/or add a transient-voltage-suppression device to the circuit if the power lines are likely to have high voltage spikes. See Relay Fails to Turn ON Load.

RELAY FAILS TO TURN ON LOAD

To troubleshoot a solid-state relay that fails to turn ON a load, apply the procedure:
1. Measure the input voltage. The relay should be ON. Troubleshoot the circuit ahead of the relay's input if the voltage is less than the relay's rated pick-up voltage. The circuit ahead of the relay is the problem if the voltage is greater than the relay's rated pick-up voltage. The higher voltage may have destroyed the relay. The relay may be a secondary problem caused by the primary problem of excessive applied voltage. Correct the high-voltage problem before replacing the relay. The relay or output circuit is the problem if the input voltage is within the pick-up limits of the relay.
2. Measure the voltage at the output of the relay. The relay is probably the problem if the relay is not switching the voltage. See Step 3. The problem is in the output circuit if the relay is switching the voltage. Check for an open circuit in the load.
3. Insert an ammeter in series with the input leads of the relay. Measure the current. The relay should be ON. The relay input is open if no current is flowing. Replace the relay. The relay is bad if the current flow is within the relay's rating. Replace the relay. The control circuit is the problem if current is flowing but is less than that required to operate the relay.

Erratic Relay Operation

Erratic relay operation is the proper operation of a relay at times, and the improper operation of the relay at other times. Erratic relay operation is caused by mechanical problems (loose connections), electrical problems (incorrect voltage), or environmental problems (high temperature). See Troubleshooting Erratic Relay Operation.

TROUBLESHOOTING ERRATIC RELAY OPERATION

To troubleshoot erratic relay operation, apply the procedure:
1. Check all wiring and connections for proper wiring and tightness. Loose connections cause many erratic problems. No sign of burning should be present at any terminal. Burning at a terminal usually indicates a loose connection.
2. Ensure that the input control wires are not next to the output line or load wires. The noise carried on the output side may cause unwanted input signals.

170 TROUBLESHOOTING ELECTRICAL/ELECTRONIC SYSTEMS

3. The relay may be half-waving if the load is a chattering AC motor or solenoid. *Half-waving* occurs when a relay fails to turn OFF because the current and voltage in the circuit reach zero at different times. Half-waving is caused by the phase shift inherent in inductive loads. The phase shift makes it difficult for some solid-state relays to turn OFF. Connecting an RC or another snubber circuit across the output load should allow the relay to turn OFF. An *RC circuit* is a circuit in which resistance (R) and capacitance (C) are used to help filter the power in a circuit.

Application — Troubleshooting Contactors and Motor Starters

Check the tightness of all terminals and busbar connections when troubleshooting control devices. Loose connections in the power circuit of contactors and motor starters cause overheating. Overheating leads to equipment malfunction or failure. Loose connections in the control circuit cause control malfunctions. Loose connections of grounding terminals lead to electrical shock and cause electromagnetic-generated interference.

The contactor or motor starter is the first device checked when troubleshooting a circuit that does not work or has a problem. The contactor or motor starter is checked first because it is the point where the incoming power, load, and control circuit are connected. Basic voltage readings are taken at a contactor or motor starter to determine where the problem lies. Since a motor starter is a contactor with added overload protection, the same basic procedure used to troubleshoot a starter works for contactors as well.

Troubleshoot the power circuit and the control circuit if the control circuit does not correctly operate a motor. The two circuits are dependent on each other but are considered two separate circuits because they are usually at different voltage levels and always at different current levels. See Troubleshooting Motor Starters.

TROUBLESHOOTING MOTOR STARTERS

To troubleshoot a motor starter, apply the procedure:
1. Inspect the motor starter and overload assembly. Service or replace motor starters that show heat damage, arcing, or wear. Replace motor starters that show burning.
2. Reset the overload relay if there is no visual indication of damage. Replace the overload relay if there is visual indication of damage.
3. Observe the motor starter for several minutes if the motor starts after resetting the overload relay. The overload relay continues to open if an overload problem still exists.
4. Check the voltage going into the starter if resetting the overload relay does not start the motor. Check circuit voltage ahead of the starter if the voltage reading is 0 V. The voltage is acceptable if the voltage reading is within 10% of the motor's voltage rating. The voltage is unacceptable if the voltage reading is not within 10% of the voltage rating of the motor.
5. Energize the starter and check the starter contacts if the voltage into the starter is present and at the correct level. The starter contacts are good if the voltage reading is acceptable. Open the starter, turn the power OFF, and replace the contacts if there is no voltage reading.
6. Check the overload relay if voltage is coming out of the starter contacts. Turn the power OFF and replace the overload relay if the voltage reading is 0 V. The problem is downstream from the starter if the voltage reading is acceptable and the motor is not operating.

Troubleshooting Guides

A troubleshooting guide is used when troubleshooting contactors and motor starters. The guide states a problem, its possible cause(s), and corrective action(s) that may be taken. See Contactor and Motor Starter Troubleshooting Guide.

CONTACTOR AND MOTOR STARTER TROUBLESHOOTING GUIDE

Problem	Possible Cause	Corrective Action
Humming noise	Magnet pole faces misaligned	Realign. Replace magnet assembly if realignment is not possible.
	Too low voltage at coil	Measure voltage at coil. Check voltage rating of coil. Correct any voltage that is 10% less than coil rating.
	Pole face obstructed by foreign object, dirt, or rust	Remove any foreign object and clean as necessary. Never file pole faces.
Loud buzz noise	Shading coil broken	Replace coil assembly.
Controller fails to drop out	Voltage to coil not being removed	Measure voltage at coil. Trace voltage from coil to supply looking for shorted switch or contact if voltage is present.
	Worn or rusted parts causing binding	Clean rusted parts. Replace worn parts.
	Contact poles sticking	Check for burning or sticky substance on contacts. Replace burned contacts. Clean dirty contacts.
	Mechanical interlock binding	Check to ensure interlocking mechanism is free to move when power is OFF. Replace faulty interlock.
Controller fails to pull in	No coil voltage	Measure voltage at coil terminals. Trace voltage loss from coil to supply voltage if voltage is not present.
	Too low voltage	Measure voltage at coil terminals. Correct voltage level if voltage is less than 10% of rated coil voltage. Check for a voltage drop as large loads are energized.
	Coil open	Measure voltage at coil. Remove coil if voltage is present and correct but coil does not pull in. Measure coil resistance for open circuit. Replace if open.

continued

continued

CONTACTOR AND MOTOR STARTER TROUBLESHOOTING GUIDE

Problem	Possible Case	Corrective Action
	Coil shorted	Shorted coil may show signs of burning. The fuse or breakers should trip if coil is shorted. Disconnect one side of coil and reset if tripped. Remove coil and check resistance for short if protection device does not trip. Replace shorted coil. Replace any coil that is burned.
	Mechanical obstruction	Remove any obstructions.
Contacts badly burned or welded	Too high inrush current	Measure inrush current. Check load for problem if higher-than-rated load current. Change to larger controller if load current is correct but excessive for controller.
	Too fast load cycling	Change to larger controller if load cycled ON and OFF repeatedly.
	Too large overcurrent protection device	Size overcurrent protection to load and controller.
	Short circuit	Check fuses or breakers. Clear any short circuit.
	Insufficient contact pressure	Check to ensure contacts are making good connection.
Nuisance tripping	Incorrect overload size	Check size of overload against rated load current. Size up if permissible per NEC®.
	Lack of temperature compensation	Correct setting of overload if controller and load are at different ambient temperatures.
	Loose connections	Check for loose terminal connection.

TROUBLESHOOTING RELAYS AND STARTERS

9

Name _____ Date _____ **Activities**

◯ Activity 9-1. Troubleshooting Electromechanical Relays

Using the Lubricant Circuit Wiring and Line Diagram and the meter readings, answer questions 1-3.

1. A solenoid-operated valve is used to dispense lubricant. Lubricant flows to the brush when the solenoid is energized. Lubricant stops flowing when the solenoid is de-energized.

_____ A. _____ NO relay contacts are used in the circuit.

_____ B. _____ NC relay contacts are used in the circuit.

_____ C. _____ and _____ are the pin numbers of the relay coil.

_____ D. _____ and _____ are the terminal strip numbers wired to the relay coil.

_____ E. _____ and _____ are the relay pin numbers in parallel with the Open Pushbutton.

_____ F. The relay contact in parallel with the pushbutton is wired to _____ and _____ terminal strip numbers.

_____ G. _____ and _____ are the relay pin numbers that control the solenoid valve.

_____ H. _____ and _____ are the relay pin numbers that control the red light.

_____ I. _____ and _____ are the relay pin numbers that control the green light.

173

174 TROUBLESHOOTING ELECTRICAL/ELECTRONIC SYSTEMS

2. A machine operator reports that no lubricant is dispensed when the Open Pushbutton is pressed and the red light is ON. The operator also reports that the lubricant tank is full, the green light turns OFF, and the red light turns ON when the circuit is tested by pressing the Open Pushbutton.

PROBLEM 1	
Meter	Reading
1	120 V (at all times)
2	119.5 V (at all times)
3	.5 V (at all times)

_____ **A.** The problem is _____.

3. The machine operator reports that the red light does not turn ON when the green light turns OFF. The operator also reports that the lubricant system is working, the green light turns OFF, and the red light does not turn ON when the circuit is tested by pressing the Open Pushbutton.

PROBLEM 2	
Meter	Reading
1	120 V (at all times)
2	120 V (when Close Pushbutton pressed) / 0 V (when Open Pushbutton pressed)
3	120 V (when Open Pushbutton pressed) / 0 V (when Close Pushbutton pressed)

_____ **A.** The problem is _____.

◯ Activity 9-2. Troubleshooting Solid-State Relays

Using the meter readings, determine the problem. *Note:* Reading 1 is shaded.

1. Reading 1 is taken when Pushbutton 1 or Pushbutton 2 is pressed. Reading 2 is taken when neither pushbutton is pressed.

_____ **A.** The problem is _____.

2. Reading 1 is taken when Pushbutton 1 and Pushbutton 2 are pressed. Reading 2 is taken when neither pushbutton is pressed.

   _____ A. The problem is _____.

3. Reading 1 is taken when the pushbutton is not pressed. Reading 2 is taken when the pushbutton is pressed.

   _____ A. The problem is _____.

176 TROUBLESHOOTING ELECTRICAL/ELECTRONIC SYSTEMS

4. Reading 1 is taken when the pushbutton is pressed. Reading 2 is taken when the pushbutton is not pressed.

_____ **A.** The problem is _____.

5. Reading 1 is taken when Pushbutton 1 or Pushbutton 2 is pressed. Reading 2 is taken when neither pushbutton is pressed.

_____ **A.** The problem is _____.

Activity 9-3. Troubleshooting Contactors and Motor Starters

1. Connect Meter 1 so that it reads Line 1 and Line 2 voltage into the motor starter. Connect Meter 2 so that it reads Line 2 and Line 3 voltage into the motor starter. Connect Meter 3 so that it reads Line 1 voltage out of the starter's power contact. Connect Meter 4 so that it reads Line 2 voltage out of the starter's power contact. Connect Meter 5 so that it reads Line 1 voltage out of the starter's heaters. Connect Meter 6 so that it reads Line 2 and Line 3 voltage out of the starter's power contacts and heaters.

2. Connect Meter 1 so that it reads Line 1 and Line 3 voltage into the motor starter. Connect Meter 2 so that it reads Line 3 voltage out of the starter's power contact. Connect Meter 3 so that it reads Line 3 voltage out of the starter's power contact and the contactor's power contact. Connect Meter 4 so that it reads Line 3 voltage out of the starter's heater.

178 TROUBLESHOOTING ELECTRICAL/ELECTRONIC SYSTEMS

> **3.** Connect Meter 1 so that it reads the line voltage into the motor starter. Connect Meter 2 so that it reads the line voltage at the motor's starting windings when the motor is connected for the forward direction of rotation. Connect Meter 3 so that it reads the line voltage at the motor's running windings when the motor is connected for the forward direction of rotation.

> **4.** Connect Meter 1 so that it reads the line voltage into the motor starter. Connect Meter 2 so that it reads the line voltage at the motor's armature when the motor is connected for the forward direction of rotation. Connect Meter 3 so that it reads the line voltage at the motor's armature when the motor is connected for the reverse direction of rotation. *Note:* The meters must be connected for the correct polarity at the meter test leads.

MOTOR ELECTRICAL PROBLEMS

10 Applications

▢ Application — Phase Unbalance

Phase unbalance is the unbalance that occurs when power lines are out-of-phase. Phase unbalance of a 3ϕ power system occurs when 1ϕ loads are applied, causing one or two of the lines to carry more or less of the load. The loads of 3ϕ power systems are balanced by electricians during installation. An unbalance begins to occur as additional 1ϕ loads are added to the system. This unbalance causes the 3ϕ lines to move out-of-phase so the lines are no longer 120 electrical degrees apart. See Phase Unbalance.

PHASE UNBALANCE

Phase unbalance causes 3ϕ motors to run at temperatures higher than their listed ratings. The greater the phase unbalance, the greater the temperature rise. High temperatures produce insulation breakdown and other related problems.

A 3ϕ motor operating in an unbalanced circuit cannot deliver its rated horsepower. For example, a phase unbalance of 3% causes a motor to work at 90% of its rated power. This requires the motor to be derated. See Phase Unbalance Derating Factor.

PHASE UNBALANCE DERATING FACTOR

▪ Application — Voltage Unbalance

Voltage unbalance is the unbalance that occurs when the voltages at different motor terminals are not equal. One winding overheats, causing thermal deterioration of the winding if the voltage is not balanced. Voltage unbalance results in a current unbalance. Line voltage should be checked for voltage unbalance periodically and during all service calls. Whenever more than 2% voltage unbalance is measured, take the steps:
- Check the surrounding power system for excessive loads connected to one line.
- Adjust the load or motor rating by reducing the load on the motor or oversizing the motor if the voltage unbalance cannot be corrected.
- Notify the power company.

Voltage unbalance is found by applying the procedure:
1. Measure the voltage between each incoming power line. The readings are taken from L1 to L2, L1 to L3, and L2 to L3.
2. Add the voltages.
3. Divide by 3 to find the voltage average.
4. Subtract the voltage average from the voltage with the largest deviation to find the voltage deviation.
5. To find voltage unbalance, apply the formula:

$$V_u = \frac{V_d}{V_a} \times 100$$

where
V_u = voltage unbalance (%)
V_d = voltage deviation (in volts)
V_a = voltage average (in volts)
100 = constant

Example: Calculating Voltage Unbalance

Calculate the voltage unbalance of a feeder system with the following voltage readings: L1 to L2 = 442 V; L1 to L3 = 474 V; L2 to L3 = 456 V. See Measuring Voltage Unbalance.

1. Measure incoming voltage. Incoming voltage is 442 V, 474 V, and 456 V.
2. Add voltages. 442 V + 474 V + 456 V = **1372 V**
3. Find voltage average.
   $V_a = \frac{V}{3}$
   $V_a = \frac{1372}{3}$
   $V_a = $ **457 V**
4. Find voltage deviation.
   $V_d = V - V_a$
   $V_d = 474 - 457$
   $V_d = $ **17 V**
5. Find voltage unbalance.
   $V_u = \frac{V_d}{V_a} \times 100$
   $V_u = \frac{17}{457} \times 100$
   $V_u = .0372 \times 100$
   $V_u = $ **3.72%**

❶ MEASURE INCOMING VOLTAGE — DISCONNECT OFF

❷ ADD VOLTAGES
442
474
456
―――
1372 V

❸ FIND VOLTAGE AVERAGE
$V_a = \frac{V}{3}$
$V_a = \frac{1372}{3}$
$V_a = $ **457 V**

❹ FIND VOLTAGE DEVIATION
$V_d = V - V_a$
$V_d = 474 - 457$
$V_d = $ **17 V**

❺ FIND VOLTAGE UNBALANCE
$V_u = \frac{V_d}{V_a} \times 100$
$V_u = \frac{17}{457} \times 100$
$V_u = .0372 \times 100$
$V_u = $ **3.72%**

MEASURING VOLTAGE UNBALANCE

A troubleshooter can observe the blackening of one or two of the stator windings which occurs when a motor has failed due to voltage unbalance. The winding with the largest voltage unbalance is the darkest. See Voltage Unbalance Motor Damage.

Application — Single Phasing

Single phasing is the operation of a motor designed to operate on three phases which is operating on two phases because one phase is lost. Single phasing occurs when one of the 3ϕ lines leading to a 3ϕ motor does not deliver voltage to the motor. Single phasing is the maximum condition of voltage unbalance.

Single phasing occurs when one phase opens on either the primary or secondary power distribution system. This happens when one fuse blows, when there is a mechanical failure within the switching equipment, or when lightning takes out one of the lines.

Single phasing can go undetected on most systems because a 3ϕ motor running on 2ϕ continues to run in most applications. The motor usually runs until it burns out. When single phasing, the motor draws all its current from two lines.

Measuring the voltage at a motor does not usually detect a single phasing condition. The open winding in the motor generates a voltage almost equal to the phase voltage that is lost. In this case, the open winding acts as the secondary of a transformer, and the two windings connected to power act as the primary.

Single phasing is reduced by using the proper size dual-element fuse and by using the correct heater sizes. In motor circuits, or other types of circuits in which a single phasing condition cannot be allowed to exist for even a short period of time, an electronic phase-loss monitor is used to detect phase loss. When a phase loss is detected, the monitor activates a set of contacts to drop out the starter coil.

A troubleshooter can observe the severe blackening of one of the 3ϕ windings which occurs when a motor has failed due to single phasing. The coil that experienced the voltage loss indicates obvious and fast damage, which includes the blowing out of the insulation on the one winding. See Single-Phasing Motor Damage.

VOLTAGE UNBALANCE MOTOR DAMAGE

SINGLE-PHASING MOTOR DAMAGE

> Single phasing is distinguished from voltage unbalance by the severity of the damage. Voltage unbalance causes less blackening (but usually over more coils) and little or no distortion. Single phasing causes severe burning and distortion to one phase coil.

Application — Improper Phase Sequence

Improper phase sequence is the changing of the sequence of any two phases (phase reversal) in a 3φ motor control circuit. Improper phase sequence reverses the motor rotation. Reversing motor rotation can damage driven machinery or injure personnel.

Phase reversal can occur when modifications are made to a power distribution system or when maintenance is performed on electrical conductors or switching equipment. The NEC® requires phase reversal protection on all personnel transportation equipment, such as moving walkways, escalators, and ski lifts. See Improper Phase Sequence.

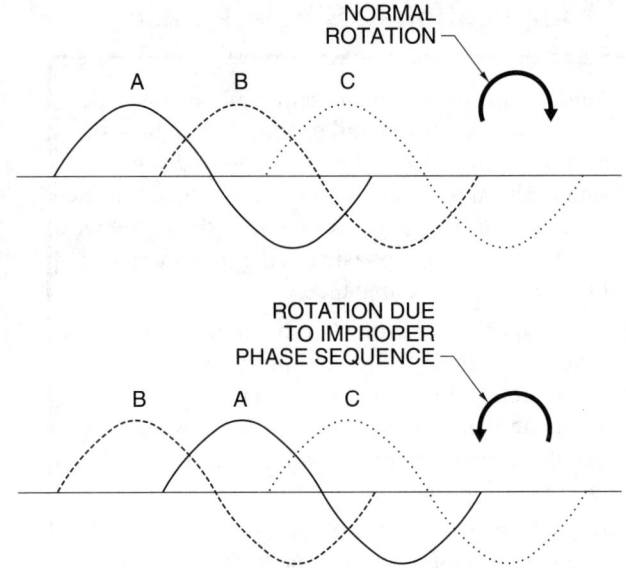

IMPROPER PHASE SEQUENCE

Application — Voltage Surges

A *voltage surge* is any higher-than-normal voltage that temporarily exists on one or more of the power lines. Lightning is a major cause of large voltage surges. A lightning surge on a power line comes from a direct hit or induced voltage. The lightning energy moves in both directions on the power lines, much like a rapidly moving wave.

This traveling surge causes a large voltage rise in an extremely short period of time. The large voltage is impressed on the first few turns of the motor windings, destroying the insulation and burning out the motor.

A troubleshooter can observe the burning and opening of the first few turns of the windings which occur when a motor has failed due to a voltage surge. The rest of the windings appear normal, with little or no damage. See Voltage Surge Motor Damage.

Lightning arresters with the proper voltage rating and connection to an excellent ground assure maximum voltage surge protection. Surge protectors are also available. Surge protectors are placed on the equipment or throughout the distribution system.

VOLTAGE SURGE MOTOR DAMAGE

Voltage surges can also occur from normal switching of high-rated power circuits. Voltage surges occurring from switching high-rated power circuits are of much less magnitude than lightning strikes and normally do not cause any problems in motors. A surge protector should be used on circuits with computer equipment to protect sensitive electronic components.

Application — AC Voltage and Frequency Variations

Voltage Variations

Motors are rated for operation at specific voltages. Motor performance is affected when the supply voltage varies from a motor's rated voltage. A motor operates satisfactorily with a voltage variation of ±10% from the voltage rating listed on the motor nameplate. See Voltage Variation Characteristics.

Frequency Variations

Motors are rated for operation at specific frequencies. Motor performance is affected when the frequency varies from a motor's rated frequency. A motor operates satisfactorily with a frequency variation of ±5% from the frequency rating listed on the motor nameplate. See Frequency Variation Characteristics.

VOLTAGE VARIATION CHARACTERISTICS

Performance Characteristics	10% above Rated Voltage	10% below Rated Voltage
Starting current	+10% to +12%	−10% to −12%
Full-load current	−7%	+11%
Motor torque	+20% to +25%	−20% to −25%
Motor efficiency	Little change	Little change
Speed	+1%	−1.5%
Temperature rise	−3°C to −4°C	+6°C to +7°C

FREQUENCY VARIATION CHARACTERISTICS

Performance Characteristics	5% above Rated Frequency	5% below Rated Frequency
Starting current	−5% to −6%	+5% to +6%
Full-load current	−1%	+1%
Motor torque	−10%	+11%
Motor efficiency	Slight increase	Slight decrease
Speed	+5%	−5%
Temperature rise	Slight decrease	Slight increase

Application — DC Voltage Variations

DC motors should be operated on pure DC power. *Pure DC power* is power obtained from a battery or DC generator. DC power is also obtained from rectified AC power. Most industrial DC motors obtain power from a rectified AC power supply. DC power obtained from a rectified AC power supply varies from almost-pure DC power to half-wave DC power.

Half-wave rectified power is obtained by placing a diode in one of the AC power lines. Full-wave rectified power is obtained by placing a bridge rectifier (four diodes) in the AC power line. Rectified DC power is filtered by connecting a capacitor in parallel with the output of the rectifier circuit. See DC Power Types.

DC motor operation is affected by a change in voltage. The change may be intentional, as in a speed-control application, or the change may be caused by variations in the power supply.

DC POWER TYPES

DC motor operation is affected by a change in voltage. The change may be intentional as in a speed-control application, or the change may be caused by variations in the power supply.

Typically, the power supply voltage should not vary by more than 10% of a motor's rated voltage. Motor speed, current, torque, and temperature are affected if the DC voltage varies from the motor rating. See DC Motor Performance Characteristics.

DC MOTOR PERFORMANCE CHARACTERISTICS

Performance Characteristics	Voltage 10% below Rated Voltage		Voltage 10% above Rated Voltage	
	Shunt	Compound	Shunt	Compound
Starting torque	−15%	−15%	+15%	+15%
Speed	−5%	−6%	+5%	+6%
Current	+12%	+12%	−8%	−8%
Field temperature	Increases	Decreases	Increases	Increases
Armature temperature	Increases	Increases	Decreases	Decreases
Commutator temperature	Increases	Increases	Decreases	Decreases

Application — Allowable Motor Starting Time

A motor must accelerate to its rated speed within a limited time period. The longer a motor takes to accelerate, the higher the temperature rise in the motor. The larger the load, the longer the acceleration time. The maximum recommended acceleration time depends on the motor's frame size. Large motor frames dissipate heat faster than small motor frames. See Maximum Acceleration Time.

Overcycling

Overcycling is the process of turning a motor ON and OFF repeatedly. Motor starting current is usually 6 to 8 times the full-load running current of the motor. Most motors are not designed to start more than 10 times per hour. Overcycling occurs when a motor is at its operating temperature and still cycles ON and OFF. This further increases the temperature of the motor, destroying the motor insulation. See Motor Overcycling.

Totally enclosed motors better withstand overcycling than open motors because they hold heat longer. When a motor application requires a motor to be cycled often, take the following steps:
- Use a motor with a 50°C rise instead of the standard 40°C.
- Use a motor with a 1.25 or 1.35 service factor instead of a 1.00 or 1.15.
- Provide additional cooling by forcing air over the motor.

MAXIMUM ACCELERATION TIME

Frame Number	Maximum Acceleration Time (in seconds)
48 and 56	8
143–286	10
324–326	12
364–505	15

MOTOR OVERCYCLING

Application — Megohmmeter Tests

A *megohmmeter* is a device that detects insulation deterioration by measuring high resistance values under high test voltage conditions. A megohmmeter detects motor insulation deterioration before a motor fails. A megohmmeter is an ohmmeter capable of measuring very high resistances by using high voltages. Typical megohmmeter test voltages range from 50 V to 5000 V. A megohmmeter is used to perform motor insulation tests to prevent electrical shock and other causes of motor insulation failure, which include excessive moisture, dirt, heat, cold, corrosive vapors or solids, vibration, and aging.

A megohmmeter measures the resistance of different windings or the resistance from a winding to ground. An ohmmeter measures the resistance of common windings and components in a motor circuit. See Megohmmeter Connections and Ohmmeter Connections.

MEGOHMMETER CONNECTIONS

OHMMETER CONNECTIONS

Several megohmmeter readings should be taken over a long period of time because the resistance of good insulation varies greatly. Megohmmeter readings are typically taken when the motor is installed and semiannually thereafter. A motor is in need of service if the megohmmeter reading is below the minimum acceptable resistance. See Recommended Minimum Resistance.

Note: A motor with good insulation may have readings of 10 to 100 times the minimum acceptable resistance. Service the motor if the resistance reading is less than the minimum value.

RECOMMENDED MINIMUM RESISTANCE*	
Minimum Acceptable Resistance	Motor Voltage Rating (from nameplate)
100,000 Ω	Less than 208
200,000 Ω	208–240
300,000 Ω	240–600
1 MΩ	600–1000
2 MΩ	1000–2400
3 MΩ	2400–5000

* values for motor windings at 40°C

Caution: A megohmmeter uses very high voltage for testing (up to 5000 V). Avoid touching the meter leads to the motor frame. Always follow the manufacturer's recommended procedures and safety rules. After performing insulation tests with a megohmmeter, connect the motor windings to ground through a 5 kΩ, 5 W resistor. The winding should be connected for 10 times the motor testing time in order to discharge the energy stored in the insulation.

186 TROUBLESHOOTING ELECTRICAL/ELECTRONIC SYSTEMS

Application — Insulation Spot Test

An *insulation spot test* is a test that checks motor insulation over the life of the motor. An insulation spot test is taken when the motor is placed in service and every six months thereafter. The test should also be taken after a motor is serviced. See Insulation Spot Test.

To perform an insulation spot test, apply the procedure:
1. Connect a megohmmeter to measure the resistance of each winding lead to ground. Record the readings after 60 seconds. Service the motor if a reading does not meet the minimum acceptable resistance. Record the lowest meter reading on an insulation spot test graph if all readings are above the minimum acceptable resistance. The lowest reading is used because a motor is only as good as its weakest point.
2. Discharge the motor windings.
3. Repeat Steps 1 and 2 every six months.

Interpret the results of the test to determine the condition of the insulation. Point A represents the motor insulation condition when the motor was placed in service. Point B represents the effects of aging, contamination, etc., on the motor insulation. Point C represents motor insulation failure. Point D represents motor insulation condition after being rewound.

INSULATION SPOT TEST

Application — Dielectric Absorption Test

A *dielectric absorption test* is a test that checks the absorption characteristics of humid or contaminated insulation. The test is performed over a 10-minute period. See Dielectric Absorption Test. To perform a dielectric absorption test, apply the procedure:
1. Connect a megohmmeter to measure the resistance of each winding lead to ground. Service the motor if a reading does not meet the minimum acceptable resistance. Record the lowest meter reading on a dielectric absorption test graph if all readings are above the minimum acceptable resistance. Record the readings every 10 seconds for the first minute and every minute thereafter for 10 minutes.
2. Discharge the motor windings.

Interpret the results of the test to determine the condition of the insulation. The slope of the curve shows the condition of the insulation. Good insulation (Curve A) shows a continual increase in resistance. Moist or cracked insulation (Curve B) shows a relatively constant resistance.

DIELECTRIC ABSORPTION TEST

A polarization index is obtained by dividing the value of the 10-minute reading by the value of the 1-minute reading. The polarization index is an indication of the condition of the insulation. A low polarization index indicates excessive moisture or contamination. See Minimum Acceptable Polarization Index Values.

For example, if the 1-minute reading of Class B insulation is 80 MΩ and the 10-minute reading is 90 MΩ, the polarization index is 1.125 (90 MΩ ÷ 80 MΩ = 1.125). The insulation contains excessive moisture or contamination.

MINMUM ACCEPTABLE POLARIZATION INDEX VALUES	
Insulation	Value
Class A	1.5
Class B	2.0
Class C	2.0

Application — Insulation Step Voltage Test

An *insulation step voltage test* is a test that creates electrical stress on internal insulation cracks to reveal aging or damage not found during other motor insulation tests. The insulation step voltage test is performed only after an insulation spot test. See Insulation Step Voltage Test.

To perform an insulation step voltage test, apply the procedure:
1. Set the megohmmeter to 500 V and connect to measure the resistance of each winding lead to ground. Take each resistance reading after 60 seconds. Record the lowest reading.
2. Place the meter leads on the winding that has the lowest reading.
3. Set the megohmmeter on increments of 500 V starting at 1000 V and ending at 5000 V. Record each reading after 60 seconds.
4. Discharge the motor windings.

Interpret the results of the test to determine the condition of the insulation. The resistance of good insulation that is thoroughly dry (Curve A) remains approximately the same at different voltage levels. The resistance of deteriorated insulation (Curve B) decreases substantially at different voltage levels.

INSULATION STEP VOLTAGE TEST

MOTOR ELECTRICAL PROBLEMS

10

Name _____ Date _____

Activities

○ Activity 10-1. Phase Unbalance

Balancing a system requires connecting the loads so that each main power line draws about the same amount of current. Determine the ammeter readings.

Circuit Device	Load Condition
Range	OFF
Refrigerator	OFF
Electric Heating Unit 1	OFF
Electric Heating Unit 2	OFF
Electric Heating Unit 3	OFF
Dryer	OFF
Furnace	OFF
Air Conditioner	OFF
Motor 1	ON
Motor 2	ON
Motor 3	ON
Motor 4	ON

_____ 1. Ammeter 1 reading equals _____ A.

_____ 2. Ammeter 2 reading equals _____ A.

_____ 3. Ammeter 3 reading equals _____ A.

_____ 4. Ammeter 4 reading equals _____ A.

_____ 5. Ammeter 5 reading equals _____ A.

LOAD RATINGS
- RANGE = 40 A
- REFRIGERATOR = 5 A
- ELECTRIC HEATING UNIT 1 = 20 A
- ELECTRIC HEATING UNIT 2 = 25 A
- ELECTRIC HEATING UNIT 3 = 15 A
- ELECTRIC DRYER = 35 A
- FURNACE = 8 A
- AIR CONDITIONER = 15 A

LOAD RATINGS
- MOTOR 1 = 10 A
- MOTOR 2 = 50 A
- MOTOR 3 = 25 A
- MOTOR 4 = 100 A

Activity 10-2. Voltage Unbalance

Determine the average voltage, voltage deviation, and voltage unbalance.

_____ 1. The average voltage equals _____ V.

_____ 2. The voltage deviation equals _____ V.

_____ 3. The voltage unbalance equals _____ %.

_____ 4. The average voltage equals _____ V.

_____ 5. The voltage deviation equals _____ V.

_____ 6. The voltage unbalance equals _____ %.

Activity 10-3. Single Phasing

Single phasing is measured when the load is a 3φ motor and the motor is ON. An uneven voltage is measured at a running motor because the open winding has an induced voltage applied to it from the powered windings. Each motor has lost power in one incoming power line. Determine the voltage unbalance and identify the incoming power line that has lost power.

_____ 1. The voltage unbalance equals _____ %.

_____ 2. The incoming power line that has lost power is Line _____.

_____ 3. The voltage unbalance equals _____ %.

_____ 4. The incoming power line that has lost power is Line _____.

Motor Electrical Problems **191**

◯ Activity 10-4. Improper Phase Sequence

A phase sequence tester is used to determine which 3φ phase power lines are powered and the direction the motor shaft will rotate. The tester is connected to the supply voltage before the motor is connected. Lights are used to indicate the circuit's condition. Three lights indicate the presence of line power and two lights indicate motor shaft rotation. The motor shaft rotation (CW or CCW) is determined when viewing the shaft end of the motor. Determine if the phase sequence is correct or incorrect.

_____ 1. The phase sequence is _____.

_____ 2. The phase sequence is _____.

◯ Activity 10-5. Voltage Surges

A surge protector may be added to the power lines to limit any voltage surges to a set limit. Determine the maximum peak voltage that the surge protector allows as measured on the scopemeter.

_____ 1. The maximum peak voltage of Scopemeter 1 equals _____.

_____ 2. The maximum peak voltage of Scopemeter 2 equals _____.

Activity 10-6. AC Voltage and Frequency Variations

A motor's torque and current are reduced when the voltage at the motor's terminals is less than the motor's rated nameplate voltage. Determine the amount of torque and current based on the measured terminal voltage.

MOTOR REDUCED VOLTAGE STARTING CHARACTERISTICS

Motor Terminal Voltage (% nameplate voltage)	Torque Reduction (%)	Current Reduction (%)
100	0	0
95	11	6
90	20	11
85	30	16
80	39	22
75	47	27
70	55	32
65	60	37
60	68	42
55	73	48
50	78	54

1. The derated torque equals _____ lb-ft.

2. The derated current equals _____ A.

3. Derated torque equals _____ lb-ft.

4. Derated current equals _____ A.

Activity 10-7. DC Voltage Variations

Match each waveform to the scopemeter on which it should appear.

Activity 10-8. Allowable Motor Starting Time

A tachometer or ammeter may be used to measure the acceleration time of a motor. A tachometer measures the speed of a motor to determine when the motor has reached its rated speed. An ammeter measures the current draw of a motor. All motors draw high inrush current when starting. A motor is at full speed when the motor current draw has leveled.

The true acceleration condition of a motor is measured when both a tachometer and ammeter are used. When a motor is operating properly, a tachometer indicates that the motor has constant acceleration to its rated speed, and an ammeter indicates that the high starting current has leveled to its rated (or less) value.

1. Using the actual readings, calculate the percent of rated speed and percent of rated current for each second of acceleration. Draw the graph of the acceleration condition of the motor based on its percent rated speed and current.

2. Does the chart indicate a problem?

Time*	Readings		Percent	
	rpm	Amps	rpm	Amps
1	172	50		
2	345	47		
3	517	44		
4	690	41		
5	862	38		
6	1035	35		
7	1207	32		
8	1380	29		
9	1552	26		
10	1725	23		
11	1725	20		
12	1725	20		

* in sec

$$\% \text{ RATED RPM} = \frac{\text{MEASURED RPM}}{\text{RATED RPM}} \times 100$$

$$\% \text{ RATED AMPS} = \frac{\text{MEASURED AMPS}}{\text{RATED AMPS}} \times 100$$

Activity 10-9. Megohmmeter Tests

1. Connect Megohmmeter 1 to measure the resistance of the armature to ground circuit. Connect Megohmmeter 2 to measure the resistance of the series winding to ground.

Activity 10-10. Insulation Spot Test

1. A 120 V split-phase motor was placed in service in January 1988. Semiannual resistance readings were taken. Develop the Insulation Spot Test Graph from the readings. Mark the point at which the motor requires service.

INSULATION SPOT TEST	
Test Taken	Resistance Reading (in MΩ)
1-88	300
7-88	300
1-89	250
7-89	225
1-90	100
7-90	100
1-91	90
7-91	10
1-92	.1
7-92	.05
1-93	.03

INSULATION SPOT TEST GRAPH

Activity 10-11. Dielectric Absorption Test

Semiannual resistance readings were taken during a dielectric absorption test at 240 V with Class A insulation.

1. Develop the Dielectric Absorption Test Graph.

_____ 2. The polarization index of the insulation is _____.

_____ 3. Is the motor insulation good or bad?

DIELECTRIC ABSORPTION TEST	
Test Taken	Resistance Reading (in MΩ)
At 10 seconds	200
At 20 seconds	200
At 30 seconds	210
At 40 seconds	220
At 50 seconds	250
At 60 seconds	250
At 2 minutes	225
At 3 minutes	200
At 4 minutes	195
At 5 minutes	190
At 6 minutes	180
At 7 minutes	180
At 8 minutes	175
At 9 minutes	175
At 10 minutes	175

DIELECTRIC ABSORPTION TEST GRAPH

Activity 10-12. Insulation Step Voltage Test

Resistance readings were taken during an insulation step voltage test.

1. Develop the Insulation Step Voltage Test Graph.

_____ 2. Is the motor insulation good or bad?

INSULATION STEP VOLTAGE TEST	
Applied Voltage (in V)	Resistance Reading (in MΩ)
500	200
1000	200
1500	225
2000	225
2500	220
3000	220
3500	210
4000	200
4500	195
5000	190

INSULATION STEP VOLTAGE TEST GRAPH

MOTOR MECHANICAL PROBLEMS

Applications — 11

Application — Heat Problems

Excessive heat is a major cause of motor failure and a sign of other motor problems. Heat destroys motor insulation. When motor insulation is destroyed, the windings are shorted, and the motor is no longer functional.

As the heat in a motor increases beyond the temperature rating of the insulation, the life of the insulation is shortened. The higher the temperature, the sooner the insulation fails. The temperature rating of motor insulation is listed as the insulation class. See Motor Insulation Class.

MOTOR INSULATION CLASS		
Class	°C	°F
A	105	221
B	130	266
F	155	311
H	180	356

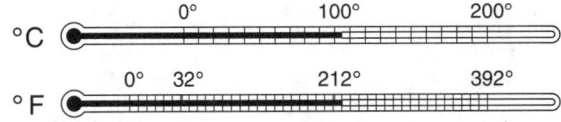

The insulation class is given in Celsius (°C) and/or Fahrenheit (°F). A motor nameplate normally lists the insulation class of the motor. Heat buildup in a motor can be caused by the conditions:
- Incorrect motor type or size for the application
- Improper cooling, normally from dirt buildup
- Excessive load, normally from improper use
- Excessive friction, normally from misalignment or vibration
- Electrical problems, normally voltage unbalance, phase loss, or a voltage surge

Improper Ventilation

All motors produce heat as they convert electrical energy to mechanical energy. This heat must be removed to prevent destruction of motor insulation. Motors are designed with air passages that permit a free flow of air over and through the motor. This airflow removes the heat from the motor. Anything that restricts airflow through a motor causes the motor to operate at a higher temperature than it is designed for.

Airflow through a motor may be restricted by the accumulation of dirt, dust, lint, grass, pests, rust, etc. If a motor becomes coated with oil from leaking seals or from overlubrication, airflow is restricted much faster. See Improper Ventilation.

IMPROPER VENTILATION

Overheating can also occur if a motor is placed in an enclosed area. A motor overheats due to the recirculation of heated air when a motor is installed in a location that does not permit the heated air to escape. Vents added at the top and bottom of the enclosed area allow a natural flow of heated air.

Overloads

An *overload* is the application of excessive load to a motor. Motors attempt to drive the connected load when the power is ON. The larger the load, the more power required. All motors have a limit to the load they can drive. For example, a 5 HP, 460 V, 3φ motor should draw no more than 7.6 A. See NEC® Table 430-150.

Overloads should not harm a properly protected motor. Any overload present longer than the built-in time delay of the protection device is detected and removed. Properly sized heaters in the motor starter assure that an overload is removed before any damage is done. See Motor Overloading.

A troubleshooter can observe the even blackening of all motor windings, which occurs when a motor has failed due to overloading. The even blackening is caused by the motor's slow destruction over a long period of time. No obvious damage or isolated areas of damage to the insulation are visible.

Current readings are taken at a motor to determine an overload problem. A motor is working to its maximum if it is drawing rated current. A motor is overloaded if it is drawing more than rated current. The motor size may be increased or the load on the motor decreased if overloads are a problem. See Motor Current Readings.

MOTOR OVERLOADING

RATED CURRENT OF MOTOR	METER READING		
	MOTOR UNDERLOADED	MOTOR FULLY LOADED	MOTOR OVERLOADED
20 A	12 A	20 A	22 A

↑ NAMEPLATE LISTED VALUE   ↑ 0 TO 95% OF LISTED VALUE   ↑ 95 TO 105% OF LISTED VALUE   ↑ 105% + OF LISTED VALUE

MOTOR CURRENT READINGS

▨ Application — Altitude Correction

Temperature rise of motors is based on motor operation at altitudes of 3300′ or less. A motor with a service factor of 1.0 is derated when it operates at altitudes above 3300′. A motor with a service factor above 1.0 is derated based on the altitude and service factor. See Motor Altitude Deratings.

MOTOR ALTITUDE DERATINGS				
Altitude Range (in ft)	Service Factor			
	1.0	1.15	1.25	1.35
3300–9000	93%	100%	100%	100%
9000–9900	91%	98%	100%	100%
9900–13,200	86%	92%	98%	100%
13,200–16,500	79%	85%	91%	94%
Over 16,500	Consult manufacturer			

Application — Motor Mounting and Positioning

Motors that are not mounted properly are more likely to fail from mechanical problems. A motor must be mounted on a flat, stable base. A flat, stable base helps reduce vibration and misalignment problems. An adjustable motor base aids in proper mounting and alignment.

To ensure a long life span, a motor should be mounted so that it is kept as clean as possible. To reduce the chance of damaging material reaching a motor, a belt cleaner should be used in any application in which the belts are likely to bring damaging material to the motor.

Adjustable Motor Base

An adjustable motor base makes the installation, tensioning, maintenance, and replacement of belts easier. An *adjustable motor base* is a mounting base that allows a motor to be easily moved over a short distance. See Adjustable Motor Base. An adjustable motor base simplifies the installation of the motor and the tightening of belts and chains.

Mounting Direction

The position of the driven machine usually determines whether a motor is installed horizontally or vertically. Standard motors are designed to be mounted with the shaft horizontal. The horizontal position is the best operating position for the motor bearings. A specially designed motor is used for vertical mounting. Motors designed to operate vertically are more expensive and require more preventive maintenance.

ADJUSTABLE MOTOR BASE

Motor Belt Tension and Mounting

Belt drives provide a quiet, compact, and durable form of power transmission and are widely used in industrial applications. A belt must be tight enough not to slip, but not so tight as to overload the motor bearings.

Belt tension is usually checked by placing a straightedge from pulley to pulley and measuring the amount of deflection at the midpoint, or by using a tension tester. Belt deflection should equal $\frac{1}{64}''$ per inch of span. For example, if the span between the center of a drive pulley and the center of a driven pulley is 16″, the belt deflection is $\frac{1}{4}''$ (16 × $\frac{1}{64}''$ = $\frac{1}{4}''$). Belt tension is usually adjusted by moving the drive component away from or closer to the driven component. See Belt Tension.

BELT TENSION

When using a belt drive for a horizontal application, belt sag should be on top. Maximum belt contact is made with the drive pulley when the taut part of the belt is on the bottom. Whenever possible, align pulley centers on the same horizontal plane. See Horizontal Pulley Drive.

Vertical belt drives cause more problems than horizontal belt drives. If an application requires a vertical belt drive, mount the drive pulley on top. Maximum belt contact is made with the drive pulley when the taut part of the belt is pulling up. See Vertical Pulley Drive.

Whenever possible, align pulley centers on the same vertical plane. If the pulley centers cannot be aligned on the same vertical plane, an angle of 45° or less between the pulley centers and the vertical plane is permissible.

Belt Cleaner

In many motor conveyor applications, the motor's belts are likely to come in contact with material used in production, such as chemicals, dirt, sand, wood pulp, food, etc. If material adheres to a motor belt, a belt wears faster, misalignment and vibration occur, and the motor is required to deliver additional power.

The motor can be damaged from reduced air flow and corrosion if a foreign material is delivered to the motor. A belt cleaner should be installed to reduce material buildup in any application in which foreign material is likely to come in contact with motor belts. A belt cleaner is a specially designed device that removes foreign material on a belt and deposits the material at one location. See Belt Cleaner.

Misalignment and Vibration

Misalignment of a motor and driven load is a major cause of motor failure. If a motor and driven load are misaligned, premature failure of the motor bearings, load, or both may occur.

Equipment shafts should be properly aligned on all new installations and checked during periodic maintenance inspections. Misalignment is normally corrected by placing shims under the feet of the motor or driven equipment. A coupling designed to allow some misalignment is used if misalignment cannot be corrected. Couplings used in misaligned applications include rubber-in-shear, flexible spring, and all-metal flex link types. See Motor Couplings.

Misalignment may be angular or parallel. *Angular misalignment* is misalignment when two shafts are not parallel. *Parallel misalignment* is misalignment when two shafts are parallel but not on the same line. See Angular Misalignment and Parallel Misalignment.

Motor Coupling Rating

Motor couplings are rated according to the amount of torque they can handle. Couplings are rated in pound-inches (lb-in.) or pound-feet (lb-ft). The coupling torque rating must be correct for the application to prevent the coupling from bending or breaking. A bent coupling causes misalignment and vibration. A broken coupling prevents the motor from doing work.

ANGULAR MISALIGNMENT

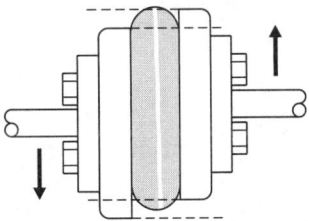

PARALLEL MISALIGNMENT

Application — Loose Connections

All motors produce vibration as they rotate. The vibration can loosen mechanical and electrical connections. Loose mechanical connections generally cause noise and are easily detected. Loose electrical connections do not cause noise, but do cause a voltage drop to the motor and excess heat. Always check mechanical and electrical connections when troubleshooting a motor.

Application — Motor Defect

A motor can fail due to a motor defect or motor damage. A *motor defect* is an imperfection created during the manufacture of a motor that impairs its use. The defect is usually caught by the manufacturer if it impairs initial motor operation. If the defect manifests itself after the motor has been in operation for some time, the troubleshooter determines that the problem is a defect in the motor. Motors with defects should be replaced and the manufacturer should be notified.

A troubleshooter can observe the effect when a motor has failed due to a defect, which is usually confined to a small area of the motor. Typical defects that may occur in a motor include windings grounded in the slot, windings grounded at the edge of the slot, windings shorted phase-to-phase, and shorted connections. See Motor Defects.

Motor damage is any damage that occurs to a properly manufactured motor. The damage may occur before or during installation and during operation. A sound maintenance schedule and proper operation of a motor minimize the occurrence of motor damage.

MOTOR DEFECTS

MOTOR MECHANICAL PROBLEMS

Name _____ Date _____

Activities

⭘ Activity 11-1. Heat Problems

Using the Ambient Temperature Correction Chart, calculate the horsepower rating.

1. The motor rating adjusted for ambient temperature is _____ HP.

TYPE	AC	PHASE	1
HP	5	CYCLE	60
VOLTS	115	AMPS	40
RPM	1725	TEMP RISE	40°C
SF	1.0	CODE	D
TIME RATE	CONT		

Thermometer: 50°C

AMBIENT TEMPERATURE CORRECTION CHART

Graph: Correction Factor (0.75 to 1.45) vs Ambient Temperature (-40°C/-40°F to 60°C/140°F)

2. The motor rating adjusted for ambient temperature is _____ HP.

TYPE	AC	PHASE	3
HP	60	CYCLE	60
VOLTS	460	AMPS	77
RPM	1150	TEMP RISE	50°C
SF	1.25	CODE	H
TIME RATE	CONT		

Thermometer: 35°C

3. The motor rating adjusted for ambient temperature is _____ HP.

TYPE	DC COMPOUND		
HP	3	CYCLE	60
VOLTS	240	AMPS	12.2
RPM	1725	TEMP RISE	40°C
SF	1.15		
TIME RATE	CONT		

Thermometer: 60°C

203

204 TROUBLESHOOTING ELECTRICAL/ELECTRONIC SYSTEMS

4. The motor rating adjusted for ambient temperature is _____ HP.

5. The motor rating adjusted for ambient temperature is _____ HP.

◯ Activity 11-2. Altitude Correction

Using Motor Altitude Deratings, calculate the horsepower rating.

MOTOR ALTITUDE DERATINGS				
Altitude Range (in ft)	Service Factor			
	1.0	1.15	1.25	1.35
3300–9000	93%	100%	100%	100%
9000–9900	91%	98%	100%	100%
9900–13,200	86%	92%	98%	100%
13,200–16,500	79%	85%	91%	94%
Over 16,500	Consult manufacturer			

1. The altitude adjusted rating equals _____ HP.

ALTITUDE = 14,400 FT

2. The altitude adjusted rating equals _____ HP.

ALTITUDE = 11,000 FT

TYPE	AC	PHASE	3
HP	20	CYCLE	60
VOLTS	230	AMPS	54
RPM	1725	TEMP RISE	40° C
SF	1.25	CODE	F
TIME RATE	CONT		

3. The altitude adjusted rating equals _____ HP.

ALTITUDE = 5500 FT

TYPE	AC	PHASE	3
HP	40	CYCLE	60
VOLTS	230	AMPS	104
RPM	1725	TEMP RISE	40° C
SF	1.0	CODE	H
TIME RATE	CONT		

○ Activity 11-3. Motor Mounting and Positioning

Determine the proper direction (CW or CCW) of rotation for the drive (small) pulleys for problems 1 through 4.

1. The direction of rotation equals _____.

2. The direction of rotation equals _____.

3. The direction of rotation equals _____.

4. The direction of rotation equals _____.

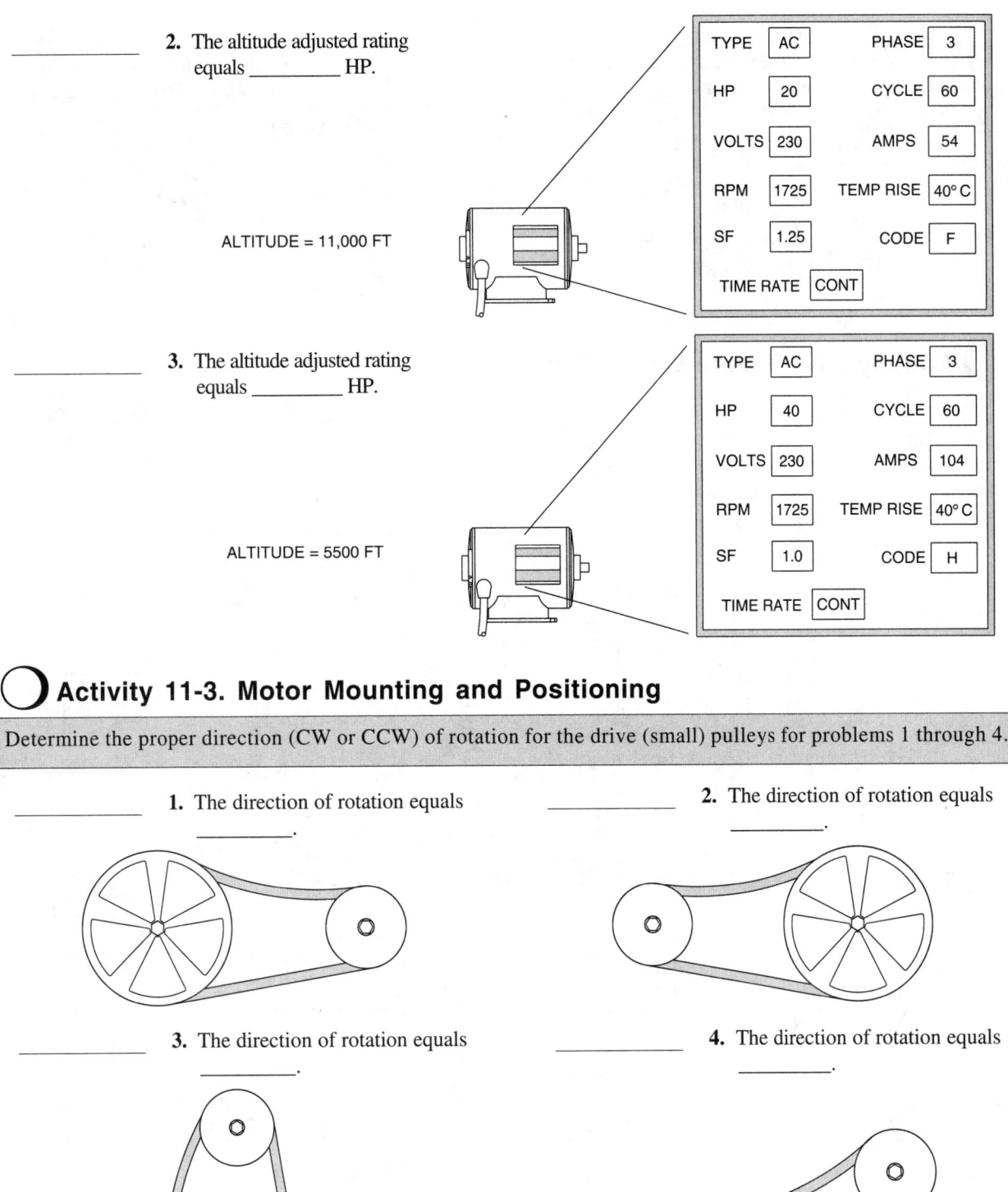

206 TROUBLESHOOTING ELECTRICAL/ELECTRONIC SYSTEMS

_____ **5.** For proper belt tightness, the belt tension tester should not read more than _____ ".

_____ **6.** For proper belt tightness, the belt tension tester should not read more than _____ ".

◐ Activity 11-4. Loose Connections

A loose connection may cause a circuit to have a low voltage condition at the motor. Determine the point in each circuit that has the problem.

_____ **1.** The problem is located at _____.

2. The problem is located at _____.

Activity 11-5. Motor Defect

Each motor failed in the first month of service. The motors were replaced and sent to the maintenance shop for inspection. Company policy is to send a motor back to the manufacturer for further inspection if it appears that a motor has failed because of a manufacturing defect. List each motor failure as defect or non-defect motor failure.

_____ 1. _____ motor failure

_____ 2. _____ motor failure

_____ 3. _____ motor failure

_____ 4. _____ motor failure

TROUBLESHOOTING DC MOTORS

12 Applications

▢ Application — DC Motors

A *direct current (DC) motor* is a motor that uses direct current connected to a field and armature to produce rotation. A *field* is the stationary windings, or magnets, of a DC motor. Field windings develop a magnetic field in a DC motor. An *armature* is the rotating part of a DC motor that is mounted on the motor shaft. See DC Motor.

When current passes through the armature located in the magnetic field, a mechanical force is exerted on the armature. A *commutator* is a series of copper segments connected to the armature. *Brushes* are the sliding contacts that provide contact between the external power source and the commutator. Brushes are made of carbon or graphite. See DC Motor Brushes.

The amount of force exerted on the motor shaft depends on the factors:
- Intensity (strength) of the magnetic field
- Current passing through the armature inside the magnetic field
- Length (or number) of the armature inside the magnetic field

The force on the motor shaft is increased by increasing any of these factors. Normally, the intensity of the field and the current are changed to increase force. The length of the armature is fixed when the motor is manufactured.

The armature windings, commutator, and brushes are arranged so that the flow of current is in one direction in the windings on one side of the armature and is in the opposite direction in the windings on the opposite side of the armature. DC power is delivered to the armature through the brushes, which ride on the commutator sections. The commutator is mounted on the same shaft as the armature and rotates with it.

To increase the torque of the motor, the armature has a large number of coils connected to a large number of commutator sections. The field coils may be permanent magnets or electromagnets.

DC MOTOR

DC MOTOR BRUSHES

DC Motor Types

There are four basic types of DC motors:
- DC series motor
- DC shunt motor
- DC compound motor
- DC permanent-magnet motor

These motors have the same external appearance, but are different in their internal construction and output performance. The selection of the type of DC motor to use is based on the mechanical requirements of the applied load.

DC SERIES MOTOR

DC Series Motor

A *DC series motor* is a motor with the field connected in series with the armature. The field carries the load current passing through the armature. The field has comparatively few turns of heavy-gauge conductor. The conductors extending from the series coil are marked S1 and S2. The conductors extending from the armature are marked A1 and A2. See DC Series Motor.

A DC series motor is used as a traction motor because it produces the highest torque of all DC motors. The DC series motor can develop 500% of its full-load torque on starting. Typical applications include traction bridges, hoists, gates, and starting motors in automobiles.

DC Shunt Motor

A *DC shunt motor* is a motor with the field connected in shunt (parallel) with the armature. The field has numerous turns of wire, and the current in the field is independent of the armature. The field windings extending from the shunt field of a DC shunt motor are marked F1 and F2. The armature windings are marked A1 and A2. See DC Shunt Motor.

A DC shunt motor has excellent speed control. A DC shunt motor is used where constant or adjustable speed is required and starting conditions are moderate. Typical applications include fans, blowers, centrifugal pumps, conveyors, elevators, woodworking machinery, and metalworking machinery.

DC SHUNT MOTOR

DC Compound Motor

A *DC compound motor* is a motor with a field connected in both series and shunt with the armature. In a DC compound motor, the field coil is a combination of the series field (S1 and S2) and shunt field (F1 and F2). The series field is connected in series with the armature. The shunt field is connected in parallel with the series field and armature. See DC Compound Motor.

A DC compound motor has high torque and constant speed. A DC compound motor is used when starting torque and fairly constant speed are required. Typical applications include punch presses, shears, bending machines, and hoists.

DC Permanent-Magnet Motor

A *DC permanent-magnet motor* is a motor that uses magnets for the field winding. The DC permanent-magnet motor has molded magnets mounted in a steel shell. The permanent magnets are the field coils. DC power is supplied only to the armature.

A DC permanent-magnet motor is used in automobiles to control power seats, power windows, and windshield wipers. A DC permanent-magnet motor produces relatively high torque at low speeds and provides some self-braking when removed from power. DC permanent-magnet motors are not designed to run continuously because they overheat rapidly. Overheating destroys the permanent magnets.

DC COMPOUND MOTOR

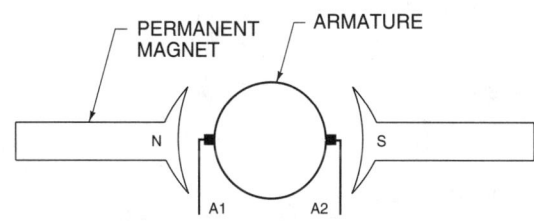

DC PERMANENT-MAGNET MOTOR

▊ Application — Troubleshooting DC Motors

Direct current motors are used in applications that require high torque. To produce high torque, the outside power supply is connected to both the armature and field. A commutator and brushes are used to supply power to the rotating field. Because of their brushes, DC motors generally require more repair than motors that do not use brushes. The brushes should be checked first when troubleshooting DC motors.

Brushes wear faster than any other component of a DC motor. The brushes ride on the fast-moving commutator. Bearings and lubrication are used to reduce friction when two moving surfaces touch. No lubrication is used between the moving brushes and the commutator because the brushes must carry current from the armature. Sparking occurs as the current passes from the commutator to the brushes. Sparking causes heat, burning, and wear of electric parts.

212 TROUBLESHOOTING ELECTRICAL/ELECTRONIC SYSTEMS

The brushes and commutator are subject to wear. The brushes are designed to wear as the motor ages. Replacing worn brushes is easier and less expensive than servicing or replacing a worn commutator. Most DC motors are designed so that the brushes and the commutator can be inspected without disassembling the motor. Some motors require disassembly for close inspection of the brushes and commutator.

Observe the brushes as a motor operates. The brushes should be riding smoothly on the commutator with little or no sparking. There should be no brush noise, such as chattering. Brush sparking, chattering, or a rough commutator indicates service is required.

Troubleshooting Brushes

The condition of the brushes and their holders is extremely important for good motor operation. The brushes should be checked every time the motor is serviced. See Troubleshooting Brushes.

TROUBLESHOOTING BRUSHES

To troubleshoot brushes, apply the procedure:
1. Turn the handle of the safety switch or combination starter OFF. Lockout and tag the starting mechanism per company policy.
2. Measure the voltage at the motor terminals to ensure the power is OFF.
3. Check the brush movement and tension. Remove the brushes. The brushes should move freely in the brush holder. The spring tension should be approximately the same on each brush.

4. Check the length of the brushes. Brushes should be replaced when they have worn down to about half of their original size. Replace all brushes if any brush is less than half its original length. Never replace only one brush.

   Brush compositions include high-grade carbon, electrographite, natural graphite, and carbon graphite. Each composition has its advantages and disadvantages. Always replace brushes with brushes of the same composition. Check manufacturer's recommendations for type of brushes to use.
5. Check the position of the brush holder in relation to the commutator. The brush holder should be $\frac{1}{16}''$ to $\frac{1}{8}''$ from the commutator. The commutator may be damaged if the brush holder is too close. The brush may break if the brush holder is too far away.
6. Check for proper brush pressure. Brush pressure is critical to proper operation. Too little pressure causes the brushes to arc excessively and groove the commutator. Too much pressure causes the brushes to chatter and wear faster than normal.

   Brush pressure varies with the composition of the brush. Brush pressure is usually about 1.5 to 5 psi of surface area. The original spring should provide the proper pressure if it is in good condition. Replace the spring with one of the same type if it is not in good condition. Brush pressure is checked using a brush-pressure tester. Follow the manufacturer's procedures.

   When checking brush pressure, remove the endbell on the side in which the commutator is located. Pull back on the gauge, noting the pressure at which the piece of paper is free to move. Divide this reading by the contact area of the brush to get actual brush pressure in psi. Check the measured pressure to ensure it falls within the manufacturer's listed range.

Troubleshooting Commutators

Brushes wear faster than the commutator. After brushes have been changed once or twice, the commutator usually needs servicing. Any markings on the commutator, such as grooves or ruts, or discolorations other than a polished, brown color where the brushes ride, indicate a problem. See Troubleshooting Commutators.

To troubleshoot commutators, apply the procedure:
1. Make a visual check of the commutator. The commutator should be smooth and concentric. A uniform dark copper oxide carbon film should be present on the surface of the commutator. This naturally occurring film acts like a lubricant by prolonging the life of the brushes and reducing wear on the commutator surface.
2. Check the mica insulation between the commutator segments. The mica insulation separates and insulates each commutator segment. The mica insulation should be undercut (lowered below the surface) approximately $\frac{1}{32}''$ to $\frac{1}{16}''$, depending on the size of the motor. The larger the motor, the deeper the undercut. Replace or service the commutator if the mica is raised.

TROUBLESHOOTING COMMUTATORS

Troubleshooting for Grounded, Open, or Short Circuits

A DC motor is tested for a grounded, open, or short circuit by using a test light. A *grounded circuit* is a circuit in which current leaves its normal path and travels to the frame of the motor. A grounded circuit is caused when insulation breaks down or is damaged and touches the metal frame of the motor.

An *open circuit* is a circuit that no longer provides a path for current to flow. An open circuit is caused when a conductor or connection has physically moved apart from another conductor or connection.

A *short circuit* is a circuit in which current takes a shortcut around the normal path of current flow. A short circuit is caused when the insulation of two conductors from different parts of a circuit touch. Short circuits are usually a result of insulation breakdown.

A test light is preferred for a quick check of a motor. A test light gives good results for obvious problems. Additional testing equipment, such as ohmmeters and megohmmeters is required for less obvious problems. For example, a megohmmeter is used to test for insulation breakdown. See Troubleshooting Circuit Faults.

TROUBLESHOOTING CIRCUIT FAULTS

To troubleshoot for a grounded circuit, open circuit, or short circuit, apply the procedure:
1. Check for a grounded circuit. Connect one lead of the test-light to the frame of the motor. Touch the other test-light lead from one motor lead to the other. A grounded circuit is present if the test light turns ON. Service and repair the motor.
2. Check for an open circuit. Connect the two test-light leads to the motor field and armature circuits as follows:

COMPOUND MOTORS	SHUNT MOTORS	SERIES MOTORS
A1 to A2	A1 to A2	A1 to A2
F1 to F2	F1 to F2	S1 to S2
S1 to S2		

The circuits are complete if the test light turns ON. The circuits are open if the test light does not turn ON. Service and repair the motor.

3. Check for a short circuit between windings. Connect the two test-light leads to the motor field and armature circuits as follows:

COMPOUND MOTORS	SHUNT MOTORS	SERIES MOTORS
A1 to F1, A2 to F2	A1 to F1	A1 to S1
A1 to F2, A2 to S1	A1 to F2	A1 to S2
A1 to S1, A2 to S2	A2 to F1	A2 to S1
A1 to S2, F1 to S1	A2 to F2	A2 to S2
A2 to F1, F1 to S2		

The circuit is shorted if the test light turns ON. The circuit is not shorted if the test light does not turn ON. Service and repair the motor.

Troubleshooting for Grounded or Shorted Commutator Windings

A commutator is grounded whenever one or more of its segments (bars) makes contact with the iron core of the commutator. An armature winding is also grounded whenever one or more windings makes contact with the iron core. See Troubleshooting Commutator Windings.

TROUBLESHOOTING COMMUTATOR WINDINGS

To troubleshoot for a grounded commutator or armature winding, apply the procedure:
1. Check for a commutator segment grounded to the shaft. Connect one lead of a test-light to the shaft. Touch the other test lead to each commutator segment. The test light turns ON if voltage passes through the commutator segments. The commutator or the winding is grounded if the test light turns ON. Service and repair the commutator.
2. Check for a short between adjacent commutator segments. Connect one test-light lead to a commutator segment. Touch the other test-light lead to each adjacent commutator segment. The test light turns ON if voltage passes through the commutator segments. The adjacent segments are shorted if the test light turns ON. Service and repair the commutator. There should also be no sparking or arcing between any segments. This indicates a partial short. Service and repair the commutator.

Troubleshooting Guides

Troubleshooting guides for motors state a problem, its possible cause(s), and corrective action(s) that may be taken. These guides may be used to quickly determine potential problems and possible courses of action. See Direct Current Motor Troubleshooting Guide.

DIRECT CURRENT MOTOR TROUBLESHOOTING GUIDE

Problem	Possible Cause	Corrective Action
Motor will not start.	Blown fuse or open CB	Test the OCPD. If voltage is present at the input, but not the output of the OCPD, the fuse is blown or the CB is open. Check the rating of the OCPD. It should be at least 125% of the motor's FLC.
	Motor overload on starter tripped.	Allow overloads to cool. Reset overloads. If reset overloads do not start the motor, test the starter.
	No brush contact	Check brushes. Replace, if worn.
	Open control circuit between incoming power and motor	Check for cleanliness, tightness, and breaks. Use a voltmeter to test the circuit starting with the incoming power and moving to the motor terminals. Voltage generally stops at the problem area.
Fuse, CB, or overloads retrip after service.	Excessive load	If the motor is loaded to excess or is jammed, the circuit OCPD will open. Disconnect the load from the motor. If the motor now runs properly, check the load. If the motor does not run and the fuse or CB opens, the problem is with the motor or control circuit. Remove the motor from the control circuit and connect it directly to the power source. If the motor runs properly, the problem is in the control circuit. Check the control circuit. If the motor opens the fuse or CB again, the problem is in the motor. Replace or service the motor.
	Motor shaft does not turn.	Disconnect the motor from the load. If the motor shaft still does not turn, the bearings are frozen. Replace or service the motor.
Brushes chip or break.	Brush material is too weak or the wrong type for motor's duty rating.	Replace with better grade or type of brush. Consult manufacturer if problem continues.
	Brush face is overheating and losing brush bonding material.	Check for an overload on the motor. Reduce the load as required. Adjust brush holder arms.
	Brush holder is too far from commutator.	Too much space between the brush holder and the surface of the commutator allows the brush end to chip or break. Set correct space between brush holder and commutator.
	Brush tension is incorrect.	Adjust brush tension so the brush rides freely on the commutator.
Brushes spark.	Worn brushes	Replace worn brushes. Service the motor if rapid brush wear, excessive sparking, chipping, breaking, or chattering is present.
	Commutator is concentric.	Grind commutator and undercut mica. Replace commutator if necessary.
	Excessive vibration	Balance armature. Check brushes. They should be riding freely.
Rapid brush wear	Wrong brush material, type, or grade	Replace with brushes recommended by manufacturer.
	Incorrect brush tension	Adjust brush tension so the brush rides freely on the commutator.
Motor overheats.	Improper ventilation	Clean all ventilation openings. Vacuum or blow dirt out of motor with low pressure, dry, compressed air.
	Motor is overloaded.	Check the load for binding. Check shaft straightness. Measure motor current under operating conditions. If the current is above the listed current rating, remove the motor. Remeasure the current under no-load conditions. If the current is excessive under load but not when unloaded, check the load. If the motor draws excessive current when disconnected, replace or service the motor.

TROUBLESHOOTING DC MOTORS

12

Name _____ Date _____   **Activities**

◯ Activity 12-1. DC Motors

The voltage and current draw of a motor are measured when testing a motor. Complete the wiring of the motor. Connect the ammeter to measure the current flowing through the positive power line. Connect the voltmeter to measure the voltage applied to the motor.

1. Connect the series motor.

218 TROUBLESHOOTING ELECTRICAL/ELECTRONIC SYSTEMS

2. Connect the DC-compound motor.

3. Connect the DC-shunt motor.

4. Complete the wiring of the motor. Connect Ammeter 1 to measure the current flowing through the shunt field. Connect Ammeter 2 to measure the current flowing through the armature.

Activity 12-2. Troubleshooting DC Motors

Determine if the test light should be ON or OFF when the motor is good.

_____ 1. Test Light 1 should be _____ when the motor is good.

_____ 2. Test Light 2 should be _____ when the motor is good.

_____ 3. Test Light 3 should be _____ when the motor is good.

220 TROUBLESHOOTING ELECTRICAL/ELECTRONIC SYSTEMS

_____ 4. Test Light 4 should be _____ when the motor is good.

_____ 5. Test Light 5 should be _____ when the motor is good.

_____ 6. Test Light 6 should be _____ when the motor is good.

_____ 7. Test Light 7 should be _____ when the motor is good.

_____ 8. Test Light 8 should be _____ when the motor is good.

_____ 9. Test Light 9 should be _____ when the motor is good.

_____ 10. Test Light 10 should be _____ when the motor is good.

TROUBLESHOOTING AC MOTORS

Applications — Chapter 13

Application — AC Motors

An *alternating current (AC) motor* is a motor that uses alternating current to produce rotation. AC motors have a rotor and stator. A *rotor* is the rotating part of an AC motor. The rotor rotates the motor shaft and delivers the work. A *stator* is the stationary part of an AC motor to which the power lines are connected. The stator produces a rotating magnetic field. See AC Motor.

AC motors are 1φ or 3φ. Single-phase motors include shaded-pole, split-phase, capacitor-start, capacitor-run, and capacitor start-and-run motors. Single-phase motors require a way of starting the motor. The method used to start a 1φ motor determines the motor's name.

AC MOTOR

Shaded-Pole Motors

A *shaded-pole motor* is an AC motor that uses a shaded stator pole for starting. Shading the stator poles is the simplest method used to start a 1φ motor. See Shaded-Pole Motor.

Shading the poles is accomplished by applying a short-circuited conductor on one side of each stator pole. The shaded pole is normally a solid single turn of copper wire placed around a portion of the main pole laminations.

The shaded pole delays the magnetic flux in the area of the pole that is shaded. Shading causes the magnetic flux at the pole area to be positioned about 90° apart from the magnetic flux of the main stator pole. The offset flux causes the rotor to move from the main pole toward the shaded pole. This movement determines the starting direction of the motor.

SHADED-POLE MOTOR

221

Shaded-pole motors are commonly 1/20 HP or less and have low starting torque. Common applications of shaded-pole motors include small appliance cooling fans found in computers and stereos.

Split-Phase Motors

A *split-phase motor* is an AC motor that can run on one or more phases. A split-phase motor has a main winding and an auxiliary winding. The main winding is the running winding. The auxiliary winding is the starting winding. The two windings are placed in the stator slots and positioned 90° apart. See Split-Phase Motor.

SPLIT-PHASE MOTOR

The running winding is made of larger wire and has a greater number of turns than the starting winding. When the motor is first connected to power, the inductive reactance of the running winding is higher and the resistance is lower than the starting winding. *Inductive reactance* is the opposition to the flow of alternating current in a circuit due to inductance.

The starting winding is made of relatively small wire and has a fewer number of turns than the running winding. When the motor is first connected to power, the inductive reactance of the starting winding is lower and the resistance is higher than the running winding.

When power is first applied, both the running winding and the starting winding are energized. The running winding current lags the starting winding current because of its different inductive reactance. This produces a phase difference between the starting and running windings. A 90° phase difference is required to produce maximum starting torque, but the phase difference is commonly much less. A rotating magnetic field is produced because the two windings are out of phase.

The rotating magnetic field starts the rotor turning. With the running and starting windings out of phase, the current changes in magnitude and direction, and the magnetic field moves around the stator. This movement forces the rotor to rotate with the rotating magnetic field.

Centrifugal Switch

To minimize energy loss and prevent heat build-up in the starting winding once the motor is started, a centrifugal switch is used to remove the starting winding when the motor reaches a set speed. A *centrifugal switch* is a switch that opens to disconnect the starting winding when the rotor reaches a certain preset speed and to reconnect the starting winding when the speed falls below a preset value. In most motors the centrifugal switch is located inside the enclosure on the shaft. See Centrifugal Switch.

CENTRIFUGAL SWITCH

As the motor shaft rotates and the speed of the motor accelerates, the switch is activated by centrifugal force. *Centrifugal force* is the force that moves rotating bodies away from the center of rotation. For some motors the centrifugal switch is often located outside the enclosure for easy repair.

The centrifugal switch is connected in series with the starting winding. The centrifugal switch automatically de-energizes the starting winding at a set speed. The set speed is usually about 60% to 80% of the running speed. After the starting winding is removed, the motor continues to operate on the running winding only.

The split-phase motor is one of the oldest and most common motor types. A split-phase motor is used in applications such as fans, business machines, machine tools, and centrifugal pumps. A split-phase motor is generally available in sizes ranging from 1/30 HP to 1/2 HP.

Thermal Protection

Many motors contain a thermal switch in addition to a centrifugal switch. A *thermal switch* is a switch that operates its contacts when a preset temperature is reached. A thermal switch is located inside a motor and is used to protect the motor windings from burnout. See Thermal Protection.

A thermal switch is activated by high temperatures and automatically removes both the starting and running windings from the power source. A higher-than-normal temperature may be caused by improper ventilation, too high of a motor load, high ambient temperature, or a mechanical problem that prevents rotation.

During normal operation, a thermal switch is in the closed position. As more current than normal is allowed to pass through the motor windings, the switch begins to heat. At a preset temperature, the switch opens. This automatically removes the motor windings from power. When the motor windings are removed from power, they begin to cool. The thermal switch closes as it cools, and the motor is automatically restarted.

The thermal switch constantly recycles the motor ON and OFF if the problem that caused the original overheating is not corrected.

Reversing

The direction of the connections for either the starting or the running windings are reversed to change the direction of rotation of a split-phase motor. Reversing the starting winding is the industrial standard for reversing the direction of rotation of a 1φ motor.

THERMAL PROTECTION

Capacitor Motors

A *capacitor motor* is a 1φ motor with a capacitor connected in series with the stator windings to produce phase displacement in the starting winding. The capacitor is added to provide a higher starting torque at lower starting current than is delivered by a split-phase motor. Applications of capacitor motors include refrigerators, air conditioners, air compressors, and some power tools.

In a capacitor motor, the capacitor causes the starting winding current to lead the applied voltage by about 40°. Since the starting and running windings are about 90° out of phase, the motor's operating characteristics are improved, a higher starting torque is produced, and the motor has a better power factor with a lower current draw. Three types of capacitor motors are the capacitor-start, capacitor-run, and capacitor start-and-run motors.

Capacitor-Start Motors

A *capacitor-start motor* is a motor that has the capacitor connected in series with the starting winding. The capacitor is connected only during starting. The capacitor and starting winding are disconnected from the power by a centrifugal switch at a set speed. See Capacitor-Start Motor.

A capacitor-start motor develops considerably more locked rotor torque per ampere than a split-phase motor because of the capacitor in the circuit. *Locked rotor torque* is the torque a motor produces when the rotor is stationary and full power is applied to the motor. Capacitor-start motors are used to drive power tools, pumps, and small machines.

CAPACITOR-START MOTOR

Capacitor-Run Motors

A *capacitor-run motor* leaves the starting winding and capacitor in the circuit at all times. The starting winding is not removed as the motor speed increases because there is no centrifugal switch. A capacitor-run motor has a lower full-load speed than a capacitor-start motor because the capacitor remains in the circuit at all times. See Capacitor-Run Motor.

The advantage of leaving the capacitor in the circuit is that the motor has more running torque than a capacitor-start motor or split-phase motor. This allows a capacitor-run motor to be used for loads that require a high running torque. Capacitor-run motors are used to drive shaft-mounted fans and blowers ranging in size from $\frac{1}{16}$ HP to $\frac{1}{3}$ HP.

CAPACITOR-RUN MOTOR

Capacitor Start-and-Run Motors

A *capacitor start-and-run motor* uses two capacitors. A capacitor start-and-run motor starts with one value capacitor in series with the starting winding and runs with a different value capacitor in series with the running winding. This motor is also known as a dual-capacitor motor. See Capacitor Start-and-Run Motor.

A capacitor start-and-run motor has the same starting torque as a capacitor-start motor. A capacitor-start-and-run motor has more running torque than a capacitor-start motor or capacitor-run motor because the capacitance is better matched for starting and running.

In a typical two-capacitor motor, one capacitor is used for starting the motor and the other capacitor remains in the circuit while the motor is running. A larger-value capacitor is used for starting and a smaller-value capacitor is used for running. Capacitor start-and-run motors are used to run refrigerators and air compressors.

CAPACITOR START-AND-RUN MOTOR

Three-Phase Motors

Three-phase motors require a rotating magnetic field. A 3ϕ motor does not require any additional components to produce this rotating magnetic field. The rotating magnetic field is set up automatically in the stator when the motor is connected to 3ϕ power lines. The coils in the stator are connected to form three separate windings (phases). Each phase contains one-third of the total number of individual coils in the motor. These windings (or phases), are the A phase, B phase, and C phase. See 3ϕ Motor Phases.

Each phase is positioned 120° from the other phases in the motor. A rotating magnetic field is produced in the stator when each phase reaches its peak value, 120° apart from the other phases.

Electrical degrees and mechanical degrees differ. In electric motors and generators, the distance traveled past one pole represents 180 electrical degrees. One revolution has 360 mechanical degrees.

3ϕ MOTOR PHASES

For example, during one revolution in a 4-pole motor, a rotor passes from a north pole, through a south pole, through a north pole, through a south pole, and back to the original north pole, completing 720 electrical degrees (4 poles × 180 electrical degrees = 720 electrical degrees). See Mechanical Versus Electrical Degrees.

Single-Voltage, 3ϕ Motors

The windings must be connected to the proper voltage to develop a rotating magnetic field in a 3ϕ motor. This voltage level is determined by the manufacturer and stamped on the motor nameplate.

A *single-voltage, 3ϕ motor* is a motor that operates at only one voltage level. Single-voltage, 3ϕ motors are less expensive to manufacture than dual-voltage, 3ϕ motors, but are limited to locations having the same voltage as the motor. Typical single-voltage, 3ϕ motor ratings are 230 V, 460 V, and 575 V. Other single-voltage, 3ϕ motor ratings are 200 V, 208 V, 220 V, and 280 V.

Wye-Connected, 3ϕ Motors

A *wye-connected, 3ϕ motor* has one end of each of the three phases internally connected to the other phases. The remaining end of each phase is brought out externally and connected to the incoming power source. See Wye-Connected, 3ϕ Motor.

The leads, which are brought out externally, are labeled terminals one, two, and three (T1, T2, and T3). When connecting a wye-connected, 3ϕ motor to the 3ϕ power lines, the power lines and motor terminals are connected L1 to T1, L2 to T2, and L3 to T3.

MECHANICAL VERSUS ELECTRICAL DEGREES

Mechanical Degrees In One Revolution	Motor Poles × 180	Electrical Degrees In One Revolution
360	2 × 180	= 360
360	4 × 180	= 720
360	6 × 180	= 1080
360	8 × 180	= 1440

WYE-CONNECTED, 3ϕ MOTOR

Delta-Connected, 3ϕ Motors

A *delta-connected, 3ϕ motor* has each phase wired end-to-end to form a completely closed loop circuit. At each point where the phases are connected, leads are brought out externally to form T1, T2, and T3. See Delta-Connected, 3ϕ Motor.

T1, T2, and T3, are connected to the three power lines. L1 is connected to T1, L2 to T2, and L3 to T3. The 3ϕ line supplying power to the motor must have the same voltage and frequency rating as the motor.

Dual-Voltage, 3ϕ Motors

A *dual-voltage, 3ϕ motor* is a motor that operates at more than one voltage level. Most 3ϕ motors are manufactured so they can be connected to either of two voltages. The purpose in making motors for two voltages is to enable the same motor to be used with two different power line voltages.

A typical dual-voltage, 3ϕ motor rating is 230/460 V. Other common dual-voltage, 3ϕ motor ratings are 240/480 V and 208-230/460 V. The dual-voltage rating of a motor is listed on the nameplate of the motor. If both voltages are available, the higher voltage is usually preferred because the motor uses the same amount of power, given the same horsepower output, for either high or low voltage. As the voltage is doubled, the current draw on the power lines is cut in half. With the reduced current, the wire size is reduced and the material cost is decreased.

DELTA-CONNECTED, 3ϕ MOTOR

Dual-Voltage, Wye-Connected, 3ϕ Motors

In a dual-voltage, wye-connected, 3ϕ motor, each phase coil (A, B, and C) is divided into two equal parts. By dividing the phase coils in two, nine terminal leads are available. These motor leads are marked terminals one through nine (T1-T9). The nine terminal leads may be connected for high or low voltage.

To connect a dual-voltage, wye-connected, 3ϕ motor for high voltage, connect L1 to T1, L2 to T2, and L3 to T3 at the motor starter. Using wire nuts, tie T4 to T7, T5 to T8, and T6 to T9. By making these connections, the individual coils in each phase are connected in series. The applied voltage is divided equally among the coils because the coils are connected in series. See Dual-Voltage, Wye-Connected, 3ϕ Motor.

DUAL-VOLTAGE, WYE-CONNECTED, 3ϕ MOTOR

To connect a dual-voltage, wye-connected, 3ϕ motor for low voltage, connect L1 to T1 and T7, L2 to T2 and T8, and L3 to T3 and T9 at the motor starter. Using a wire nut, tie T4, T5, and T6 together. By making these connections, the individual coils in each phase are connected in parallel. The applied voltage is present across each set of coils because the coils are connected in parallel.

Dual-Voltage, Delta-Connected, 3ϕ Motors

In a dual-voltage, delta-connected, 3ϕ motor, each phase coil (A, B, and C) is divided into two equal parts. By dividing the phase coils in two, nine terminal leads are available. These motor leads are marked terminals one through nine (T1-T9). The nine terminal leads can be connected for high or low voltage.

To connect a dual-voltage, delta-connected, 3ϕ motor for high voltage, connect L1 to T1, L2 to T2, and L3 to T3 at the motor starter. Using wire nuts, tie T4 to T7, T5 to T8, and T6 to T9. By making these connections, the individual coils in each phase are connected in series. Since the coils are connected in series, the applied voltage divides equally among the coils. See Dual-Voltage, Delta-Connected, 3ϕ Motor.

To connect a dual-voltage, delta-connected, 3ϕ motor for low voltage, connect L1 to T1, L2 to T2, and L3 to T3 at the motor starter. Using wire nuts, tie T1 to T7 and T6, T2 to T8 and T4, and T3 to T9 and T5. By making these connections, the individual coils in each phase are connected in parallel. Since the coils are connected in parallel, the applied voltage is present across each set of coils.

DUAL-VOLTAGE, DELTA-CONNECTED, 3ϕ MOTOR

Reversing

The direction of rotation of 3ϕ motors is reversed by interchanging any two of the 3ϕ power lines to the motor. The industrial standard is to interchange T1 and T3. This standard holds true for all 3ϕ motors.

Interchanging T1 and T3 is a standard for safety reasons. When first connecting a motor, the direction of rotation is not usually known until the motor is started. It is common practice to temporarily connect the motor to determine the direction of rotation before making permanent connections. Motor leads of temporary connections are not taped. By always interchanging T1 and T3, T2 may be permanently connected to L2, creating an insulated barrier between T1 and T3.

Application — Troubleshooting AC Motors

Most problems with 1ϕ motors involve the centrifugal switch, thermal switch, or capacitor(s). A motor is usually serviced and repaired if the problem is in the centrifugal switch, thermal switch, or capacitor. A motor is usually replaced if the motor is more than 10 years old and less than 1 HP. A motor is almost always replaced if the motor is less than 1/8 HP. Three-phase motors usually operate for many years without any problems because 3ϕ motors have fewer components that may malfunction than other motor types.

The motor is serviced or replaced if a 3ϕ motor is the problem. Servicing usually requires that the motor be sent to a motor repair shop for rewinding. A motor is replaced if the motor is less than 1 HP and more than five years old. A motor may be serviced or replaced if the motor is more than 1 HP but less than 5 HP. A motor is usually serviced if the motor is more than 5 HP.

Troubleshooting Shaded-Pole Motors

Shaded-pole motors that fail are usually replaced. The reason for the motor failure should be discovered. For example, replacing a motor because it failed due to a jammed load, does not solve the problem, etc. See Troubleshooting Shaded-Pole Motors.

TROUBLESHOOTING SHADED-POLE MOTORS

To troubleshoot a shaded-pole motor, apply the procedure:
1. Turn the power to the motor OFF. Visually inspect the motor. Replace the motor if it is burned, the shaft is jammed, or if there is any sign of damage.
2. Check the stator winding. The stator winding is the only electrical circuit that may be tested without taking the motor apart. Measure the resistance of the stator winding. Set the ohmmeter to the lowest scale for taking the reading. The winding is open if the ohmmeter indicates an infinity reading. Replace the motor. The winding is shorted if the ohmmeter indicates a zero reading. Replace the motor. The winding may still be good if the ohmmeter indicates a low-resistance reading. Check the winding with a megohmmeter before replacing.

Troubleshooting Split-Phase Motors

Some split-phase motors include a thermal switch that automatically turns the motor OFF when it overheats. Thermal switches may have a manual reset or an automatic reset. Caution should be taken with any motor that has an automatic reset should the motor automatically restart at any time. See Troubleshooting Split-Phase Motors.

TROUBLESHOOTING SPLIT-PHASE MOTORS

To troubleshoot a split-phase motor, apply the procedure:
1. Turn power to motor OFF. Visually inspect the motor. Replace the motor if it is burned, the shaft is jammed, or if there is any sign of damage.
2. Check to determine if the motor is controlled by a thermal switch. Reset the thermal switch and turn the motor ON if the thermal switch is manual.
3. Use a voltmeter to check for voltage at the motor terminals if the motor does not start. The voltage should be within 10% of the motor's listed voltage. Troubleshoot the circuit leading to the motor if the voltage is not correct. Turn the power to the motor OFF so the motor may be tested if the voltage is correct.
4. Turn the handle of the safety switch or combination starter OFF. Lockout and tag the starting mechanism per company policy.
5. With power OFF, connect the ohmmeter to the same motor terminals from which the incoming power leads were disconnected. The ohmmeter reads the resistance of the starting and running windings. Their combined resistance is less than the resistance of either winding alone because the windings are in parallel. A short is present if the meter reads zero. An open circuit is present if the meter reads infinity. Replace the motor. *Note:* The motor size is too small for a repair to be cost efficient.
6. Visually inspect the centrifugal switch for signs of burning or broken springs. Service or replace the switch if any obvious signs of problems are present. Check the switch using an ohmmeter if no obvious signs of problems are present. Manually operate the centrifugal switch. (The endbell on the switch side may have to be removed.) The resistance on the ohmmeter decreases if the motor is good. A problem exists if the resistance does not change. Continue to check the motor to determine the problem.

Thermal Switch Testing

A thermal switch removes the starting and running motor windings from the circuit at a preset temperature. When the windings cool, the contacts close. When a thermal switch is not operating properly, a motor either does not operate or operates without terminal protection. See Thermal Switch Testing.

THERMAL SWITCH TESTING

To test a thermal switch, apply the procedure:
1. Remove the motor from the power lines and let cool.
2. Remove the end of the motor that includes the thermal switch.
3. Remove one of the leads running from the thermal switch to the motor windings.
4. In a motor containing a two-terminal thermal switch, check for continuity (very low resistance) using an ohmmeter. Set the ohmmeter on the lowest resistance scale and check across the switch contacts. The contacts are open and the switch is defective if a high-resistance reading is obtained.

   In a motor containing a three-terminal thermal switch, check for continuity across the switch contacts using an ohmmeter set on the lowest scale. The contacts are open and the switch is defective if a high-resistance reading is obtained. Check for continuity across the heater element. The heater element is open and defective if a high-resistance reading is obtained.

   In a motor containing a four-terminal thermal switch, check for continuity across the switch contacts and the heater element using an ohmmeter set on the lowest scale. The contacts or the heater element is open and defective if a high-resistance reading is obtained. Replace the defective component.

Troubleshooting Capacitor Motors

Troubleshooting capacitor motors is similar to troubleshooting split-phase motors. The only additional device to be considered is the capacitor. Capacitors have a limited life and are often the problem in capacitor motors. Capacitors may have a short circuit, an open circuit, or may deteriorate to the point that they must be replaced.

Deterioration may also change the value of a capacitor, which may cause additional problems. When a capacitor short-circuits, the winding in the motor may burn out. When a capacitor deteriorates or opens, the motor has poor starting torque. Poor starting torque may prevent the motor from starting, which usually trips the overloads.

All capacitors are made with two conducting surfaces separated by dielectric material. *Dielectric material* is a medium in which an electric field is maintained with little or no outside energy supply. Dielectric material is used to insulate the conducting surfaces of a capacitor. Capacitors are either oil or electrolytic. Oil capacitors are filled with oil and sealed in a metal container. The oil serves as the dielectric material.

More motors use electrolytic capacitors than oil capacitors. Electrolytic capacitors are formed by winding two sheets of aluminum foil separated by pieces of thin paper impregnated with an electrolyte. An *electrolyte* is a conducting medium in which the current flow occurs by ion migration. The electrolyte is used as the dielectric material. The aluminum foil and electrolyte are encased in a cardboard or aluminum cover. A vent hole is provided to prevent a possible explosion in the event the capacitor is shorted or overheated. AC capacitors are used with capacitor motors. Capacitors that are designed to be connected to AC have no polarity. See Troubleshooting Capacitor Motors.

TROUBLESHOOTING CAPACITOR MOTORS

To troubleshoot a capacitor motor, apply the procedure:
1. Turn the handle of the safety switch or combination starter OFF. Lockout and tag the starting mechanism per company policy.
2. Using a voltmeter, measure the voltage at the motor terminals to ensure the power is OFF.
3. Capacitors are located on the outside frame of the motor. Remove the cover of the capacitor.
   Caution: A good capacitor will hold a charge, even when power is removed.
4. Visually check the capacitor for leakage, cracks, or bulges. Replace the capacitor if present.
5. Remove the capacitor from the circuit and discharge it. To safely discharge a capacitor, place a 20,000 Ω, 5 W resistor across the terminals for five seconds.
6. After the capacitor is discharged, connect the ohmmeter leads to the capacitor terminals. The ohmmeter indicates the general condition of the capacitor. A capacitor is either good, shorted, or open.

- Good Capacitor. The reading changes from zero resistance to infinity. When the reading reaches the halfway point, remove one of the leads and wait 30 seconds. When the lead is reconnected, the reading should change back to the halfway point and continue to infinity. This demonstrates that the capacitor can hold a charge. The capacitor cannot hold a charge and must be replaced when the reading changes back to zero resistance.
- Shorted Capacitor. The reading changes to zero and does not move. The capacitor is bad and must be replaced.
- Open Capacitor. The reading does not change from infinity. The capacitor is bad and must be replaced.

Motor Capacitor Failure

Single-phase motor capacitors fail because of excessive temperature, excessive voltage, excessive duty cycles, internal corrosion, or an open fuse. A substitute capacitor is placed in the circuit to check the motor if a capacitor is suspected to be defective. The original capacitor is defective if motor operation improves. Capacitors normally operate in ambient temperatures up to 80°C (176°F). A capacitor's life is shortened at high temperatures. Operating at low temperatures does not harm a capacitor. The capacitance of a capacitor decreases at temperatures below 0°C.

Excessive voltage causes arcing in the capacitor that leads to permanent damage. Excessive voltage is applied by connecting a motor to a supply voltage higher than the capacitor voltage rating or by a faulty centrifugal switch. The voltage rating of a capacitor is normally listed on the capacitor. The capacitor voltage rating must be equal to or greater than the supply voltage applied to the motor.

Caution: Never replace a capacitor with one that has a lower voltage rating than the original. Never use a capacitor with a voltage rating higher than 10% of the original. Using a capacitor with too high or too low of a voltage rating increases the amperage and wattage draw of the motor. This increase can burn out the motor windings. See Capacitor Ratings. See Capacitor Ratings in Appendix.

CAPACITOR RATINGS

Typical Ratings*	Dimensions**		Model Number***
	Diameter	Length	
88-106	1 7/16	2 3/4	EC8815
108-130	1 7/16	2 3/4	EC10815
130-156	1 7/16	2 3/4	EC13015
145-174	1 7/16	2 3/4	EC14515

* in µF
** in inches
*** model numbers vary by manufacturer

Faulty centrifugal-switch contacts chatter before opening. The voltage on the capacitor is several times higher than the supply voltage when centrifugal-switch contacts chatter. The high voltage is induced by the collapsing magnetic field of the starting winding.

A capacitor is damaged by normal voltage if the motor is started many times or if the load applied to the motor requires a long starting time. A starting capacitor is normally in a circuit for less than 3 seconds. Extra acceleration time causes excessive heat buildup in the capacitor, which reduces capacitor life. A capacitor with an open seal allows moisture absorption. Moisture absorption leads to corrosion that destroys the film inside the capacitor.

Many capacitors have an internal fuse that opens when excessive voltage or current is applied to the capacitor. This internal fuse is also opened by improper servicing of the motor. A capacitor with an open internal fuse shows a resistance reading of infinity when checked with an ohmmeter.

Troubleshooting 3ϕ Motors

The extent of troubleshooting a 3ϕ motor depends on the motor's application. Testing is normally limited to checking the voltage at the motor if a motor is used in an application that is critical to an operation or production. The motor is assumed to be the problem if the voltage is present and correct. Unless it is very large, the motor is normally replaced at this time so production may continue. Further tests may be made to determine the exact problem if time is not a critical factor. See Troubleshooting 3ϕ Motors.

TROUBLESHOOTING 3ϕ MOTORS

To troubleshoot a 3ϕ motor, apply the procedure:
1. Using a voltmeter, measure the voltage at the motor terminals. The motor must be checked if the voltage is present and at the correct level on all three phases. The incoming power supply must be checked if the voltage is not present on all three phases.
2. Turn the handle of the safety switch or combination starter OFF if voltage is present but the motor is not operating. Lockout and tag the starting mechanism per company policy.

3. Disconnect the motor from the load.
4. After the load is disconnected, turn power ON to try restarting the motor. Check the load if the motor starts.
5. Turn the motor OFF and lockout the power if the motor does not start.
6. With an ohmmeter, check the motor windings for any opens or shorts. Take a resistance reading of the T1-T4 coil. This coil must have a resistance reading. The coil is shorted when the reading is zero. The coil is open when the reading is infinity. Since the coil winding is made of wire only, the resistance is low. There is resistance on a good coil winding, however. The larger the motor, the smaller the resistance reading.

After the resistance of one coil has been found, the basic electrical laws of series and parallel circuits are applied. When measuring the resistance of two coils in series, the total resistance is twice the resistance of one coil. When measuring the resistance of two coils in parallel, the total resistance is one half the resistance of one coil.

Troubleshooting Guides

Troubleshooting guides for motors state a problem, its possible cause(s), and corrective action(s) that may be taken. These guides may be used to quickly determine potential problems and possible courses of action. See Troubleshooting Guides.

SHADED-POLE MOTOR TROUBLESHOOTING GUIDE		
Problem	Possible Cause	Corrective Action
Motor will not start.	Blown fuse or open CB	Test the OCPD. If voltage is present at the input, but not the output of the OCPD, the fuse is blown or the CB is open. Check the rating of the OCPD. It should be at least 125% of the motor's FLC.
	Motor overload on starter tripped.	Allow overloads to cool. Reset overloads. If reset overloads do not start the motor, test the starter.
	Low or no voltage applied to motor	Check the voltage at the motor terminals. The voltage must be present and within 10% of the motor nameplate voltage. If voltage is present at the motor but the motor is not operating, remove the motor from the load the motor is driving. Reapply power to the motor. If the motor runs, the problem is with the load. If the motor does not run, the problem is with the motor. Replace or service the motor.
	Open control circuit between incoming power and motor	Check for cleanliness, tightness, and breaks. Use a voltmeter to test the circuit starting with the incoming power and moving to the motor terminals. Voltage generally stops at the problem area.
Fuse, CB, or overloads retrip after service.	Excessive load	If the motor is loaded to excess or jammed, the circuit OCPD will open. Disconnect the load from the motor. If the motor now runs properly, check the load. If the motor does not run and the fuse or CB opens, the problem is with the motor or control circuit. Remove the motor from the control circuit and connect it directly to the power source. If the motor runs properly, the problem is in the control circuit. Check the control circuit. If the motor opens the fuse or CB again, the problem is in the motor. Replace or service the motor.
Excessive noise	Unbalanced motor or load	An unbalanced motor or load causes vibration, which causes noise. Realign the motor and load. Check for excessive end play or loose parts. If the shaft is bent, replace the rotor or motor.
	Dry or worn bearings	Dry or worn bearings cause noise. Bearings may be dry due to dirty oil, oil not reaching the shaft, or motor overheating. Oil bearings as recommended. If noise remains, replace the bearings or motor.
	Excessive grease	Ball bearings that have excessive grease may cause bearings to overheat. Overheated bearings cause noise. Remove excess grease.

SPLIT-PHASE MOTOR TROUBLESHOOTING GUIDE

Problem	Possible Cause	Corrective Action
Motor will not start.	Thermal cutout switch is open.	Reset the thermal switch. **Caution:** Resetting the thermal switch may automatically start the motor.
	Blown fuse or open CB	Test the OCPD. If voltage is present at the input, but not the output of the OCPD, the fuse is blown or the CB is open. Check the rating of the OCPD. It should be at least 125% of the motor's FLC.
	Motor overload on starter tripped.	Allow overloads to cool. Reset overloads. If reset overloads do not start the motor, test the starter.
	Low or no voltage applied to motor	Check the voltage at the motor terminals. The voltage must be present and within 10% of the motor nameplate voltage. If voltage is present at the motor but the motor is not operating, remove the motor from the load the motor is driving. Reapply power to the motor. If the motor runs, the problem is with the load. If the motor does not run, the problem is with the motor. Replace or service the motor.
	Open control circuit between incoming power and motor	Check for cleanliness, tightness, and breaks. Use a voltmeter to test the circuit starting with the incoming power and moving to the motor terminals. Voltage generally stops at the problem area.
	Starting winding not receiving power	Check the centrifugal switch to make sure it connects the starting winding when the motor is OFF.
Fuse, CB, or overloads retrip after service.	Blown fuse or open CB	Test the OCPD. If voltage is present at the input, but not the output of the OCPD, the fuse is blown or the CB is open. Check the rating of the OCPD. It should be at least 125% of the motor's FLC.
	Motor overload on starter tripped.	Allow overloads to cool. Reset overloads. If reset overloads do not start the motor, test the starter.
	Low or no voltage applied to motor	Check the voltage at the motor terminals. The voltage must be present and within 10% of the motor nameplate voltage. If voltage is present at the motor but the motor is not operating, remove the motor from the load the motor is driving. Reapply power to the motor. If the motor runs, the problem is with the load. If the motor does not run, the problem is with the motor. Replace or service the motor.
	Open control circuit between incoming power and motor	Check for cleanliness, tightness, and breaks. Use a voltmeter to test the circuit starting with the incoming power and moving to the motor terminals. Voltage generally stops at the problem area.
	Motor shaft does not turn.	Disconnect the motor from the load. If the motor shaft still does not turn, the bearings are frozen. Replace or service the motor.
Motor produces electric shock.	Broken or disconnected ground strap	Connect or replace ground strap. Test for proper ground.
	Hot power lead at motor connecting terminals is touching motor frame.	Disconnect the motor. Open the motor terminal box and check for poor connections, damaged insulation, or leads touching the frame. Service and test motor for ground.
	Motor winding shorted to frame	Remove, service, and test motor.
Motor overheats.	Starting windings are not being removed from circuit as motor accelerates.	When the motor is turned OFF, a distinct click should be heard as the centrifugal switch closes.
	Improper ventilation	Clean all ventilation openings. Vacuum or blow dirt out of motor with low-pressure, dry, compressed air.
	Motor is overloaded.	Check the load for binding. Check shaft straightness. Measure motor current under operating conditions. If the current is above the listed current rating, remove the motor. Remeasure the current under no-load conditions. If the current is excessive under load but not when unloaded, check the load. If the motor draws excessive current when disconnected, replace or service the motor.

continued

continued

	Dry or worn bearings	Dry or worn bearings cause noise. The bearings may be dry due to dirty oil, oil not reaching the shaft, or motor overheating. Oil the bearings as recommended. If noise remains, replace the bearings or the motor.
	Dirty bearings	Clean or replace bearings.
Excessive noise	Excessive end play	Check end play by trying to move the motor shaft in and out. Add end-play washers as required.
	Unbalanced motor or load	An unbalanced motor or load causes vibration, which causes noise. Realign the motor and load. Check for excessive end play or loose parts. If the shaft is bent, replace the rotor or motor.
	Dry or worn bearings	Dry or worn bearings cause noise. The bearings may be dry due to dirty oil, oil not reaching the shaft, or motor overheating. Oil the bearings as recommended. If noise remains, replace the bearings or the motor.
	Excessive grease	Ball bearings that have excessive grease may cause the bearings to overheat. Overheated bearings cause noise. Remove any excess grease.

THREE-PHASE MOTOR TROUBLESHOOTING GUIDE

Problem	Possible Cause	Corrective Action
Motor will not start.	Wrong motor connections	Most 3φ motors are dual-voltage. Check for proper motor connections.
	Blown fuse or open CB	Test the OCPD. If voltage is present at the input, but not the output of the OCPD, the fuse is blown or the CB is open. Check the rating of the OCPD. It should be at least 125% of the motor's FLC.
	Motor overload on starter tripped.	Allow overloads to cool. Reset overloads. If reset overloads do not start the motor, test the starter.
	Low or no voltage applied to motor	Check the voltage at the motor terminals. The voltage must be present and within 10% of the motor nameplate voltage. If voltage is present at the motor but the motor is not operating, remove the motor from the load the motor is driving. Reapply power to the motor. If the motor runs, the problem is with the load. If the motor does not run, the problem is with the motor. Replace or service the motor.
	Open control circuit between incoming power and motor	Check for cleanliness, tightness, and breaks. Use a voltmeter to test the circuit starting with the incoming power and moving to the motor terminals. Voltage generally stops at the problem area.
Fuse, CB, or overloads retrip after service.	Power not applied to all three lines	Measure voltage at each power line. Correct any power supply problems.
	Blown fuse or open CB	Test the OCPD. If voltage is present at the input, but not the output of the OCPD, the fuse is blown or the CB is open. Check the rating of the OCPD. It should be at least 125% of the motor's FLC.
	Motor overload on starter tripped.	Allow overloads to cool. Reset overloads. If reset overloads do not start the motor, test the starter.
	Low or no voltage applied to motor	Check the voltage at the motor terminals. The voltage must be present and within 10% of the motor nameplate voltage. If voltage is present at the motor but the motor is not operating, remove the motor from the load the motor is driving. Reapply power to the motor. If the motor runs, the problem is with the load. If the motor does not run, the problem is with the motor. Replace or service the motor.

continued

continued

	Open control circuit between incoming power and motor	Check for cleanliness, tightness, and breaks. Use a voltmeter to test the circuit starting with the incoming power and moving to the motor terminals. Voltage generally stops at the problem area.
	Motor shaft does not turn.	Disconnect the motor from the load. If the motor shaft still does not turn, the bearings are frozen. Replace or service the motor.
Motor overheats.	Motor is single phasing.	Check each of the 3ϕ power lines for correct voltage.
	Improper ventilation	Clean all ventilation openings. Vacuum or blow dirt out of motor with low-pressure, dry, compressed air.
	Motor is overloaded.	Check the load for binding. Check shaft straightness. Measure motor current under operating conditions. If the current is above the listed current rating, remove the motor. Remeasure the current under no-load conditions. If the current is excessive under load but not when unloaded, check the load. If the motor draws excessive current when disconnected, replace or service the motor.

TROUBLESHOOTING AC MOTORS   13

Name _____ Date _____   **Activities**

◯ Activity 13-1. AC Motors

The voltage and current draw of a motor are measured when testing a motor. Complete the wiring of the motor. Connect the clamp-on ammeter to measure the current flowing through L1 and the voltmeter to measure the voltage applied to the motor.

1. Connect the split-phase motor.

2. Connect the dual-voltage, split-phase motor.

3. Connect the dual-voltage, wye-connected, 3ϕ motor.

HIGH VOLTAGE (SERIES)

SUPPLY VOLTAGE

Activity 13-2. Troubleshooting AC Motors

Determine whether the meter reading indicates an open winding, shorted winding, faulty centrifugal switch, faulty capacitor, faulty (or open) thermal switch, or a normal reading.

_____ 1. Meter 1 reading indicates _____.

_____ 2. Meter 2 reading indicates _____.

_____ 3. Meter 3 reading indicates _____.

_____ 4. Meter 4 reading indicates _____.

_____ 5. Meter 5 reading indicates _____.

_____ 6. Meter 6 reading indicates _____.

7. Meter 7 reading indicates _____.
_____
8. Meter 8 reading indicates _____.
_____
9. Meter 9 reading indicates _____.
_____

244 TROUBLESHOOTING ELECTRICAL/ELECTRONIC SYSTEMS

_____ **10.** Meter 10 reading indicates _____.

_____ **11.** Meter 11 reading indicates _____.

_____ **12.** Meter 12 reading indicates _____.

POWER DISTRIBUTION — Applications

Application — Power Conditioning and Protection

Many electrical loads such as solenoids, lights, motors, and heating elements may withstand temporary voltage fluctuations (a few milliseconds) without sustaining damage. A *voltage fluctuation* is any change in the voltage level above or below the specified normal range of the power supply. Electronic and other voltage-sensitive equipment may be damaged by temporary voltage fluctuations. A voltage fluctuation in the power supply of a computer may cause memory loss, chip damage, and electronic circuit malfunctions. Solenoids, lights, motors, and heating elements may be damaged if the voltage fluctuation is excessive or lasts longer than a few seconds.

High and Low Voltages

The voltage in a circuit may be consistently, temporarily, or momentarily too high or too low. Voltage may be consistently too high if a circuit is near the beginning of a power distribution system. Voltage may be consistently too low if a circuit is near the end of a power distribution system.

Voltage may be temporarily too high in the office area of a plant when the production lines are shut down. Voltage may be temporarily too low during certain parts of the day or during a brown-out. A *brown-out* is the deliberate reduction of the voltage level by a power company to conserve power during peak usage times.

Voltage may be momentarily too high or too low at any time. A *voltage surge* is a higher-than-normal voltage that temporarily exists on one or more power lines. Voltage surges are normally caused by the switching OFF of heavy electrical loads. A *voltage dip* is a momentary low voltage. Voltage dips are normally caused by the switching ON of heavy electrical loads.

Voltage dips are often followed by voltage surges as voltage regulators overadjust for the voltage dip. *Voltage variance* is the difference in voltage between a voltage surge and a voltage dip. The voltage surge following a voltage dip increases the voltage variance. See Voltage Problems.

A recording meter or a multimeter with a min/max mode is used to measure voltage surges and voltage dips. Input voltage is measured over a 24-hour period when a problem recurs in a piece of electronic equipment. A voltage stabilizer should be added when the voltage fluctuates more than ±8%.

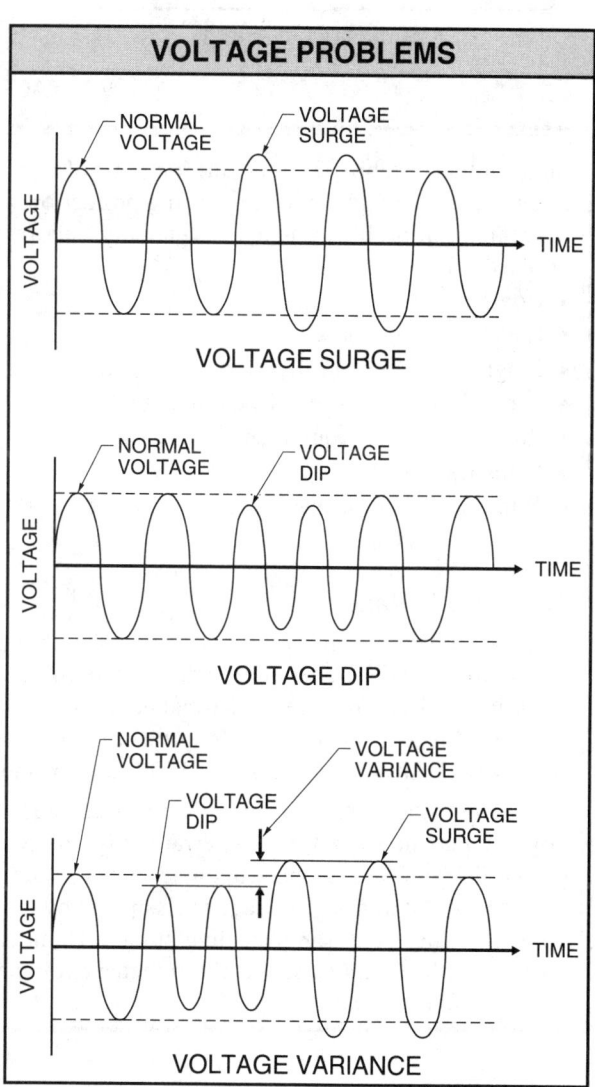

Voltage Stabilizers

Most electronic devices are designed to operate at a constant voltage level. A voltage stabilizer is required for any application containing an electronic circuit that requires a constant voltage level. A *voltage stabilizer* (regulator) is a device that provides precise voltage regulation to protect equipment from voltage dips and voltage surges. See Voltage Stabilizer.

VOLTAGE STABILIZER

Most voltage stabilizers hold the output voltage within ±1%. A voltage stabilizer is placed between the incoming power and the load requiring protection when the input voltage varies to ±15%. A voltage stabilizer prevents such problems as loss of memory, incorrect data output, and shutdowns. Voltage stabilizers should be used on:

- Computers
- Local area networks
- Programmable controllers
- X-ray and other medical equipment
- Radar and sonar equipment
- Industrial robots
- Numerical control machines
- Air traffic control equipment
- Electronic testing and monitoring equipment
- Photographic equipment
- Induction heaters
- Electronic cash registers and banking equipment
- Electronic security systems
- Communications equipment

Transient Voltages

Most power received from a utility company is changed by the addition of transient voltages on the power lines. *Transient voltages* are temporary, unwanted voltages in an electrical circuit. Transient voltages occur due to lightning strikes, unfiltered electrical equipment, and electricity being switched ON and OFF. Transient voltages may range in value from a few volts to several thousand volts and may occur at any phase angle on the AC sine wave. Transient voltages on a 120 V power line may reach 5000 V or more.

Transient voltages differ from voltage surges and voltage dips in that transient voltages are normally much larger in magnitude and are very erratic. Lightning is the most common source of transient voltages on outdoor power distribution systems. Lightning induces a surge that travels in two directions from the point of origin. In most cases, the transient voltage is dissipated by the utility companies' grounding systems after it has traveled 10 to 20 utility poles (or grounding points). Damage is unlikely if equipment is located farther than 20 utility poles from the lightning strike. Unprotected equipment may be severely damaged if the lightning strike is close to equipment in use.

Transient voltages caused by lightning strikes must be considered in all electrical applications since lightning may strike at almost any place. Low-level transient voltages are produced on indoor power lines when most loads are switched OFF. Transient voltages may be continuously produced when using:
- Single-phase switched mode power supplies
- Fluorescent lighting using electronic ballasts
- Variable-speed motor drives
- Soft-start motor starters
- Electronically controlled welding equipment
- Motors that include brushes (DC and universal)

Transient voltages include voltage spikes and electrical noise, or both. A *voltage spike* is an increase in voltage (typically several thousand volts) that lasts for a very short time (microseconds to milliseconds). Voltage spikes are shown as sharp peaks on the AC sine wave. *Electrical noise* is any unwanted disturbance that interferes with a signal. Electrical noise distorts the normally pure AC sine wave. Electrical noise is shown as jagged lines on the AC sine wave. An oscilloscope or scopemeter may be used to detect voltage spikes and electrical noise. See Transient Voltages.

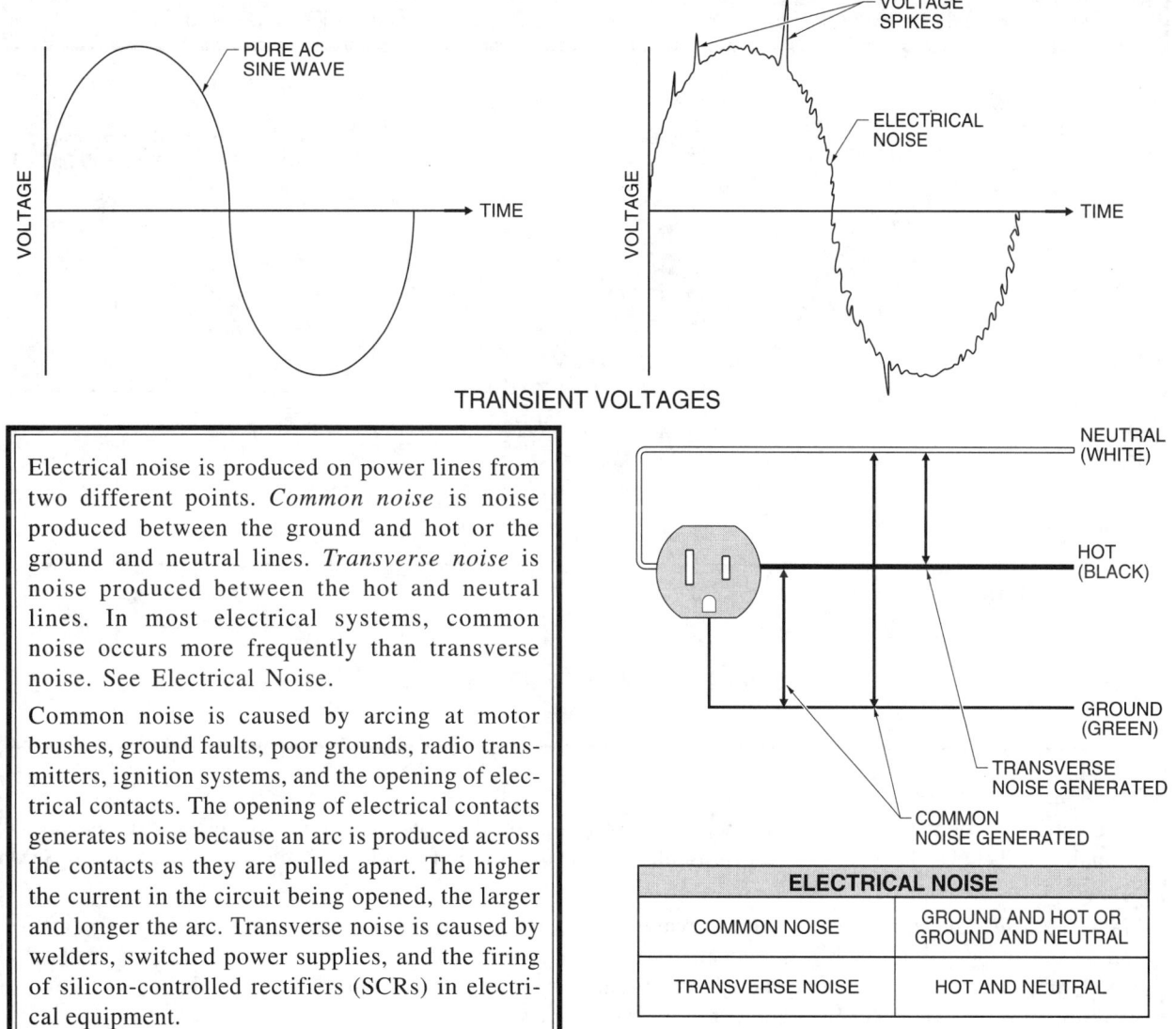

TRANSIENT VOLTAGES

Electrical noise is produced on power lines from two different points. *Common noise* is noise produced between the ground and hot or the ground and neutral lines. *Transverse noise* is noise produced between the hot and neutral lines. In most electrical systems, common noise occurs more frequently than transverse noise. See Electrical Noise.

Common noise is caused by arcing at motor brushes, ground faults, poor grounds, radio transmitters, ignition systems, and the opening of electrical contacts. The opening of electrical contacts generates noise because an arc is produced across the contacts as they are pulled apart. The higher the current in the circuit being opened, the larger and longer the arc. Transverse noise is caused by welders, switched power supplies, and the firing of silicon-controlled rectifiers (SCRs) in electrical equipment.

ELECTRICAL NOISE	
COMMON NOISE	GROUND AND HOT OR GROUND AND NEUTRAL
TRANSVERSE NOISE	HOT AND NEUTRAL

ELECTRICAL NOISE

Voltage Spike and Electrical Noise Protection

Computers, electronic circuits, and specialized equipment require protection against voltage spikes and electrical noise. Protection against voltage spikes and electrical noise includes:

- Proper wiring, grounding, and shielding of the power lines. Proper wiring, grounding, and shielding include the use of a proper grounding system (See NEC® 250.) and also ensure the proper use of the ground and neutral conductors. The neutral conductor should always carry the normal load current. The ground conductor should only carry fault current. The ground conductor should never carry the normal current drawn by a load. The amount of current carried on the hot, neutral, and ground conductors is measured with a clamp-on ammeter. The current measured on the neutral conductor should equal the current measured on the hot conductor under normal operating conditions. The current measured on the ground conductor should always equal zero under normal operating conditions.
- Proper sizing of neutral and ground conductors.
- Ensuring that all wire connections are tight. Loose connections cause noise and arcing which may damage computer and other electronic circuits.
- Proper selecting and applying of surge suppressors and noise suppressors. A *surge suppressor* is a protection device that limits voltage surges which may be present on power lines. Surge suppressors limit voltage surges by discharging or bypassing the damaging voltage. Surge suppressors may be installed separately or may be included with a noise suppressor. See Surge and Noise Suppressor.

SURGE AND NOISE SUPPRESSOR

System Protection

An electrical system is protected at several points along the distribution system. An electrical system is normally protected at the main entrance into a plant, at the distribution panel, and at the individual circuits. See System Protection.

Main entrance protection devices protect against major surges and reduce transient voltages in the main entrance to an electrical distribution system. Main-entrance protection should include surge suppressors, such as lightning arresters that are large enough to dissipate a lightning strike or major utility fault. A major utility fault occurs when two power lines come in contact during a storm.

Power Distribution **249**

SYSTEM PROTECTION

Distribution panel protection devices protect the plant distribution system from internally-generated transient voltages. Distribution panel protection should be used in industrial environments that include industrial controls, motors, automated equipment, and fluorescent lighting. Distribution panel protection should also be used in computer rooms, hospitals, laboratories, banks, and other areas that include electronic equipment. See Distribution Panel Protection.

DISTRIBUTION PANEL PROTECTION

Plug-in protection devices protect individual equipment from voltage surges, dips, and transient voltages. Plug-in protection devices are used in residential and commercial applications to protect computers, TVs, VCRs, word processors, printers, and other electronic equipment. See Plug-In Protection Devices.

PLUG-IN PROTECTION DEVICES

Application — Isolated Ground Receptacles

When a standard receptacle is installed in a metal outlet box, the receptacle ground is connected to the common grounding system. This common grounding system typically includes all metal wiring boxes, conduits, gas pipes, water pipes, and the noncurrent-carrying metal parts of most electrical equipment. When a piece of electrical equipment is plugged into the receptacle, its ground becomes part of the larger grounding system. See Standard Receptacle Grounding.

The common grounding system may act as a large antenna and conduct electrical noise. This electrical noise may cause interference in sensitive electrical equipment such as computers, medical, security, and communication equipment.

An isolated grounded receptacle may be used to minimize problems in sensitive applications or in areas of high electrical noise. An *isolated grounded receptacle* is a receptacle that minimizes electrical noise by providing a separate grounding path. Isolated grounded receptacles are identified by an orange triangle on the face of the receptacle and/or an orange colored faceplate. See Isolated Receptacle Grounding.

In the isolated grounding system, a separate ground conductor is run with the circuit conductors. The grounding system must be installed per the NEC®.

STANDARD RECEPTACLE GROUNDING

ISOLATED RECEPTACLE GROUNDING

Application — Transformer Load Cycle

Transformers are used to deliver power to a set number of loads. For example, a transformer may be used to deliver power to a school. As loads in the school are switched ON and OFF, the power delivered by the transformer changes. At certain times (night), the power output required from the transformer may be low. At other times (during school hours), the power output required from the transformer may be high. Peak load is the maximum output requirement of a transformer. See Transformer Load Cycle.

A transformer is overloaded when it is required to deliver more power than it is rated for. A transformer is not damaged when overloaded for a short time period. A transformer is not damaged because the heat storage capacity of a transformer ensures a relatively slow increase in internal transformer temperature.

To help dissipate heat in a transformer, most power transformers are self-cooled or oil immersed. Transformer manufacturers list the length of time a transformer may safely be overloaded at a given peak level. For example, a transformer that is overloaded 15 times its rated current has a permissible overload time of 5.5 sec. See Temporary Transformer Overloading.

TRANSFORMER LOAD CYCLE

TEMPORARY TRANSFORMER OVERLOADING

Application — Wye and Delta Transformer Connections

Transformers are connected in wye and delta configurations. A *wye configuration* is a transformer connection that has one end of each transformer coil connected together. The remaining end of each coil is connected to the incoming power lines (primary side) or used to supply power to the load(s) (secondary side). A *delta configuration* is a transformer connection that has each transformer coil connected end-to-end to form a closed loop. Each connecting point is connected to the incoming power lines or used to supply power to the load(s). The voltage output and type available for the load(s) is determined by whether the transformer is connected in a wye or delta configuration. See Wye and Delta Configurations.

252 TROUBLESHOOTING ELECTRICAL/ELECTRONIC SYSTEMS

Application — Transformer Tap Connections

A transformer with taps is used to compensate for voltage differences in a control circuit. *Taps* are connecting points that are provided along the transformer coil. Taps are usually provided at 2½% increments along one end of the transformer coil. See Determining Transformer Tap Connections.

DETERMINING TRANSFORMER TAP CONNECTIONS

To determine the proper transformer tap connections, apply the procedure:
1. Measure the incoming voltage on the primary side of the transformer.
2. Determine the secondary voltage. The secondary voltage is the voltage rating of the loads that are connected to the transformer.
3. Determine the connections using the manufacturer's specifications or transformer nameplate.
4. Turn the power OFF, lock out, and tag the incoming power to the transformer. Ensure the power is OFF by testing the circuit with a voltmeter.
5. Connect the transformer per manufacturer's specifications or transformer nameplate. Check each connection twice.
6. Turn the power ON.
7. Measure the secondary voltage of the transformer.
   Repeat Steps 1-6 if the secondary voltage is not correct.

Application — Transformer Sizing and Balancing

A correctly-sized transformer provides years of trouble-free service. Heat destroys a transformer by damaging the insulation on the windings. Heat in a transformer is produced by the current flow through the windings. The greater the current flow, the greater the heat. Overcurrent in a transformer is usually the result of overloading a transformer. Overloading a transformer occurs when a transformer is undersized or loads are unbalanced.

Sizing Single-Phase Transformers

When servicing or replacing a burned-out 1φ transformer, the electrical loads and available supply voltage are considered. A transformer is selected based on the electrical load and supply voltage requirements. See Sizing 1φ Transformers.

SIZING 1φ TRANSFORMERS

To size a 1φ transformer, apply the procedure:
1. Determine the total voltage required by the loads if more than one load is connected. The secondary side of the transformer must have a rating equal to the voltage of the loads.
2. Determine the amperage rating or kVA capacity required by the load(s). Add all loads that are (or may be) ON concurrently.

3. Check load(s) frequency on the nameplate. The frequency of the supply voltage and the electrical load(s) must be the same. Transformers change voltage levels but not frequency.
4. Check the supply voltage to the primary side of the transformer. The primary side of a transformer must have a rating equal to the supply voltage. Consider each voltage if there is more than one source voltage available. Use a transformer that has primary taps if there is a variation in the supply voltage. The transformer must have a kVA capacity of at least 10% greater than that required by the loads. A conversion table may be used to determine proper kVA capacity when the load rating is given in amperes. See Single-Phase Full-Load Currents.

SINGLE-PHASE FULL-LOAD CURRENTS*

kVA	120 V	208 V	240 V	277 V	380V	480 V
.050	.4	.2	.2	.2	.1	.1
.100	.8	.5	.4	.3	.2	.2
.150	1.2	.7	.6	.5	.4	.3
.250	2.0	1.2	1	.9	.6	.5
.500	4.2	2.4	2.1	1.8	1.3	1
.750	6.3	3.6	3.1	2.7	2	1.6
1	8.3	4.8	4.2	3.6	2.6	2.1
1½	12.5	7.2	6.2	5.4	3.9	3.1
2	16.7	9.6	8.3	7.2	5.2	4.2
3	25	14.4	12.5	10.8	7.9	62
5	41	24	20.8	18	13.1	10.4
7½	62	36	31	27	19.7	15.6
10	83	48	41	36	26	20.8
15	125	72	62	54	39	31

* full-load current in amperes

Example: Determining 1ϕ Transformer kVA — Conversion Table

A 240 V circuit has a total load rating of 41 A. Determine the kVA capacity.

A 240 V circuit with a total load rating of 41 A has a kVA capacity of 10 kVA (from conversion table). To calculate kVA capacity when voltage and current are known, apply the formula:

$$kVA_{CAP} = \frac{E \times I}{1000}$$

where
kVA_{CAP} = transformer capacity (in kVA)
E = voltage (in volts)
I = current (in amperes)
1000 = constant

Note: When computing kVA, multiply amperage by voltage. Never use the wattage rating of the loads.

To calculate amperage when kVA capacity and voltage are known, apply the formula:

$$I = \frac{kVA_{CAP} \times 1000}{E}$$

Example: Determining 1ϕ Transformer kVA — Calculation

A 120 V circuit has loads of 5 A and 15 A. Determine the capacity (in kVA) required by the secondary of the transformer.

$$kVA_{CAP} = \frac{E \times I}{1000}$$

$$kVA_{CAP} = \frac{120 \times 20}{1000}$$

$$kVA_{CAP} = \frac{2400}{1000}$$

$$kVA_{CAP} = \mathbf{2.4 \text{ kVA}}$$

Sizing Three-Phase Transformers

When servicing or replacing a burned-out 3ϕ transformer, the electrical loads and available supply voltage are considered. A transformer is selected based on its electrical load and supply voltage requirements. See Sizing 3ϕ Transformers.

To size a 3ϕ transformer, apply the procedure:
1. Determine the total voltage required by the loads if more than one load is connected. The secondary side of the transformer must have a rating equal to the voltage of the loads.
2. Determine the amperage rating or kVA capacity required by the load(s). Add all loads that are (or may be) on concurrently.
3. Check load(s) frequency on the nameplate. The frequency of the supply voltage and the electrical load(s) must be the same. Transformers change voltage levels, not frequency.
4. Determine the type of 3ϕ voltage available. This includes three-wire no ground or three-wire with ground (four-wire).
5. Check the supply voltage to the primary side of the transformer.

SIZING 3ϕ TRANSFORMERS

The primary side of a transformer must have a rating equal to the supply voltage. Consider each voltage when there is more than one source of voltage available. Use a transformer that has primary taps when there is a variation in the supply voltage. The transformer must have a kVA capacity of at least 10% greater than that required by the loads.

A conversion table is used to determine the kVA capacity of 3ϕ circuits when the load rating is given in amperes. See Three-Phase Full-Load Currents.

THREE-PHASE FULL-LOAD CURRENTS*

kVA	208 V	240 V	480 V	600 V
3	8.3	7.2	3.6	2.9
4	12.5	10.8	5.4	4.3
6	16.6	14.4	7.2	5.8
9	25	21.6	10.8	8.6
15	41	36	18	14.4
22	62	54	27	21.6
30	83	72	36	28
45	124	108	54	43

* full-load current in amperes

To calculate kVA capacity when voltage and current are known, apply the formula:

$$kVA_{CAP} = \frac{E \times 1.732 \times I}{1000}$$

where
kVA_{CAP} = transformer capacity (in kVA)
E = voltage (in volts)
1.732 = constant (for 3ϕ power)
I = current
1000 = constant

Note: When calculating kVA, multiply amperage by voltage. Never use the wattage rating of the loads.

To calculate 3ϕ amperage when kVA capacity and voltage are known, apply the formula:

$$I = kVA_{CAP} \times \frac{1000}{E} \times 1.732$$

Note: When 3ϕ problems are calculated, the following values* may be substituted to eliminate one mathematical step:

For 208 V × 1.732, use 360
For 230 V × 1.732, use 398
For 240 V × 1.732, use 416
For 440 V × 1.732, use 762
For 460 V × 1.732, use 797
For 480 V × 1.732, use 831

* $\sqrt{3} = 1.732$

Example: Determining Three-Phase Transformer kVA — Calculation

A 480 V, 3ϕ circuit has two loads of 18 A each. Determine the capacity (in kVA) required by the secondary of the transformer. See Transformer Capacity Calculation.

$$kVA_{CAP} = \frac{E \times 1.732 \times I}{1000}$$

$$kVA_{CAP} = \frac{831 \times 36}{1000}$$

$$kVA_{CAP} = \frac{29,916}{1000}$$

$$kVA_{CAP} = \mathbf{29.916 \text{ kVA}}$$

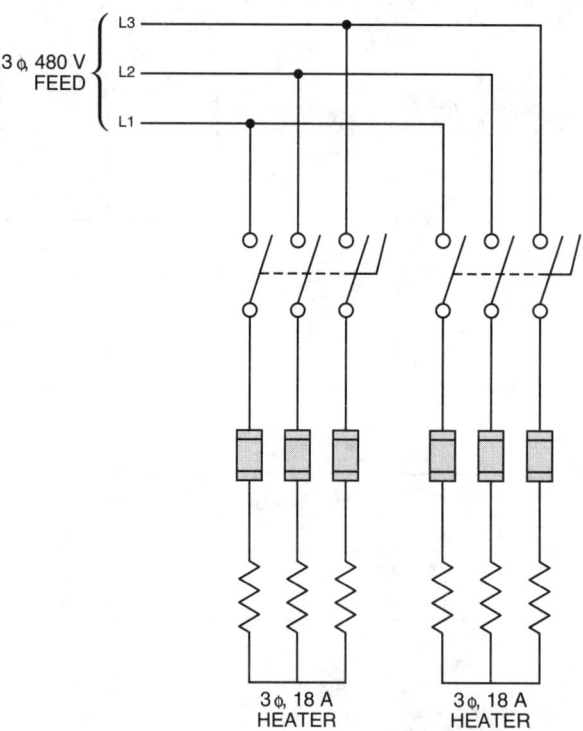

TRANSFORMER CAPACITY CALCULATION

Standard Ambient Temperature Compensation

Temperature rise in a transformer is the temperature of the windings above the existing ambient temperature. Transformers list their maximum temperature rise. Normal ambient temperature is 40°C.

A transformer must be derated if the ambient temperature exceeds 40°C. Transformer derating charts are used to derate transformers in high ambient temperatures. See Transformer Deratings.

TRANSFORMER DERATINGS

Maximum Ambient Temperature (°C)	Maximum Transformer Loading (%)
40	100
45	96
50	92
55	88
60	81
65	80
70	76

Example: Standard Ambient Temperature Compensation

Calculate the rating of a 2.5 kVA transformer installed in an ambient temperature of 50°C.

kVA = Rated kVA × Maximum Load
kVA = 2.5 × .92 (See Transformer Deratings.)
kVA = **2.3 kVA**

Special Ambient Temperature Compensation

Standard ambient temperature is the average temperature of the air that cools a transformer over a 24-hour period. Standard ambient temperature assumes that maximum temperature does not exceed 10°C above average ambient temperature.

A transformer is derated above the standard values when the maximum temperature exceeds the average temperature by more than 10°C. A transformer is derated by 1½% for each 1°C above 40°C when the maximum ambient temperature exceeds 10°C above the average temperature. See Standard Compensation Rating and Special Compensation Rating.

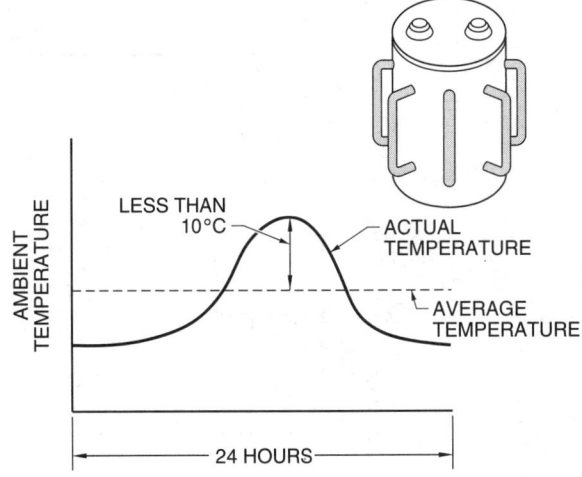

STANDARD COMPENSATION RATING

Example: Special Ambient Temperature Compensation

Calculate the rating of a 2.5 kVA transformer installed in an average ambient temperature of 50°C. The maximum temperature exceeds 10°C above average temperature.

The transformer is derated by 1½% for each degree above 40°C. A 50°C average temperature is 10°C above 40°C, so the transformer must be derated 15% (10°C × 1½% = 15%).

kVA = Rated kVA × Maximum Load
kVA = 2.5 × .85 (15% derating)
kVA = **2.125 kVA**

SPECIAL COMPENSATION RATING

Transformer Load Balancing

When connecting loads to a transformer, the loads should be connected so that the transformer is as electrically balanced as possible. *Electrical balance* occurs when loads on a transformer are placed so that each coil of the transformer carries the same amount of current. See Transformer Load Balancing.

A clamp-on ammeter is used to check for an unbalanced transformer. The ammeter shows the amount of current draw on each line. The measured current draw on each hot (ungrounded) power line should be nearly equal. The loads should be placed in balance if any current reading is higher than the others and the transformer is at or near its capacity.

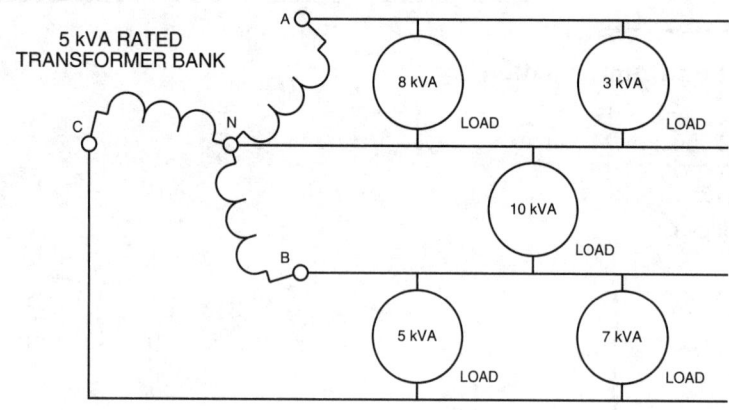

THREE-PHASE

CORRECT LOADING

A TO N = 11 kVA
B TO N = 10 kVA
C TO N = 12 kVA

EACH TRANSFORMER WINDING LOADED LESS THAN 15 kVA/ϕ LIMIT

$$\frac{45 \text{ kVA}}{3\phi} = \text{kVA}/\phi \text{ LIMIT}$$

SINGLE-PHASE

CORRECT LOADING

ELECTRICALLY BALANCE EACH SECONDARY WINDING RATED AND LOADED TO 5 kVA

TRANSFORMER LOAD BALANCING

POWER DISTRIBUTION

Name _____ Date _____

Activities

Activity 14-1. Power Conditioning and Protection

Surge suppressor manufacturers provide voltage specifications for each surge suppressor. The voltage specifications show the let-through voltage of a surge suppressor as shown on an oscilloscope or scopemeter. *Let-through voltage* is the maximum voltage surge allowed to pass through a surge suppressor. The more sensitive the equipment, the less the let-through voltage value should be. The equipment to be protected determines the best surge suppressor to use because cost increases as let-through voltage decreases. Determine the let-through voltage if the surge suppressor allows the voltage spikes to pass.

_____ **1.** Maximum peak let-through voltage equals _____ V.

_____ **2.** Maximum peak let-through voltage equals _____ V.

260 TROUBLESHOOTING ELECTRICAL/ELECTRONIC SYSTEMS

_____ **3.** Maximum peak let-through voltage equals _____ V.

_____ **4.** Maximum peak let-through voltage equals _____ V.

_____ **5.** Maximum peak let-through voltage equals _____ V.

Power Distribution **261**

◯ Activity 14-2. Isolated Ground Receptacles

1. Draw arrows to indicate the normal path of current flow through the receptacle's conductors when an electrical device is plugged into the receptacle.

A.

B.

2. Draw arrows to indicate the path to ground for electrical noise entering the grounded frame of an electrical device plugged into the receptacle.

A.

B.

◯ Activity 14-3. Transformer Load Cycle

Using Temporary Transformer Overloading, determine the permissible overload time for each transformer.

_____ **1.** Permissible overload time equals _____.

TEMPORARY TRANSFORMER OVERLOADING

2. Permissible overload time equals _____.

3. Permissible overload time equals _____.

4. Permissible overload time equals _____.

Power Distribution

Activity 14-4. Wye and Delta Transformer Connections

1. Connect Transformer Bank 1 in a delta-delta configuration. Connect the primary side of each transformer to the main power lines to form a basic delta transformer bank. Connect the secondary side of each transformer to form a delta transformer bank with the center tap of Transformer 1 used to produce a 1φ, low-voltage (120 VAC) neutral. Connect each load to the secondary side of the transformer bank so that each load is connected to the proper voltage. Connect the loads so that the transformer bank is as balanced as possible.

240 V, 3φ
50 A MOTOR

240 V, 1φ
10 A MOTOR

120 V, 1φ
15 A MOTOR

120 V, 5 A
LIGHT

120 V, 5 A
LIGHT

120 V, 5 A
LIGHT

2. Connect Transformer Bank 2 in a delta-wye configuration. Connect the primary side of each transformer to the main power lines to form a basic delta transformer bank. Connect the secondary side of each transformer to form a wye transformer bank with a common neutral. Connect each load to the secondary side of the transformer bank so that each load is connected to the proper voltage. Connect the loads so that the transformer bank is as balanced as possible.

Power Distribution **265**

3. Connect Transformer Bank 3 in a wye-wye configuration. Connect the primary side of each transformer to the main power lines to form a basic wye transformer bank with a common neutral. Connect the secondary side of each transformer to form a wye transformer bank with a common neutral. Connect each load to the secondary side of the transformer bank so that each load is connected to the proper voltage. Connect the loads so that the transformer bank is as balanced as possible.

266 TROUBLESHOOTING ELECTRICAL/ELECTRONIC SYSTEMS

4. Connect Transformer Bank 4 in a wye-delta configuration. Connect the primary side of each transformer to the main power lines to form a basic wye transformer bank with a common neutral. Connect the secondary side of each transformer to form a delta transformer bank. Connect each load to the secondary side of the transformer bank so that each load is connected to the proper voltage. Connect the loads so that the transformer bank is as balanced as possible.

Activity 14-5. Transformer Tap Connections

Each tap on the primary side of a transformer changes the secondary voltage by about $2\frac{1}{2}\%$. Determine the new position of each tap to adjust the output voltage.

_____ 1. The new tap position for Meter Reading 1 is _____.

_____ 2. The new tap position for Meter Reading 2 is _____.

Activity 14-6. Transformer Sizing and Balancing

Determine the kVA capacity of the loads. Use Transformer Selections to determine the catalog number of the smallest transformer that could be used for each application. *Note:* Select a transformer that has at least 10% greater capacity than that required by the loads.

1.

SINGLE-PHASE LOADS				
Device	Quantity	Volts	Amps	Usage
Lamps	50	120	1	50%
¼ HP motor	2	120	5.8	100%
1 HP motor	1	240	8	100%
2 HP motor	2	240	16	50%
Heater	1	120	15	100%
Heater	1	240	30	100%
Heater	1	240	10	100%

TRANSFORMER SELECTIONS				
Catalog Number	Phase	Primary Voltage	Secondary Voltage	kVA Rating
01	1	240/480	120/240	37.5
02	1	240/480	120/240	50
03	1	240/480	120/240	75
04A	1	208	120/240	50
04B	1	208	120/240	75
05A	1	277	120/240	50
05B	1	277	120/240	75
06	3	480 delta	208/120 wye	75
07	3	480 delta	208/120 wye	112.5
08	3	480 delta	208/120 wye	150
09	3	480 delta	208/120 wye	225
10	3	480 delta	240 delta	75
11	3	480 delta	240 delta	150
12	3	600 delta	208/120 wye	75
13	3	600 delta	208/120 wye	112.5
14	3	600 delta	208/120 wye	150
15	3	600 delta	208/120 wye	225

_____ A. The kVA capacity for 120 V loads equals _____ kVA.

_____ B. The kVA capacity for 240 V loads equals _____ kVA.

_____ C. The kVA capacity for 120 V and 240 V loads equals _____ kVA.

_____ D. Transformer catalog number equals _____.

2.

THREE-PHASE LOADS				
Device	Quantity	Volts	Amps	Usage
Heater	2	208	25	50%
25 HP motor	2	208	75	50%
10 HP motor	1	208	31	100%
2 HP motor	1	208	7.5	100%

SINGLE-PHASE LOADS				
Device	Quantity	Volts	Amps	Usage
Lamps	100	120	1	75%
Heater	10	120	5	100%
½ HP motor	5	120	9.8	100%

_____ A. The kVA capacity for 3φ loads equals _____ kVA.

_____ B. The kVA capacity for 1φ loads equals _____ kVA.

_____ C. Total kVA capacity for 1φ and 3φ loads equals _____ kVA.

_____ D. Transformer catalog number with 480 primary equals _____.

_____ E. Transformer catalog number with 600 primary equals _____.

TROUBLESHOOTING MOTOR CONTROL CIRCUITS — Applications

Application — Power Loss

The first step in troubleshooting an electrical circuit is to ensure that the circuit has power. Power in a circuit is used by lights, motors, solenoids, etc. All electrical circuits are protected from short circuits, overcurrents, and overloads. Circuit overcurrent protection is provided by fuses, CBs, and overloads. These protection devices are located throughout the electrical system from the point at which power is delivered to a building to the final point at which the electricity is used. See Power Circuit Distribution System.

When power is lost in a circuit, the problem is normally caused by a fuse, CB, or an overload contact. When a protection device removes power from a circuit, a troubleshooter must reestablish power and find the reason the protection device removed power.

POWER CIRCUIT DISTRIBUTION SYSTEM

270 TROUBLESHOOTING ELECTRICAL/ELECTRONIC SYSTEMS

When the power in a circuit is lost, an overall check of the electrical system is required since the power may be lost at any one or more of the protection points in the system. See System Protection Points.

Overload Contacts

Overload contacts have a built-in time delay to allow a motor to start. When overload contacts trip, the problem normally occurs because the motor is overloaded when running. Power is reestablished by resetting the overload contacts. The reason that the overload contacts tripped should be determined, especially if the overload contacts trip again. Check for the correct size overload heaters and a motor problem.

SYSTEM PROTECTION POINTS

Fuses and CBs

Fuses open and CBs trip in the power circuit when there is a short circuit or overcurrent condition. A *short circuit* occurs when the current in a circuit takes a shortcut around the normal path of current flow. An *overcurrent* is any current above the normal current level. See Normal Current Flow and Short Circuit.

When fuses open or CBs trip in the power circuit, the problem is normally a motor starting (not running) problem. Power is reestablished by changing the fuses or resetting the CBs. The reason the fuses opened or the CBs tripped should be determined, especially if the fuses open or the CBs trip again. Check for the correct size fuses or CBs and a short circuit. Fuses open and CBs trip the moment power is turned ON when there is a short circuit. Fuses open and CBs trip a few seconds after power is turned ON when there is an overload.

Fuses open and CBs trip in the control circuit when a short circuit or overcurrent condition exists in the control circuit. A short circuit is caused when power lines (L1 and L2) are connected or one of the ungrounded (hot) power lines is connected to ground. An overcurrent is caused by an excessive current draw in the control circuit. Excessive current draw normally occurs when there is a problem with one of the loads in the control circuit or additional loads are added to the circuit. Power is reestablished by changing the fuses or resetting the CBs. The reason the fuses opened or CBs tripped should be determined, especially if the fuses open or CBs trip again. Check for the correct size control transformer and a short circuit.

▮ Application — Troubleshooting Power Circuits

A *power circuit* is the part of an electrical circuit that connects the loads to the main power lines. *Loads* are devices that convert electrical energy to mechanical energy, heat, light, or sound. The starting point for troubleshooting the power circuit depends on the problem(s) observed. Troubleshooting the power circuit normally involves determining the point in the system where power is lost. This point may be at the load, the primary substation, or any point between the two.

Building Main Problems

A fault in the building main substation, transformers, or utility feed may interrupt power to a building. Problems in one of these areas cause a loss of power to all or most of the building. Common problems in a building main include lightning strikes, short circuits, accidental grounding, and brownouts. The building main and feeder circuit breakers are designed to open the circuit during a fault to prevent fatal injuries, damage to equipment, and electrical fires. When building main power is lost, a troubleshooter must find the location of the open, determine the problem, and reestablish power. See Troubleshooting Building Mains.

TROUBLESHOOTING BUILDING MAINS

To troubleshoot a building main, apply the procedure:
1. Check the unit substation and main switchboard. Check the power coming into the building when the entire building is without power. Contact the utility company when there is no power coming into the building.
2. Ensure that the power coming into the building is at the correct voltage level and present on each power line. Care must be taken when the power is correct. There is probably a short circuit in the system if a main CB trips. Turn OFF all circuits connected to the main circuit before resetting the main CB.
3. Reset the main CB and turn the circuits ON one at a time. Fuses or CBs trip the moment power is turned ON when there is a short circuit in one of the feeder lines.

Individual Load Problems

Individual load problems occur when one of the loads in a system is not operating. Individual load problems normally indicate a problem with one section (usually the load) of a system. A system problem may exist when the problem occurs at certain times or after repairs are made. The system problem may include low voltage, transient voltages, operator error, or a material or production problem. See Troubleshooting Individual Load Problems.

Troubleshooting Motor Control Circuits

TROUBLESHOOTING INDIVIDUAL LOAD PROBLEMS

To troubleshoot a power circuit containing individual load problems, apply the procedure:
1. Check the fuses, CBs, and overload contacts closest to the load. Reset the devices and try to restart the motor when the overcurrent protection devices are tripped.
2. Measure the current draw of the motor to ensure there is no excessive current draw when the motor starts. Disconnect the load from the motor and measure the current when there is an excessive current draw. Check the motor if the current draw is still excessive. Check for an excessive load if the current returns to normal.
3. Measure the incoming supply voltage on the line side of the starter if the overloads are not tripped. Measure the voltage between L1 and L2, L1 and L3, and L2 and L3. The problem lies upstream from the starter if the voltage is not the same between L1, L2, and L3 or at the correct level. Check the fuses and CBs to which the starter is connected. See Step 6.
4. Measure the voltage after the starter power contacts and before the overload contacts if the voltage is correct. Manually close the overload contacts if they cannot be electrically engaged. Most manufacturers include a manual test lever or point on the starter. The problem lies downstream from the overload contacts when the measured voltage between L1, L2, and L3 is the same and at the correct level. The problem lies with the starter power contacts when the voltage between L1, L2, and L3 is not the same or at the correct level.
5. Measure the voltage after the starter overload contacts. This is the point at which the load is connected to the starter. The problem lies downstream from the overload contacts when the measured voltage between L1, L2, and L3 is the same and at the correct level. Check the motor starter by measuring the voltage at the motor terminals. The problem lies in the overload contacts when the measured voltage between L1, L2, and L3 is not the same and at the correct level. Replace any overload that is open, burned, or has any other problem.
6. Measure the voltage coming into the fuses or CBs when the correct voltage is at the motor starter. The problem lies in the fuses or CBs when the voltage between L1, L2, and L3 is the same and at the correct level. Test for an open fuse or CB. The problem lies upstream from the starter when the measured voltage between L1, L2, and L3 is not the same or at the correct level. Check the next main fuse or CB protecting the unit.

274 TROUBLESHOOTING ELECTRICAL/ELECTRONIC SYSTEMS

> **Multiple Load Problems**
>
> Multiple load problems occur when more than one load in a system does not operate. Multiple load problems normally indicate malfunctions that have opened the fuses or tripped the CBs which protect the system. The problem is normally a short circuit in the system, a load that has a major problem (drawing excessive current), or all of the loads together are drawing excessive current. See Troubleshooting Multiple Load Problems.

PROBLEM: MORE THAN ONE MOTOR DOES NOT START

TROUBLESHOOTING MULTIPLE LOAD PROBLEMS

> To troubleshoot a power circuit with multiple load problems, apply the procedure:
> 1. Measure the voltage into the main fuses or CB protecting the loads. Check the motor control center, secondary switchboard, or power panel that feeds the loads. The problem lies upstream from the main fuses or CB when the voltage is not the same and at the correct level. Check the next main fuse or CB feeding the unit. Check the voltage out of the fuses or CB when the voltage is correct. The problem lies in the fuses or CB when the voltage is not correct. Test for an open fuse or CB. Troubleshoot from the fuses or CB to the load when the main fuses or CB has power. Troubleshoot from the fuses or CB back to the main power into the building when the main fuses or CB does not have power.
> 2. Measure the voltage at the primary side of the control transformer. The problem lies upstream from the control transformer when the voltage is not present or is incorrect. Check the fuses or CB that protects the transformer.
> 3. Measure the voltage at the secondary of the transformer. The problem lies in the control transformer when the voltage is not present or is incorrect. Replace the transformer.
> 4. Measure the voltage exiting the fuse which protects the secondary of the transformer. The problem lies in the fuse when the voltage is not present or is incorrect. Replace the fuse.

Application — Troubleshooting Control Circuits

A line diagram illustrates the logic of the control circuit in simplest form. A line diagram is used when troubleshooting the control circuit. See Troubleshooting Control Circuits.

To troubleshoot the control circuit, apply the procedure:
1. Measure the voltage out of the control circuit transformer. Connect a voltmeter across the secondary side of the transformer to measure the power in the control circuit. Troubleshoot the power circuit feeding the transformer when the voltage at the secondary side of the transformer is incorrect. Check the fuses or CBs protecting the power to the transformer when there is no power entering the transformer. The transformer is the problem when there is power entering but no power exiting the transformer. Replace the transformer.
2. Check the voltage through the control circuit when there is power exiting the transformer. Leave one side of the voltmeter lead connected to the X2 side of the control transformer when the secondary side of the transformer has the correct voltage present. Move the other meter lead from the X1 side of the transformer through the control circuit connecting it to the input and output side of each component. Voltage should be present at each point to which the test lead is moved. The point where voltage is lost is the point where the circuit is open.
3. Test the voltage in and out of each switch. Press each button to test the natural open circuit of each selector switch and pushbutton. No voltage should exit a NO switch when the switch is open. Voltage should only exit a NO switch when the switch is closed.
4. Check the voltage delivered by the switches at the L1 side of the starter coil. Move one test lead through the control circuit until the test lead is at the L1 side of the starter coil. Voltage at this point indicates the control switches are delivering the voltage to the starter coil.
5. Check the overload contact. Move one test lead from the X2 side of the transformer to each side of the overload contacts. Leave the other with the other test lead at the L1 side of the starter coil. The problem is an open overload contact when voltage is present at the L2 side of the overload contact but not on the coil side. Check to ensure the reset function on the overload is working. The coil is the problem when the meter measures a voltage when connected directly across the starting coil but the starter is not energized. Replace the starting coil.

TROUBLESHOOTING CONTROL CIRCUITS

276 TROUBLESHOOTING ELECTRICAL/ELECTRONIC SYSTEMS

Control Panel Testing

A line diagram is used to illustrate the flow of current through the various devices in a control circuit. The line diagram is intended to demonstrate the basic operation of the circuit and not the physical wiring of each device. The actual wiring of the control circuit is found in the circuit wiring diagram. In troubleshooting, the line diagram is used to show circuit logic, and the wiring diagram is used to find the actual test points at which a meter is connected. See Troubleshooting Inside the Control Panel.

TROUBLESHOOTING INSIDE THE CONTROL PANEL

Note: To simplify the control panel wiring, only a portion of the circuit is shown wired.

A voltmeter is used to measure the voltage out of the control transformer. In the line diagram, the meter is shown connected across the secondary side of the transformer and the fuse. In the control panel, the voltmeter is actually connected to the terminal strip. To measure the voltage out of the control transformer, the meter is connected to terminal posts 1 and 5. Using a terminal strip simplifies the troubleshooting process. Start at the terminal strip when troubleshooting inside a control panel.

To troubleshoot inside the control panel, apply the procedure:
1. Use the line diagram as a reference point for determining where to connect the meter and what the expected meter readings should be.
2. Measure the voltage at the terminal strip. After finding the point on the terminal strip at which power is lost, troubleshoot the individual component(s) connected to that point.

TROUBLESHOOTING MOTOR CONTROL CIRCUITS

15

Name _____ Date _____

Activities

◯ Activity 15-1. Power Loss

Using Motor Control Circuit 1, determine the starting point for troubleshooting each problem.

1. The operator reports that the pump motor in Motor Control Circuit 1 does not operate. The pump light turns ON when the selector switch is placed in the hand position and the Run Pushbutton is pressed. The pump motor does not turn ON. The discharge motor operates correctly when the Start Pushbutton and Stop Pushbutton are pressed. The cooling motor operates when the Override Pushbutton is pressed.

_____ **A.** Troubleshooting should start at the _____.

MOTOR CONTROL CIRCUIT 1

277

278 TROUBLESHOOTING ELECTRICAL/ELECTRONIC SYSTEMS

2. The operator reports that none of the motors in Motor Control Circuit 1 are operating. None of the motors operate manually when tested by pressing the pushbuttons at the operator station. None of the indicator lights operate. The starter overload contacts and circuit breakers are not tripped. Manually operating the starting contacts by pressing down on the contact assembly starts the pump motor.

    _____ A. Troubleshooting should start at the _____.

3. The operator reports that none of the motors in Motor Control Circuit 1 operate. None of the motors operate manually when tested by pressing the pushbuttons at the operator station. None of the indicator lights operate. The starter overload contacts and circuit breakers are not tripped. Manually operating the starting contacts by pressing down on the contact assembly does not start the pump motor.

    _____ A. Troubleshooting should start at the _____.

Activity 15-2. Troubleshooting Power Circuits

1. Connect each meter to measure the required voltage. Connect Meter A so the meter checks the incoming voltage on L1 and L3 before the main disconnect. Connect Meter B so the meter checks the voltage at motor terminals two and three of Motor 2. Connect Meter C so the meter checks the voltage out of the heaters delivering L1 and L3 power to Motor 1. Connect Meter D so the meter checks Fuse 1 of Motor 2. Connect Meter E so the meter checks Fuse 2 of Motor 1. Connect Meter F so the meter checks Main Fuse 3.

Activity 15-3. Troubleshooting Control Circuits

1. Using the Operator Station Wiring Diagram, answer the questions.

    A. Meter A reads _____ V when the selector switch is in the auto position.

    B. Meter B reads _____ V when the selector switch is in the auto position.

    C. Meter B reads _____ V when the selector switch is in the hand position and the Run Pushbutton is pressed.

    D. Meter C reads _____ V when the Override Pushbutton is pressed.

OPERATOR STATION WIRING DIAGRAM

2. Connect each meter to measure the required voltage. Connect Meter A at the terminal strip to measure the voltage directly into the control circuit. Connect Meter B at the terminal strip to measure the voltage out of the Run Pushbutton when the selector switch is in the hand position. Connect Meter C to measure the voltage directly across the light that indicates the cooling motor is ON. Connect Meter D at the terminal strip to measure the voltage out of the Stop Pushbutton. Connect Meter E at the terminal strip to measure the voltage out of the Start Pushbutton.

3. Connect each meter to measure the required voltage. Using the Overhead Door Control Circuit, connect Meter A at the terminal strip so that the meter tests the supply voltage into the control circuit. Connect Meter B at the terminal strip so that the meter tests the voltage out of the Stop Pushbutton. Connect Meter C at the terminal strip so that the meter tests the voltage across the starter coil that raises the door. Connect Meter D at the terminal strip so that the meter tests the voltage across the starter coil that lowers the door.

OVERHEAD DOOR CONTROL CIRCUIT

282 TROUBLESHOOTING ELECTRICAL/ELECTRONIC SYSTEMS

4. Using the Overhead Door Control Circuit, determine the problem. *Note:* The operator reports that the door does not rise when the Up Pushbutton is pressed. Meter A reads 240 V at all times. Meter B reads 240 V when the Up Pushbutton is pressed, but there is no movement of the door. Meter C reads 240 V when the Down Pushbutton is pressed, and the door attempts to move further down.

_____ **A.** The problem is _____.

5. Using the Overhead Door Control Circuit, determine the problem. *Note:* The operator reports that the door opens only when the Up Pushbutton is held closed. The door automatically closes after the Down Pushbutton is pressed and released. Meter A reads 240 V only when the Up Pushbutton is held closed. Meter B reads 240 V only when the Up Pushbutton is held closed.

_____ **A.** The problem is _____.

LIGHTING CIRCUITS — Applications

▫ Application — Lighting Terminology

Lamp output is expressed in lumen. A *lumen (lm)* is the unit used to measure the total amount of light produced by a light source. Lamps are rated in initial lumen and mean lumen. *Initial lumen* is the amount of light produced when a lamp is new. *Mean lumen* is the average light produced after a lamp has operated about 40% of its rated life. See Lamp Ratings.

Illumination

Illumination is the effect that occurs when light falls on a surface. The unit of measure of illumination is the footcandle. A *footcandle (fc)* is the amount of light produced by a lamp (lumen) divided by the area that is illuminated.

LAMP RATINGS		
Lamp	Initial Lumen	Mean Lumen
40 W standard incandescent	480	N/A
100 W standard incandescent	1750	N/A
40 W standard fluorescent	3400	3100
100 W tungsten-halogen	1800	1675
100 W mercury-vapor	4000	3000
250 W mercury-vapor	12,000	9800
100 W high-pressure sodium	9500	8500
250 W high-pressure sodium	30,000	27,000
250 W metal-halide	20,000	17,000

The light falling on a surface may be direct or indirect light. *Direct light* is light that travels directly from a lamp to the surface being illuminated. *Indirect light* is light that is reflected from one or more objects to the surface being illuminated. In indoor applications, both direct and indirect light is considered. In outdoor applications, only direct light is considered. See Direct Light and Indirect Light.

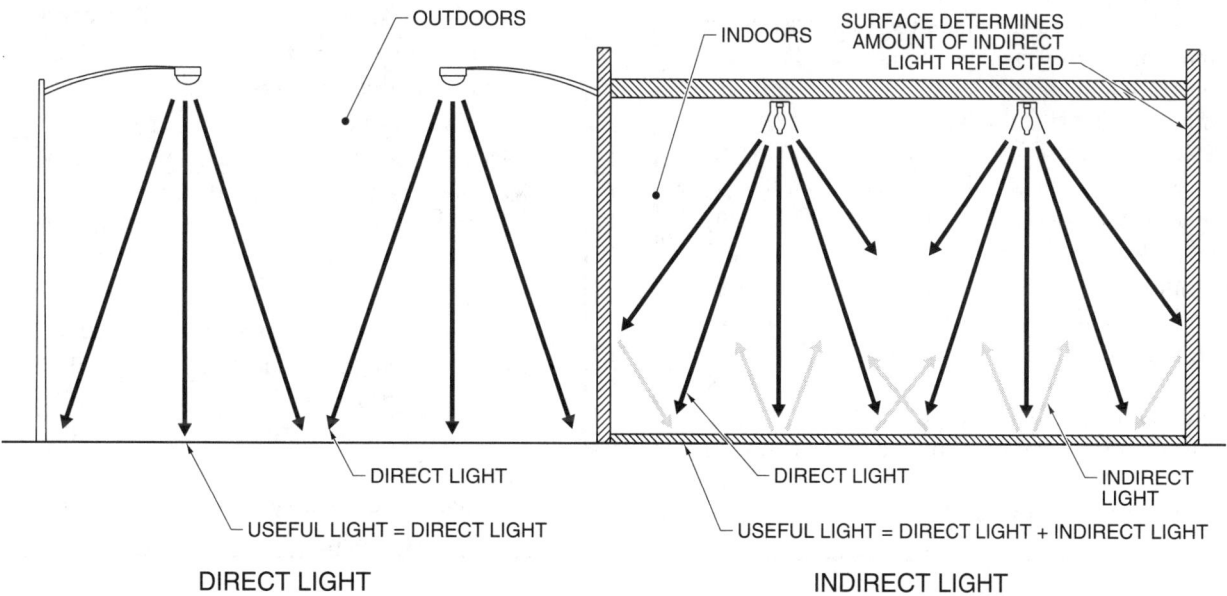

DIRECT LIGHT

INDIRECT LIGHT

The amount of illumination produced by a lamp may be measured or calculated. Footcandle level is measured using a light meter. A *light meter* is a portable instrument that measures light. A light meter measures from a fraction of a footcandle to several thousand footcandles.

A number of readings are taken at different locations and an average of the readings is used because a light meter indicates the level of illumination only at the location of the light-sensitive measuring element. When measuring the amount of illumination, both horizontal and vertical readings are taken to measure both direct and indirect light. See Light Meter Use.

LIGHT METER USE

The intensity of a light source in a given direction depends on the angle the light travels from the lamp. Some manufacturers rate a lamp's output in candela. *Candela (cd)* is the unit of luminous intensity produced by a light source in a given direction. For example, a 175 W mercury-vapor lamp's bulb may be rated at 181 cd at a 45° vertical angle, 135 cd at 60°, and 11 cd at 90°.

Illumination is calculated before lamps are installed or measured after lamps are in operation to determine the number of footcandles that are produced on a given surface. This is checked against recommended light levels. For example, the recommended light level for an accounting office is 150 fc. See Recommended Light Levels.

| RECOMMENDED LIGHT LEVELS ||
Area	Footcandles
Hospital operating table	2500
Factory assembly area	100-200
Accounting office	150
Major league baseball infield	150
Major league baseball outfield	100
School classroom	60-90
Home kitchen	50-70
Bank lobby	50
Active warehouse storage area	20-30
Inactive warehouse storage area	5
Auditorium	15
Elevator	10
Parking lot	5
Interstate roadway	1.4
Street	.9

Manufacturers use footcandle levels to aid in selecting and installing lamps. Manufacturers' charts list the expected footcandle rating for an area given the mounting height and spacing of the lamps. For example, about 30 fc is produced when lamps are mounted 25′ high and are spaced 30′ apart. Thirty footcandles satisfy the aisle floor lighting requirements of an active warehouse storage area. See Aisle Floor Footcandles.

AISLE FLOOR FOOTCANDLES

Spacing*	Mounting Height*						
	15	20	25	30	35	40	45	50
20	—	—	46	39	33	63	26	24
25	—	45	36	30	26	24	21	19
30	—	37	30	25	22	19	18	15
35	—	31	26	22	19	17	15	14
40	33	28	23	19	17	14	13	12
45	30	25	20	17	14	13	12	10

* in feet

The location of a lamp is considered when determining the amount of light produced on a surface. Manufacturers provide light distribution data that is used to determine the amount of light produced for different lamp types. See Determining Footcandles.

DETERMINING FOOTCANDLES

286 TROUBLESHOOTING ELECTRICAL/ELECTRONIC SYSTEMS

To determine the light delivered to an indoor surface, apply the procedure:
1. Determine the mounting height of the lamp fixture.
2. Determine the horizontal distance from the fixture location to the measuring point.
3. Determine the vertical angle. See Step 4 if the vertical angle is given. The vertical angle is found on Horizontal Footcandles Per 1000 Candelas. The vertical angle is determined by the sloping line that intersects the mounting height and horizontal distance. The sloping line is followed to the vertical angle.
4. Determine the candelas delivered to the measuring point. The candelas delivered are found on Light Distribution Data* for the lamp used. To determine the candelas delivered to the measuring point, apply the formula:

$$cd = rcd \times \frac{rlm}{1000}$$

where
cd = candelas
rcd = rated candela at given vertical angle (from Light Distribution Data* for lamp used)
rlm = rated lamp lumen
1000 = constant

5. Determine the footcandles delivered per 1000 candelas at the given vertical angle. The footcandles delivered per 1000 candelas at the given angle are located at the intersection of the mounting height and horizontal distance on Horizontal Footcandles Per 1000 Candelas.

To determine the fc delivered to the measuring point, convert cd to fc. To convert cd to fc, apply the formula:

$$fc = rfc \times \frac{cd}{1000}$$

where
fc = footcandles at measuring point
rfc = rated footcandles per 1000 candelas (from Horizontal Footcandles Per 1000 Candelas)
cd = candelas at measuring point
1000 = constant

Example: Determining Footcandles

A 400 W high-pressure sodium lamp rated at 50,000 initial lumen is mounted 12′ high. Determine the footcandles delivered to a surface 10′ away from the lamp. See Surface Lighting.

1. Determine the mounting height of the lamp fixture. Mounting height equals **12′**.
2. Determine the horizontal distance from the fixture location to the measuring point. Horizontal distance equals **10′**.
3. Determine the vertical angle. Vertical angle equals **40°**.
4. Determine the candelas delivered to the measuring point (from Light Distribution Data*).

$$cd = rcd \times \frac{rlm}{1000}$$

$$cd = 181 \times \frac{50,000}{1000}$$

$cd = $ **9050 cd**

5. Determine the footcandles delivered per 1000 candelas at the given vertical angle. The footcandles delivered per 1000 candelas equals **3.15** (from Horizontal Footcandles Per 1000 Candelas).

Convert cd to fc.

$$fc = rfc \times \frac{cd}{1000}$$

$$fc = 3.15 \times \frac{9050}{1000}$$

$fc = $ **28.5 fc**

LIGHT DISTRIBUTION DATA*

Angle**	Rated cd
180	0
175	0
50	179
45	184
40	181
35	177

* 400 W high-pressure sodium lamp rated at 50,000 initial lumen
** in degrees

SURFACE LIGHTING

Lamps should be spaced so that the light produced on a surface is a combination of more than one light source. This helps to reduce glare and shadows. When determining the amount of light on a surface, all lamp sources that produce light are considered. To determine the total light on a surface, the light from several properly spaced lamps is calculated. See Lamp Spacing.

LAMP SPACING

Lumen per watt (lm/W) is the number of lumen produced per watt of electrical power. Lumen per watt is the measure of the effectiveness (efficacy) of the light source in producing light from electrical energy. The higher the lm/W rating, the more economical the light source is to operate.

Lamp Efficiency

Most of the energy produced by a lamp (65% to 90%) is in the form of ultraviolet energy, infrared energy, or heat. A small percentage of the total energy produced by a lamp is visible. Lamps are rated by the electrical efficiency (lm/W) of light produced. The electrical efficiency of lamps determines the cost to light a given area. See Lamp Electrical Efficiency.

LAMP ELECTRICAL EFFICIENCY	
Lamp	Lumen Per Watt*
Incandescent	15-25
Mercury-vapor	50-60
Fluorescent	55-100
Metal-halide	80-125
High-pressure sodium	80-150
Low-pressure sodium	160-200

* exact lumen output depends on size and type of lamp used

Total lamp efficiency consists of electrical efficiency and economical efficiency. Economical efficiency depends on the initial cost of the lamp, lamp life, and lamp usage. The frequency of lamp service is important from a maintenance standpoint. Lamp service includes repair and changing burnt-out bulbs. See Lamp Comparison.

Many different lamps are available. Some lamps, such as incandescent lamps, have a low initial cost, but are more expensive to operate than others. High-pressure sodium lamps have a high initial cost but produce five times the light per watt of an incandescent lamp. See Lamp Advantages and Disadvantages.

LAMP COMPARISON

Lamp	Advantages	Disadvantages
Incandescent, tungsten-halogen	Low initial cost Simple construction No ballast required Available in many shapes and sizes Requires no warm-up or restart time Inexpensively dimmed Simple maintenance	Low electrical efficiency High operating temperature Short life Bright light source in small space Does not allow large distribution of light
Fluorescent	Available in many shapes and sizes Moderate cost Good electrical efficiency Long life Low shadowing Low operating temperature Short turn-ON delay	Not suited for high-level light in small, highly-concentrated applications Requires ballast Higher initial cost than incandescent lamps Light output and color affected by ambient temperature Expensive to dim
Low-pressure sodium, mercury-vapor, metal-halide, high-pressure sodium	Good electrical efficiency Long life High light output Slightly affected by ambient temperature	May cause color distortion Long start and restart time High initial cost High replacement cost Requires ballast Expensive or not possible to dim Problem starting in cold weather High-socket voltage required

Application — Incandescent Lamps

An *incandescent lamp* is an electric lamp that produces light by the flow of current through a tungsten filament inside a gas-filled, sealed glass bulb. An incandescent lamp has a low initial cost and is simple to install and service. An incandescent lamp has a lower electrical efficiency and shorter life than other lamps. See Incandescent Lamp.

An incandescent lamp operates as current flows through a filament. A *filament* is a conductor that has a resistance high enough to cause the conductor to heat. The filament glows white-hot and produces light. The filament is made of tungsten and limits the current to a safe operating level.

Inrush current is higher than operating current because the filament has a low resistance when cold. The high inrush current causes most incandescent lamps to fail when first switched ON.

INCANDESCENT LAMP

The air inside the bulb is removed before it is sealed to prevent oxidation of the filament. The filament burns out quickly if oxygen is present. A gas mixture (nitrogen and argon) is placed inside most incandescent bulbs to increase the life of the lamp.

Incandescent Lamp Bases

The base of an incandescent lamp holds the lamp firmly in the socket and connects electricity from the outside circuit to the filament. The ends of the filament are brought out to the base. The base simplifies the replacement of an incandescent lamp. Most incandescent lamp bases contain threads. Some smaller lamps are fitted with bayonet bases. The standard screw base has right-hand threads that allow removal of the lamp by turning counterclockwise (CCW) and installation by turning clockwise (CW). See Incandescent Lamp Bases.

INCANDESCENT LAMP BASES

Incandescent Bulb Shapes

Incandescent lamp bulbs are available in a variety of shapes and sizes. Bulb shapes are designated by letters. The most common bulb shape is the A bulb. The A bulb is used for most residential and commercial indoor lighting applications. The A bulb is normally available in sizes from 15 W to 200 W. A common variation of the A bulb is the PS bulb. The PS bulb has a long neck. The PS bulb is available in sizes from 150 W to 2000 W. The PS bulb is used for commercial, school, industrial, and some old street lighting applications. See Incandescent Bulb Shapes.

Other bulbs are used for decorative or special-purpose applications. The C bulb is designed to withstand moderate vibration and is used in sewing, washing, and other machines. The B, F, and G bulbs are used for decorative applications and are normally of low wattage. The R bulb is also used as a decorative light, but is one of high wattage. The PAR bulb is normally used as an outdoor spotlight.

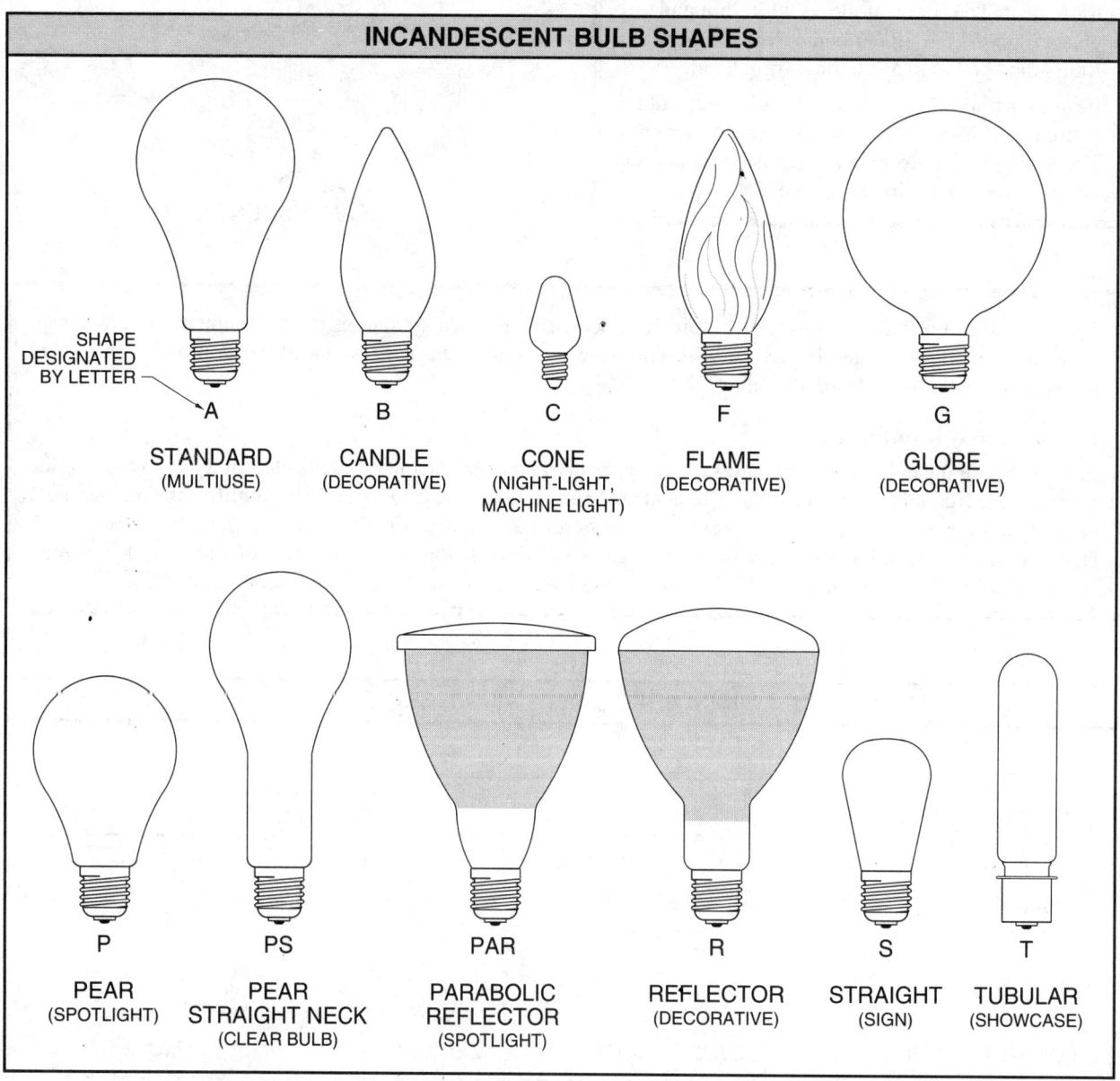

Incandescent Lamp Size

The size of an incandescent lamp is determined by the diameter of the lamp's bulb. See Common Incandescent Lamps. The diameter of a bulb is expressed in eighths of an inch ($\frac{1}{8}''$). For example, an A-21 bulb is 21 eighths ($^{21}\!/_8''$) or $2\frac{5}{8}''$ in diameter at the bulb's maximum dimension.

A scale calibrated in eighths is used to measure the size of a lamp. The scale is placed at the largest diameter of the bulb. See Sizing Incandescent Lamps.

Tungsten-Halogen Lamps

A *tungsten-halogen lamp* is an incandescent lamp that is filled with a halogen gas (iodine or bromine). The gas combines with tungsten that is evaporated from the filament as the lamp burns. Tungsten-halogen lamp wattages range from 15 W to 1500 W. Tungsten-halogen lamps are available in a limited selection of bulb sizes and wattages. See Tungsten-Halogen Lamp.

Tungsten-halogen lamps are used for display lighting, outdoor lighting, and in photocopy machines because they produce a large amount of light instantly. The lamp lasts about twice as long as a standard incandescent lamp.

Care must be taken when replacing a burned-out tungsten-halogen lamp. Ensure that all power is OFF because a high temperature is produced when the lamp is turned ON. The bulb is made of quartz because of the high temperature of the lamp. High-wattage tungsten-halogen lamps are available in T or PAR bulbs. The replacement cost is about three times the cost of an incandescent lamp.

COMMON INCANDESCENT LAMPS			
Type	Watts	Size*	Volts
A-19	40-100	$2\frac{3}{8}$	120-130
F-15	25-60	$1\frac{7}{8}$	120
PS-35	300-500	$4\frac{3}{8}$	120
PAR-38	115-150	$4\frac{3}{4}$	120-130
T-19	40-100	$2\frac{3}{8}$	120

* in in.

SIZING INCANDESCENT LAMPS

TUNGSTEN-HALOGEN LAMP

Application — Fluorescent Lamps

A *fluorescent lamp* is a low-pressure discharge lamp in which ionization of mercury-vapor transforms ultraviolet energy generated by the discharge into light. The bulb contains a mixture of inert gas (normally argon and mercury-vapor) which is bombarded by electrons from the cathode. This provides ultraviolet light. The ultraviolet light causes the fluorescent material on the inner surface of the bulb to emit visible light.

A fluorescent lamp consists of a cylindrical glass tube (bulb) sealed at both ends. Each end (base) includes a cathode that supplies the current to start and maintain the arc. Both cathodes are constructed so they allow conduction in either direction. Conduction in either direction allows the lamp to be operated on alternating current. The cathode is normally made of coiled tungsten wire and coated with a material that emits electrons. See Fluorescent Lamp Construction.

Standard fluorescent lamps vary in diameter from $\frac{5}{8}''$ to $2\frac{1}{8}''$ and in length from 6" to 96". The overall diameter of the lamp is designated by a number that indicates eighths of an inch. The most common size is the 40 W lamp, which is designated T-12 ($1\frac{1}{2}''$ in diameter) and has a length of 48". See Common Fluorescent Lamps.

292 TROUBLESHOOTING ELECTRICAL/ELECTRONIC SYSTEMS

FLUORESCENT LAMP CONSTRUCTION

COMMON FLUORESCENT LAMPS				
Type	Watts	Length*	Size*	Base
T-5	4-13	6-21	5/8	Miniature Bi-pin
T-5	13	21	5/8	Miniature Bi-pin
T-8	15-30	18-36	1	Medium Bi-pin
T-9	20-40	6-16 OD	1 1/8	4-pin
T-10	40	48	1 1/4	Medium Bi-pin
T-12	14-75	15-96	1 1/2	Medium Bi-pin
T-12	60-75	96	1 1/2	Single Pin

* in in.

Fluorescent lamp holders are designed for the various types of lamp bases. The base is used to support a fluorescent lamp and provide the required electrical connections. Fluorescent lamps that use preheat or rapid-start cathodes require four electrical contacts (pins). The four connections are made by using a bi-pin base at either end. The three standard types of bi-pin bases are miniature bi-pin, medium bi-pin, and mogul bi-pin. Instant-start fluorescent lamps require a single pin at each end of the lamp. See Fluorescent Lamp Bases.

Voltage Fluctuations

A fluorescent lamp is not connected directly to power lines. The high inrush current would destroy the lamp. A ballast is used to limit and control the current flow in a fluorescent lamp. A *ballast* is a transformer or solid-state circuit that limits current flow and supplies the starting voltage for a fluorescent lamp. See Fluorescent Lamp Fixture.

A supply voltage higher or lower than the rating of the ballast affects the life of the lamp and ballast. The voltage supplied to a fluorescent lamp should be within ±7% of the lamp's rating.

A supply voltage lower than the rated voltage but within 7% of the rated voltage normally has no serious effect on the lamp or ballast. A supply voltage deviation of more than 7% below the rated voltage shortens lamp life and reduces the light output. The reduced voltage may cause frequent flashing during starting.

A supply voltage higher than the rated voltage of the ballast increases light output and shortens lamp and ballast life. A higher-than-rated voltage also increases the lamp temperature. High temperatures break down insulation and reduce ballast life. See Fluorescent Lamp Voltage Change Characteristics.

Ballast Sound

A slight hum is present in fluorescent lamps. The hum originates from the inherent magnetic action in the core and coil assembly of the ballast. The sound is normally not a problem. The sound may be annoying if amplified. Reasons ballast sound may be amplified include:

- Improper ballast mounting. Use all mounting holes to securely mount a ballast.
- Loose parts in the fixture. Ensure no loose parts are present including loose screws and extra screws.
- Sound resonating through ceilings, walls, and furniture. Avoid mounting fluorescent lamps near air ducts, in loosely assembled metal display cabinets, or other locations that may amplify sound.

A fluorescent lamp ballast is selected based on the quietest rating for a specific location. Most ballasts are sound-rated by a letter code. An A ballast has the least hum and an F ballast has the most hum. The greater the number of fixtures used in an area, the higher the total sound output. See Ballast Sound Rating.

FLUORESCENT LAMP FIXTURE

FLUORESCENT LAMP
VOLTAGE CHANGE CHARACTERISTICS

| BALLAST SOUND RATING ||||
Sound Rating	Noise Level*	Recommended Application
A	20-24	Broadcasting booths, churches, study areas
B	25-30	Libraries, classrooms, quiet office areas
C	31-36	General office areas, commercial buildings
D	37-42	Retail stores, stock rooms
E	43-48	Light production areas, general outdoor lighting
F	Over 48	Street lighting, heavy production areas

* in decibels

Ballast Ventilation

A fluorescent lamp ballast generates heat during normal operation. A ballast using Class A insulation should have a maximum ballast coil temperature of 105°C (221°F) and a maximum ballast case temperature of 90°C (194°F) at its hottest point. Ballast life is reduced and an unsafe condition is produced when a ballast is operated at temperatures above these limits. A ballast should be positioned at a distance great enough to allow for the dissipation of the combined heating effect in multiple ballast enclosures. Per the NEC®, thermal insulation for recessed lighting must be at least 3″ from the fixture. See Ballast Ventilation.

The temperature rise of ballasts may be reduced by:
- Mounting ballasts with the maximum number of sides in direct contact with the metal channel of the fixture to help dissipate the heat.
- Providing fixture ventilation.
- Painting unpainted fixture channels with a non-metallic finish to increase thermal radiation.
- Placing the ballast in a cool location outside the fixture when fixture ventilation is restricted.
- Installing the fixture for maximum heat dissipation by conduction, convection, or radiation.

BALLAST VENTILATION

Cold Weather Operation

Lumen rating of a fluorescent lamp applies to operation in still air at an ambient temperature of 77°F. Fluorescent lamps and fluorescent lamp ballasts provide good light output down to 50°F. Further decreases in ambient temperature result in decreased light output.

Variations in humidity, line voltage, fixture design, and within the design of a lamp and ballast affect the low-temperature starting limit of a fluorescent lamp. A fluorescent lamp with a high wall-bulb temperature is recommended for efficient cold weather operation. *Wall-bulb temperature* is the operating temperature of the lamp's wall, which affects the mercury-vapor pressure inside the bulb. Variations in mercury-vapor pressure change the light output of the lamp. The colder the bulb, the less light output.

Radio and Sound Interference

Radio waves are created at the arc of fluorescent lamp electrodes. The radio waves may interfere with radio reception by direct radiation from the fluorescent lamp to the radio, feedback from the lamp through the power line to the radio, and direct radiation from the power source line to the radio. See Fluorescent Lamp Interference.

Interference caused by fluorescent lamps is eliminated by:
- Separating radio and fluorescent lamps by at least 10′ or more.
- Providing a good ground for the radio or sound system.
- Adding external circuit filtering.
- Connecting radio and fluorescent lamp fixtures to separate branch circuits.

FLUORESCENT LAMP INTERFERENCE

Fluorescent Lamp Starting

A fluorescent lamp cannot be operated directly from a standard electrical circuit. A fluorescent lamp connected directly to a standard electrical circuit does not light. All fluorescent lamps require a means of starting. The method used to start the lamp determines the circuit's name. The methods of fluorescent lamp-starting include preheat, instant-start, and rapid-start.

Preheat Circuits

A *preheat circuit* is a fluorescent lamp-starting circuit that heats the cathode before an arc is created. Preheat circuits are used in some low-wattage lamps (4 W to 20 W). Preheat circuits are common in desk lamps.

A preheat circuit uses a separate starting switch that is connected in series with the ballast. The electrodes in a preheat circuit require a few seconds to reach the proper temperature to start the lamp after the starting switch is pressed. See Preheat Circuit.

PREHEAT CIRCUIT

The opening of the starting switch breaks the path of current flow through the starting switch. This leaves the gas in the lamp as the only path for current to travel. A high-voltage surge is produced by the collapsing magnetic field as the starting switch is no longer connected to power. This high-voltage surge starts the flow of current through the gas in the lamp. Once started, the ballast limits the flow of current through the lamp.

Instant-Start Circuits

An *instant-start circuit* is a fluorescent lamp-starting circuit that provides sufficient voltage to strike an arc instantly. The instant-start circuit was developed to eliminate the starting switch and overcome the starting delay of preheat circuits. An arc strikes without preheating the cathodes when a high enough voltage is applied across a fluorescent lamp.

The high initial voltage requires a large autotransformer as an integral part of the ballast. The autotransformer delivers an instant voltage of 270 V to 600 V, depending on bulb size and voltage rating. The larger the bulb, the higher the voltage. Instant-start lamps require only one pin at each base because no preheating of the electrode is required. See Instant-Start Circuit.

A safety circuit is used with an instant-start lamp to prevent electrical shock due to the high starting voltage. When the lamp is removed, the base pin acts as a switch, interrupting the circuit to the ballast. The lamp is replaced by pushing the lamp into the spring lampholder at the high-voltage end of the fixture and then inserting it into the rigid lampholder at the low-voltage end of the fixture.

Both lamp ends must be in place before current can flow through the ballast winding. The circuit does not operate when the lamp is not in place.

INSTANT-START CIRCUIT

Rapid-Start Circuits

A *rapid-start circuit* is a fluorescent lamp-starting circuit that brings the lamp to full brightness in about 2 seconds. A rapid-start circuit is the most common circuit used in fluorescent lighting.

A rapid-start circuit uses lamps that have short, low-voltage electrodes that are automatically preheated by the lamp ballast. The rapid-start ballast preheats the cathodes by means of a heater winding. The heater winding continues to provide current to the lamp after ignition. See Rapid-Start Circuit.

Application — High-Intensity Discharge Lamps

High-Intensity Discharge Lamps

A *high-intensity discharge (HID) lamp* is a lamp that produces light from an arc tube. An *arc tube* is the light-producing element of an HID lamp. An arc tube contains metallic and gaseous vapors and electrodes inside the tube. An arc is produced in the tube between the electrodes. The arc tube is enclosed in a bulb which may contain phosphor or diffusing coating that improves color rendering, increases light output, and reduces surface brightness.

RAPID-START CIRCUIT

HID lamps include low-pressure sodium, mercury-vapor, metal-halide, and high-pressure sodium lamps. All HID lamps are electric discharge lamps. An *electric discharge lamp* is a lamp that produces light by an arc discharged between two electrodes. High vapor pressure is used to convert a large percentage of the energy produced into visible light. Arc tube pressure for most HID lamps are normally from one to eight atmospheres.

HID lamps provide an efficient, long-lasting source of light and are used for street, parking lot, and general lighting applications.

Low-Pressure Sodium Lamps

A *low-pressure sodium lamp* is an HID lamp that operates at a low vapor pressure and uses sodium as the vapor. A low-pressure sodium lamp has a U-shaped arc tube. The arc tube has both electrodes located at the same end. The arc tube is placed inside a glass bulb and contains a mixture of neon, argon, and sodium-metal. On start-up, an arc is discharged through the neon, argon, and sodium-metal. As the sodium-metal heats and vaporizes, the amber color of sodium is produced. See Low-Pressure Sodium Lamp.

LOW-PRESSURE SODIUM LAMP

A low-pressure sodium lamp gets its name because it uses sodium inside the arc tube. A low-pressure sodium lamp has the highest efficiency rating of any lamp. Some low-pressure sodium lamps deliver up to 200 lm/W of power. This is 10 times the output of an incandescent lamp.

Mercury-Vapor Lamps

A *mercury-vapor lamp* is an HID lamp that produces light by an electric discharge through mercury-vapor. Mercury-vapor lamps are used for general lighting applications. Phosphor coating is added to the inside of the bulb to improve color-rendering characteristics. See Mercury-Vapor Lamp.

A mercury-vapor lamp contains a starting electrode and two main electrodes. An electrical field is set up between the starting electrode and one main electrode when power is first applied to the lamp. The electrical field causes current to flow and an arc to strike. Current flows between the two main electrodes as the heat vaporizes the mercury.

Metal-Halide Lamps

A *metal-halide lamp* is an HID lamp that produces light by an electric discharge through mercury-vapor and metal-halides in the arc tube. A *metal-halide* is an element (normally sodium and scandium iodide) which is added to the mercury in small amounts. The metal-halide improves the light output of the lamp. A metal-halide lamp produces more lumen per watt than a mercury-vapor lamp. See Metal-Halide Lamp.

The light produced by a metal-halide lamp does not produce as much color distortion as a mercury-vapor lamp. A metal-halide lamp is an efficient source of white light. A metal-halide lamp has a shorter bulb life than the other HID lamps.

High-Pressure Sodium Lamps

A *high-pressure sodium lamp* is an HID lamp that produces light when current flows through sodium-vapor under high pressure and high temperature. A high-pressure sodium lamp is a more efficient lamp than a mercury-vapor or metal-halide lamp. The light produced from a high-pressure sodium lamp appears as a golden-white color. See High-Pressure Sodium Lamp.

MERCURY-VAPOR LAMP

METAL-HALIDE LAMP

HIGH-PRESSURE SODIUM LAMP

A high-pressure sodium lamp is constructed with a bulb and an arc tube. The arc tube is made of ceramic to withstand high temperature. The bulb is made of weather-resistant glass to prevent heat loss and protect the arc tube.

HID Lamp Bases

HID lamps are designed to be easily connected and removed from the lamp fixture. Mercury-vapor, metal-halide, and high-pressure sodium lamps use the same medium and mogul brass screw bases as incandescent lamps. Low-pressure sodium lamps normally use a bayonet type base. Many lamp bases include a date code on the base to record the month and year of installation.

HID Lamp Operating Characteristics

HID lamps take several minutes to warm up before full light output is reached. Any short interruption in the power supply may extinguish the arc. An HID lamp cannot restart immediately after being turned OFF. The lamp must cool enough to reduce the vapor pressure in the tube to a point where the arc can re-strike. See HID Lamp Operating Characteristics.

HID LAMP OPERATING CHARACTERISTICS

Lamp	Start Time*	Restart Time
Low-pressure sodium	6-12	4-12 sec
Mercury-vapor	5-6	3-5 min
Metal-halide	2-5	10-15 min
High-pressure sodium	3-4	30-60 sec

* in minutes

HID Lamp Color Rendering

Color rendering is the appearance of a color when illuminated by a light source. For example, a red color may be rendered light, dark, pinkish, or yellowish depending on the light source under which it is viewed. Color rendering of HID lamps varies depending on the lamp used.

Low-pressure sodium lamps produce yellow to yellow-orange light. The color of the lamp distorts the true color of objects viewed under the lamp. The yellow light is produced by the sodium in a low-pressure sodium lamp. Low-pressure sodium lamps are normally not used where the appearance of people and colors are important. The yellow light produces severe color distortion on most light-colored objects.

Mercury-vapor lamps with clear bulbs have poor rendering of reds. Blue colors appear purplish with most other colors appearing normal. Phosphor-coated bulbs improve color rendering. Blue colors viewed under phosphor-coated bulbs still have a slight purplish hue and yellow colors take on a greenish overtone.

Metal-halide lamps produce good overall color rendering. Red colors appear slightly muted with some pinkish overtones. High-pressure sodium lamps have good color rendering but reds, blues, greens, and violets are slightly muted.

HID Lamp Ballasts

HID lamps require ballasts to limit the current in the lamps to the correct operating level and provide the proper starting voltage to strike and maintain the arc. Each HID ballast is designed for a specific lamp, bulb size, voltage range, and line frequency.

Different lamps are not interchangeable because lamp wattage is controlled by the ballast and not the lamp. For example, when a 250 W mercury-vapor lamp is connected to a 1000 W ballast, the lamp tries to operate at 1000 W. The lamp does not operate correctly or is destroyed because the 250 W lamp is not designed to operate at 1000 W.

HID lamp sizes cannot be interchanged without changing the ballast. For example, a 100 W lamp cannot be replaced with a 175 W lamp to produce more light. Also, a 175 W lamp cannot be replaced with a 100 W lamp to produce less light.

Low-Pressure Sodium Ballasts

Low-pressure sodium lamps must be operated on a ballast designed to meet the lamp's starting and running requirements. Low-pressure sodium lamps do not have a starting electrode or ignitor. The ballast must provide an open-circuit voltage of approximately three to seven times the lamp's rated voltage to start and sustain the arc. See Low-Pressure Sodium Ballast.

Mercury-Vapor Ballasts

Mercury-vapor ballasts include reactor, high-reactance autotransformer, constant-wattage autotransformer, and two-winding constant-wattage ballasts. The ballast used normally depends on economics. See Mercury-Vapor Ballasts.

LOW-PRESSURE SODIUM BALLAST

MERCURY-VAPOR BALLASTS

REACTOR

HIGH-REACTANCE AUTOTRANSFORMER

CONSTANT-WATTAGE AUTOTRANSFORMER

TWO-WINDING CONSTANT-WATTAGE

A *reactor ballast* is a ballast that connects a coil (reactor) in series with the power line leading to the lamp. A reactor ballast is used when the incoming supply voltage meets the starting voltage requirements of the lamp. This is common when the incoming supply voltage is 240 V or 277 V. Both 240 V and 277 V mercury-vapor lamps are standard. A capacitor is added to some reactor ballasts to improve the power factor. A reactor ballast costs the least, but has poor lamp-wattage regulation. A reactor ballast should be used only when line voltage regulation is good because a 5% change in line voltage produces a 10% change in lamp wattage in a reactor ballast.

A *high-reactance autotransformer ballast* is a ballast that uses two coils (primary and secondary) to regulate both voltage and current. A high-reactance autotransformer ballast is used when the incoming supply voltage does not meet the starting requirements of the lamp. Incoming voltages of 115 V, 208 V, and 460 V require a voltage change to the lamp.

A *constant-wattage autotransformer ballast* is a high-reactance autotransformer ballast with a capacitor added to the circuit. The capacitor improves the power factor. A constant-wattage autotransformer ballast is the most commonly used ballast.

A *two-winding constant-wattage ballast* is a ballast that uses a transformer that provides isolation between the primary and secondary circuits. A two-winding constant-wattage ballast has excellent lamp-wattage regulation. A 13% change in line voltage produces only a 2% to 3% change in lamp wattage.

Metal-Halide Ballasts

A metal-halide ballast uses the same basic circuit as the constant-wattage autotransformer mercury-vapor ballast. The ballast is modified to provide high starting voltage required by metal-halide lamps. See Metal-Halide Ballast.

High-Pressure Sodium Ballasts

High-pressure sodium lamps do not have a starting electrode. The ballast must deliver a voltage pulse high enough to start and maintain the arc. This voltage pulse must be delivered every cycle and must be 4000 V to 6000 V for 1000 W lamps and 2500 V to 4000 V for smaller lamps. The starter (ignitor) is the device inside the ballast that produces the high starting voltage. See High-Pressure Sodium Ballast.

A high-pressure sodium ballast is similar to a mercury-vapor reactor ballast. The main difference is the added starter. The reactor ballast is used where the input voltage meets the lamp's requirements. A transformer or autotransformer is added to the ballast circuit when the incoming voltage does not meet the lamp's requirements.

METAL-HALIDE BALLAST

HIGH-PRESSURE SODIUM BALLAST

HID Ballast Loss

All ballasts consume a certain amount of power. Depending on the size of the lamp used, the amount of power consumed by the ballast may be a large percentage of the total lamp power.

With HID lamps, the larger the lamp, the less power the ballast loss is as a percent of total power used by the lamp. For example, a 40 W mercury-vapor lamp has a 32% ballast loss and a 1000 W mercury-vapor lamp has a 7% ballast loss. See Ballast Loss in Appendix.

HID Lamp Identification

ANSI lists a standard code that is used to identify HID lamps. This code is used among manufacturers and allows for interchangeability of lamps. See Lamp Identification in Appendix.

HID Lamp Selection

When selecting an HID lamp for an application, consider using low-pressure sodium lamps first because they are the most efficient. Low-pressure sodium lamps are good for outdoor lighting installations. The yellow-orange color is acceptable for street, highway, parking lot, and floodlight applications.

Low-pressure sodium lamps are also used for some indoor applications such as warehouse lighting and other areas where color distortion is not critical to an operation. The long start-up time is not a problem for outdoor lighting because the lights are normally turned ON by a photoelectric cell at dusk. Photoelectric-cell switches may be adjusted to turn ON at different light levels.

Consider using metal-halide lamps when the color rendering of low-pressure sodium lamps is not acceptable for the application. Metal-halide lamps are used for sport, street, highway, parking lot, and floodlighting applications. Metal-halide lamps are normally specified for most sport, indoor, and outdoor lighting applications.

Consider using mercury-vapor lamps when the initial installation cost is of major importance. Mercury-vapor lamps are used in all types of applications. Use metal-halide lamps when true color appearance is important such as in a car dealer's lot.

Consider using high-pressure sodium lamps when lower operating cost is important and some color distortion is acceptable. High-pressure sodium lamps may be used in parking lots, street lighting, shopping centers, exterior buildings, and most storage areas. See Lamp Characteristics Summary.

LAMP CHARACTERISTICS SUMMARY

Lamp	Lm/W	Rated Bulb Life*	Color Rendition	Operating Cost
Incandescent	15-25	750-1000	Excellent	Very high
Tungsten-halogen	20-25	1500-2000	Excellent	High
Fluorescent	55-100	7500-24,000	Very good	Average
Low-pressure sodium	190-200	1800	Poor	Low
Mercury-vapor	50-60	16,000-24,000	Depends on type used	Average
Metal-halide	80-125	3000-20,000	Very good	Average
High-pressure sodium	65-115	7500-14,000	Good (golden white)	Low

* in hours

LIGHTING CIRCUITS

16

Name _____ Date _____ **Activities**

◐ Activity 16-1. Lighting Terminology

Using Horizontal Footcandles Per 1000 Candelas and Light Distribution Data*, determine the footcandles delivered at each measuring point.

LIGHT DISTRIBUTION DATA*

Angle**	Rated cd	Angle**	Rated cd
180	2	85	20
175	2	80	33
170	2	75	53
165	2	70	79
160	2	65	108
155	2	60	135
150	2	55	157
145	2	50	173
140	2	45	181
135	2	40	177
130	2	35	162
125	2	30	145
120	2	25	133
115	2	20	121
110	2	15	107
105	3	10	97
100	4	5	93
95	7	0	93
90	11		

* 175 W mercury-vapor lamp rated at 8600 initial lumen
** in degrees

HORIZONTAL DISTANCE FROM FIXTURE LOCATION (FT)

HORIZOTAL FOOTCANDLES PER 1000 CANDELAS

1. A 175 W mercury-vapor lamp rated at 8600 initial lumen is used for an application.

_____ **A.** _____ fc is delivered at Measuring Point 1.

_____ **B.** _____ fc is delivered at Measuring Point 2.

303

304 TROUBLESHOOTING ELECTRICAL/ELECTRONIC SYSTEMS

_____ **C.** The vertical angle between the lamp and Measuring Point 3 equals _____°.

_____ **D.** _____ fc is delivered at Measuring Point 3.

_____ **E.** The vertical angle between the lamp and Measuring Point 4 equals _____°.

_____ **F.** _____ fc is delivered at Measuring Point 4.

2. A 400 W high-pressure sodium lamp rated at 50,000 initial lumen is used for an application.

_____ **A.** The vertical angle between the lamp and Measuring Point 5 equals _____°.

_____ **B.** _____ fc from Lamp 1 is delivered at Measuring Point 5.

_____ **C.** _____ fc from Lamp 2 is delivered at Measuring Point 5.

_____ **D.** The total fc delivered at Measuring Point 5 equals _____ fc.

Angle**	Rated cd	Angle**	Rated cd
180	0	80	18
175	0	85	28
170	0	75	42
165	0	70	63
160	0	65	928
155	0	60	128
150	0	55	160
145	0	50	179
140	0	45	184
135	0	40	181
130	2	35	177
125	5	30	176
120	7	25	174
115	8	20	166
110	8	15	153
105	8	10	142
100	7	5	138
95	9	0	137
90	11		

LIGHT DISTRIBUTION DATA*

* 400 W high-pressure sodium lamp rated at 50,000 initial lumen
** in degrees

_____ E. The vertical angle between Lamp 1 and Measuring Point 6 equals _____°.

_____ F. _____ fc from Lamp 1 is delivered at Measuring Point 6.

_____ G. The total fc delivered at Measuring Point 6 equals _____ fc.

_____ H. The vertical angle between Lamp 1 and Measuring Point 7 equals _____°.

_____ I. _____ fc from Lamp 1 is delivered at Measuring Point 7.

_____ J. _____ fc from Lamp 3 is delivered at Measuring Point 7.

_____ K. The total fc delivered at Measuring Point 7 equals _____ fc.

○ Activity 16-2. Incandescent Lamps

An incandescent lamp is the most costly lamp to operate because it is the least efficient. Energy is saved and operating costs reduced when standard incandescent lamps are replaced with energy-efficient lamps. An energy-efficient lamp delivers the same number of lumen at a slightly reduced wattage rating than a standard lamp.

Energy-efficient lamps initially cost more than standard lamps. The amount of energy used is always reduced when using energy-efficient lamps.

To determine the operating cost of an energy-efficient lamp, apply the formula:

$$oc = (W - Wr) \times rll \times \frac{er}{1000}$$

where
oc = operating cost (in $)
W = lamp rating (in watts)
Wr = watt reduction
rll = rated lamp life (in hours)
er = electric rate (in $/kWh)
1000 = constant

To determine the operating cost of a standard lamp, apply the formula:

$$oc = W \times rll \times \frac{er}{1000}$$

| STANDARD INCANDESCENT LAMPS ||||||
| Lamp Rating (in W) | Initial Cost (in $) | Rated Lamp Life (in hr) | Electric Rate (in $/kWh) |||
			#1 (low)	#2 (med)	#3 (high)
40	1.44	1500	.09	.15	.20
60	1.15	1000	.09	.15	.20
75	1.15	750	.09	.15	.20
100	1.15	750	.09	.15	.20
150	1.64	750	.09	.15	.20

| ENERGY-EFFICIENT INCANDESCENT LAMPS ||||
Lamp Rating (in W)	Initial Cost (in $)	Rated Lamp Life (in hr)	Watt Reduction
40	1.44	1500	6
60	1.29	1000	8
75	1.29	750	8
100	1.29	750	10
150	2.28	750	15

306 TROUBLESHOOTING ELECTRICAL/ELECTRONIC SYSTEMS

Using Standard Incandescent Lamps and Energy-Efficient Incandescent Lamps, answer the questions.

1. A 60 W energy-efficient incandescent lamp is used to replace a standard incandescent lamp.

   _____ A. The watt reduction of using an energy-efficient incandescent lamp is _____ W.
   _____ B. The rated lamp life of a 60 W standard incandescent lamp is _____ hr.
   _____ C. The operating cost of a standard incandescent lamp at electric rate #1 is $_____.
   _____ D. The operating cost of an energy-efficient incandescent lamp at electric rate #1 is $_____.
   _____ E. The cost savings using rate #1 is $_____.
   _____ F. The cost savings using rate #2 is $_____.
   _____ G. The cost savings using rate #3 is $_____.

2. A 100 W energy-efficient incandescent lamp is used to replace a standard incandescent lamp.

   _____ A. The watt reduction of using an energy-efficient incandescent lamp is _____ W.
   _____ B. The rated lamp life of a 100 W standard incandescent lamp is _____ hr.
   _____ C. The operating cost of a standard incandescent lamp at electric rate #1 is $_____.
   _____ D. The operating cost of an energy-efficient incandescent lamp at electric rate #1 is $_____.
   _____ E. The cost savings using rate #1 is $_____.
   _____ F. The cost savings using rate #2 is $_____.
   _____ G. The cost savings using rate #3 is $_____.

○ Activity 16-3. Fluorescent Lamps

Fluorescent lamps are available in energy-efficient lamps. The high lamp life (10 to 20 times longer than incandescent lamps) makes great savings possible. Using Standard Fluorescent Lamps and Energy-Efficient Fluorescent Lamps, answer the questions.

STANDARD FLUORESCENT LAMPS					
Lamp Rating (in W)	Initial Cost (in $)	Rated Lamp Life (in hr)	Electric Rate (in $/kWh)		
			#1 (low)	#2 (med)	#3 (high)
30	5.98	18,000	.09	.15	.20
40	2.53	20,000	.09	.15	.20
75	9.96	12,000	.09	.15	.20
110	7.83	12,000	.09	.15	.20

ENERGY-EFFICIENT FLUORESCENT LAMPS			
Lamp Rating (in W)	Initial Cost (in $)	Rated Lamp Life (in hr)*	Watt Reduction
30	7.14	18,000	5
40	6.29	15,000	6
75	10.65	12,000	15
110	9.21	12,000	15

* at 3 hours average run time per start

1. A 30 W energy-efficient fluorescent lamp is used to replace a standard fluorescent lamp.

   _____ A. The watt reduction of using an energy-efficient fluorescent lamp is _____ W.
   _____ B. The rated lamp life of a 30 W standard fluorescent lamp is _____ hr.
   _____ C. The operating cost of a standard fluorescent lamp at electric rate #1 is $_____.

_____ D. The operating cost of an energy-efficient fluorescent lamp at electric rate #1 is $_____.
_____ E. The cost savings using rate #1 is $_____.
_____ F. The cost savings using rate #2 is $_____.
_____ G. The cost savings using rate #3 is $_____.

2. A 75 W energy-efficient fluorescent lamp is used to replace a standard fluorescent lamp.

_____ A. The watt reduction of using an energy-efficient fluorescent lamp is _____ W.
_____ B. The rated lamp life of a 75 W standard fluorescent lamp is _____ hr.
_____ C. The operating cost of a standard fluorescent lamp at electric rate #1 is $_____.
_____ D. The operating cost of an energy-efficient fluorescent lamp at electric rate #1 is $_____.
_____ E. The cost savings using rate #1 is $_____.
_____ F. The cost savings using rate #2 is $_____.
_____ G. The cost savings using rate #3 is $_____.

Activity 16-4. High-Intensity Discharge Lamps

Using High Mast Lighting, Footcandles Delivered, and Minimum Recommended Light Levels, answer the questions.

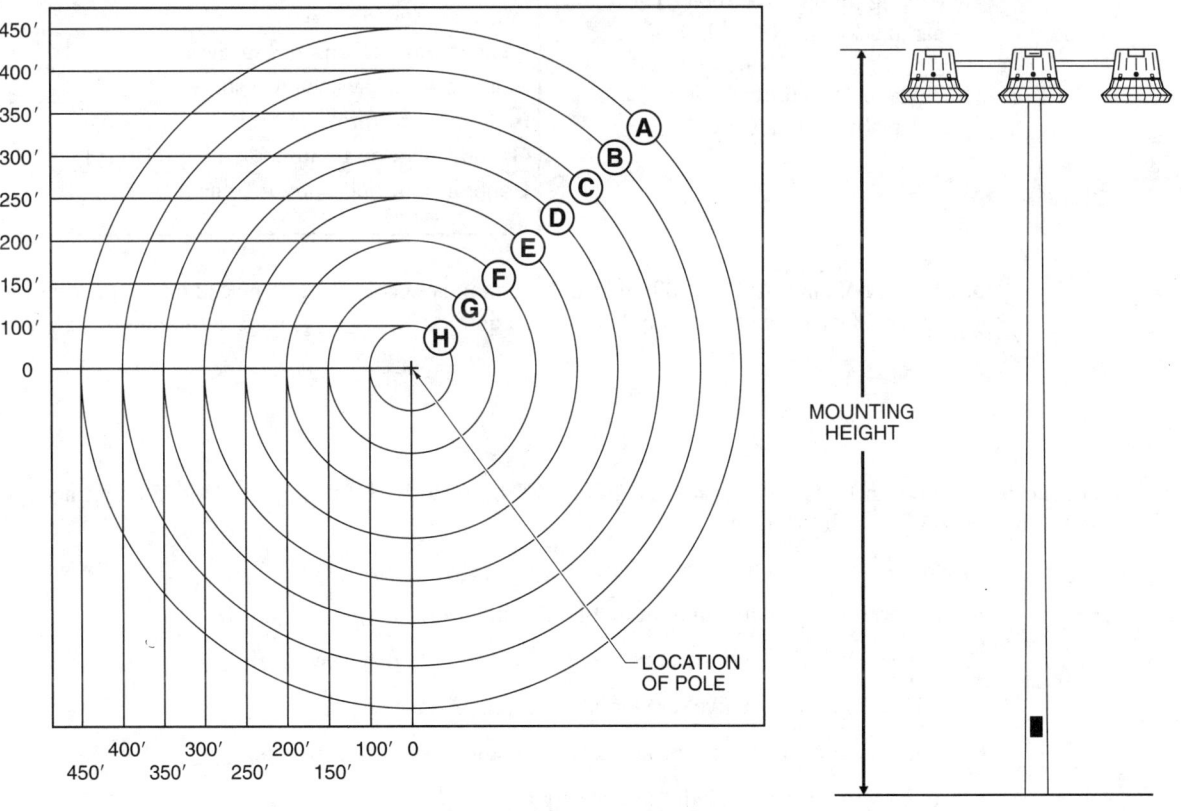

HIGH MAST LIGHTING

308 TROUBLESHOOTING ELECTRICAL/ELECTRONIC SYSTEMS

FOOTCANDLES DELIVERED												
Area	Mounting Height*											
	80			100			120			150		
	Lamp Number			Lamp Number			Lamp Number			Lamp Number		
	4	8	12	4	8	12	4	8	12	4	8	12
A	.4	3.2	6	.3	.9	1	.2	.4	.8	.1	.3	.5
B	2.2	8	15	1.2	2.2	4.2	.9	1.5	2.7	.7	1	1.8
C	10	16	31	3	5.5	10	2.3	3	7	1.5	2.3	4.5
D	21	31	62	7	12	20	5	7	14	3.2	4.5	9
E	52	76	90	15	21	42	9	12	26	6	9	17
F	100	125	165	32	50	95	22	35	68	15	21	45
G	120	160	300	65	98	200	46	68	145	30	45	86
H	160	240	480	110	160	310	70	100	220	45	68	135

* in feet

MINIMUM RECOMMENDED LIGHT LEVELS	
Area	Minimum fc (per sq ft)
Race track (auto, horse, motorcycle)	20
Outdoor ice rink	5
Prison yard	5
Quarry	5
Shopping center parking area	5
Industrial parking area	2
Lumberyard, storage (protective)	1
Railroad yard (switching point)	2
Railroad yard (protective)	.3
General park and garden lighting (display)	5
General park and garden lighting (protective)	.2

_____ 1. A 120′ pole with 8 lamps is placed in the middle of a railroad switching yard. The maximum distance from the pole that has an acceptable minimum light level is _____'.

_____ 2. A high mast light is placed in the middle of a shopping center parking lot on a 100′ pole. The parking lot is 350′ by 350′. _____ lamps are required on the pole to provide an acceptable light level.

_____ 3. A 100′ pole in a 200′ by 200′ industrial parking area could not be placed on the parking lot surface. Determine how many lamps are required. _____ lamps are required.

4. Determine the maximum distance that a mast pole 120′ high with eight lamps provides as an acceptable minimum light level for each location.

_____ A. Prison yard minimum light distance equals _____'.

_____ B. Industrial parking area minimum light distance equals _____'.

_____ C. Lumberyard storage (protective) area minimum light distance equals _____'.

_____ D. Railroad yard (protective) area minimum light distance equals _____'.

_____ E. General park and garden (protective) minimum light distance equals _____'.

_____ F. Auto racing minimum light distance equals _____'.

TROUBLESHOOTING LIGHTING CIRCUITS

Applications

Application — Troubleshooting Incandescent Lamps

Incandescent Lamps

An incandescent lamp is the simplest light used. No auxiliary equipment, such as a ballast, is required for operation. The lamp is always replaced if it is burned out. Frequent lamp replacement is cause for concern. Problems that require frequent lamp replacement are normally caused by using the wrong lamp type or a voltage problem.

Incandescent Lamp Voltage Ratings

Incandescent lamps are rated based on their voltage and power. The voltage rating is the maximum voltage that may be applied to the lamp. All incandescent lamps should be operated as close to their voltage rating as possible. Operating a lamp at its voltage rating assures the lamp is operating at its highest efficiency and life. Increasing the applied voltage increases the light output of the lamp and decreases lamp life.

All incandescent lamps produce less light as voltage is decreased. This is observed when large current-consuming appliances, such as furnaces, are connected to the same circuit as a lamp. The heavy current draw when the appliance is turned ON may cause a temporary voltage drop in the circuit. This voltage drop dims the lamp until the voltage returns to full level. Temporary voltage fluctuations do not significantly reduce the life of an incandescent lamp. Consistent operation at a voltage other than the lamp rating reduces lamp efficiency, lamp life, and increases the overall cost of lighting. Light output is reduced with a decrease in applied voltage and increased with an increase in applied voltage. See Incandescent Lamp Characteristics.

An increase in voltage causes more current to flow through the lamp and increase the brightness of the lamp. This also increases bulb temperature and shortens lamp life.

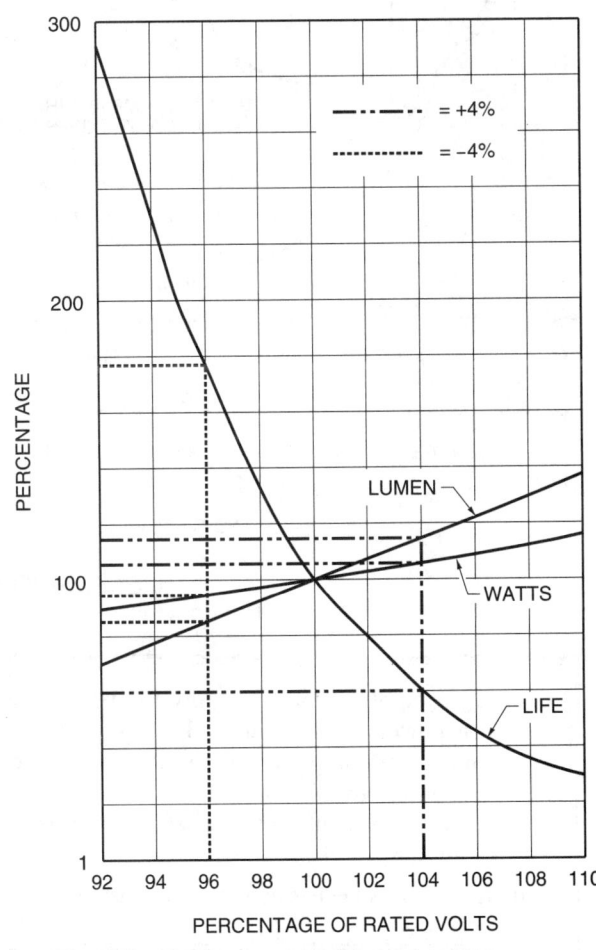

INCANDESCENT LAMP CHARACTERISTICS

310 TROUBLESHOOTING ELECTRICAL/ELECTRONIC SYSTEMS

For example, operating a 120 V lamp on 125 V (+4%) produces approximately 15% more lumen, uses 7% more watts, and shortens lamp life 40%. Operating the same 120 V lamp on 115 V (−4%) produces approximately 15% less lumen, uses 7% less watts, and increases lamp life 70%.

Measuring Applied Voltage

A voltmeter is used to measure the voltage applied to a lamp. A standard voltmeter may be used to measure the applied voltage when troubleshooting, replacing, or during lamp installation. A voltmeter with a min/max function may be used to record voltage levels over time. See Measuring Applied Voltage.

MEASURING APPLIED VOLTAGE

To measure applied voltage, apply the procedure:
1. Turn power to the lamp OFF.
2. Remove the lamp from the socket when the wires are not exposed.
3. Turn power to the lamp ON.
4. Measure the lamp voltage. Set the voltmeter to the correct voltage setting and connect the voltmeter to the lamp socket.

Warning: The socket has power. Care must be taken to ensure that no body part comes in contact with the circuit.

Standard Voltage Ratings

The standard voltage ratings of the majority of incandescent lamps are 115 V and 120 V. Incandescent lamps are also designed for use with 105 V, 110 V, and 130 V. A lamp with a high voltage rating may be used for an application when a lamp is not lasting as long as expected and the voltage is equal to or higher than the rating of the lamp.

Lamp Problems

Replacing burned-out lamps is generally routine for incandescent lamps. There may be a problem when lamp replacement becomes excessive. Consider lamp position, bulb type, service conditions, and operation time.

Bulb Blackening

Standard incandescent lamps may generally be placed in any position. The best position for most lamps is base up. Reduction of light may occur in lamps that are positioned horizontally because the bulb tends to blacken on the side that faces down.

Blackening of the bulb is normal as a lamp ages. The blackening is the result of the normal evaporation of the tungsten filament. Mount a lamp base up when bulb blackening is a problem. In this position, the hot gas inside the bulb carries the burnt particles up to the base. See Bulb Positioning.

OK

BEST

BULB POSITIONING

Rough Service Lamps

Standard incandescent lamps cannot receive much shock without burning out. A rough service lamp should be used in applications that include shocks, bumps, or rough handling. A *rough service lamp* is a lamp made of a specially processed tungsten filament that has extra supports to resist breakage. Rough service lamp applications include extension cord lights, service areas, and machine shops.

Vibration-Resistant Lamps

High-speed machinery produce high-frequency vibrations that destroy standard incandescent lamp filaments. In such applications, vibration-resistant lamps may be used. Vibration-resistant lamps have a special tungsten structure that resists the effect of vibration by allowing the filament to move.

Special lamps have decreased lumen ratings. The lower light output is a result of designing the lamp for long life, vibration, or rough service. See 100 Watt Lamp Ratings.

100 W LAMP RATINGS			
Voltage	Service Rating	Rated Lumen Per Watt	Rated Life*
120	General lighting	19.5	750
120	Long life	15	2500
120	Vibration resistant	14	1000
120	Rough service	12.5	1000

* in hours

Incandescent Lamp Testing

An incandescent lamp may be tested for continuity when it is removed from a circuit. See Incandescent Lamp Test.

To test an incandescent lamp, apply the procedure:
1. Set the meter on the lowest resistance scale.
2. Connect the meter leads to the center contact and screw base of the lamp. Polarity of the meter leads is not important. A meter reading of infinity indicates an open (bad) filament. Any resistance reading indicates a good filament.

The resistance reading varies directly with the filament temperature and wattage rating. A cold filament has a low resistance reading. The higher the lamp wattage, the lower the resistance reading. The lower the lamp wattage, the higher the resistance reading.

INCANDESCENT LAMP TEST

Group Lamp Replacement

Incandescent lamps may be replaced individually as they burn out or in groups. As a lamp ages, its life expectancy and output decrease. Lamps begin to fail after reaching about 70% of their rated life. Light output also decreases as a bulb blackens with age. Lamps are replaced in groups in applications such as schools and commercial buildings. Group replacement is performed when a lamp reaches between 70% and 85% of its rated life. This is accomplished by establishing a maintenance schedule for lamp replacement. See Incandescent Lamp Life Expectancy.

INCANDESCENT LAMP LIFE EXPECTANCY

Application — Troubleshooting Fluorescent Lamps

Bulb Discoloration

A uniform blackening throughout the length of a fluorescent bulb is normal as the lamp ages. The blackening starts gradually and is not normally detectable until the ends of the bulb blacken heavily. Heavy end blackening soon after a new lamp is installed indicates a problem. The problem may be a defective lamp, but is normally caused by excessive or low voltage. Replace any lamp with more than 2" of end blackening. Light spotting may occur on new lamps at any point on the bulb. Light spotting is due to mercury condensation in the bulb. See Fluorescent Bulb Discoloration.

Lamp Blinking

A fluorescent lamp that blinks ON and OFF is an indication of normal lamp failure that requires replacement. Other possible causes of lamp blinking include low voltage or low ambient temperature. Blinking lamps should be replaced as soon as they are detected because a blinking lamp may ruin the lamp and the starter.

Lamp Seating

Fluorescent lamps may turn ON before they are properly seated in the lamp holder. The lamp may fall out and lamp life is normally reduced when a lamp is not properly seated. Ensure that a fluorescent lamp is seated properly by aligning the base mark on the lamp with the center of the lamp holder. Both ends of the lamp must be checked for correct seating. See Fluorescent Lamp Seating.

BLACKENING OF ENDS IS NORMAL AND MAY OCCUR AT END OF LAMP LIFE

HEAVY SPOTTING MAY DEVELOP GRADUALLY DURING NORMAL LAMP LIFE

END BANDS MAY DEVELOP GRADUALLY DURING NORMAL LAMP LIFE

LIGHT SPOTTING MAY OCCUR IN NEW LAMPS AT ANY POINT ON THE BULB

FLUORESCENT BULB DISCOLORATION

FLUORESCENT LAMP SEATING

Preheat Circuit Testing

A preheat circuit requires a few seconds to heat the filament before light is produced. A starting switch is used to preheat the lamp filament. As the lamp ages, the starting switch has to be pressed to start and restart the lamp several times before the lamp stays ON. Ballast life is reduced when a hard-to-start lamp is not replaced. Replace the lamp any time a hard-to-start lamp is allowed to remain in operation for more than one week. In two-lamp circuits, premature ballast failure occurs when only one lamp is operating. Replace any inoperative lamp immediately. The ballast may require replacement on any circuit in which only one lamp is allowed to operate for longer than a few days. See Preheat Circuit Testing.

PREHEAT CIRCUIT TESTING

To test a preheat circuit, apply the procedure:
1. Turn power ON by placing the line switch in the ON position.
2. Connect a clamp-on ammeter around one of the incoming power lines.
3. Press and release the starting switch after two seconds. Record the starting current as soon as the switch is released. The starting current is the current value measured just after the starting switch is released. The starting current should be about 20% to 80% higher than the operating current when the circuit is working properly. The operating current is the current value measured after the lamp is energized.

No current reading indicates an open circuit. The same procedure may be used when an automatic starting switch is used. The starting current is read immediately after the line switch is closed. There is a problem with the starter or lamp when the lamp does not ignite properly and continues to try to start.

A lamp does not start when it is at the end of its normal life. The starter continues to try to start the lamp when the lamp is not replaced. This damages the starter. An old lamp that cycles during starting must be replaced. The starter requires replacement when a new lamp continues to cycle.

Instant-Start Circuit Testing

An instant-start circuit produces light instantly. Short lamp life is normally due to low supply voltage, improper lamp and lamp holder contact, or miswiring of the ballast. In two-lamp circuits, immediately replace any lamp that burns out. Replace both lamps when the good lamp shows signs of blackening. Look for ballast problems when there is a problem shortly after replacing a lamp in any circuit in which one lamp was allowed to operate alone for more than a few weeks.

Warning: Care must be taken to prevent electrical shock when measuring the high voltage. The voltmeter must be set to the high-voltage range. See Instant-Start Circuit Testing.

INSTANT-START CIRCUIT TESTING

To test an instant-start circuit, apply the procedure:
1. Turn the power to the circuit OFF by placing the line switch in the OFF position.
2. Remove the lamp.
3. Set the voltmeter to the 1000 V range. Connect the voltmeter leads to each end of the lamp holder. The voltmeter must complete the primary circuit at the lamp holder having two conductors. The two conductors may be shorted using a jumper wire.

Warning: Ensure that power is OFF before shorting the pins.

4. Turn the power to the circuit ON.
5. Measure the starting voltage. The ballast must provide an open circuit voltage of about three times the operating voltage to produce the arc required to light the lamp. There is a problem with the circuit when there is not a high starting voltage present. A zero voltage reading indicates an open circuit. A low voltage normally indicates a ballast problem.

316 TROUBLESHOOTING ELECTRICAL/ELECTRONIC SYSTEMS

> **Rapid-Start Circuit Testing**
>
> A rapid-start circuit produces light in less than one second. Long start times normally indicate a lamp problem. There is normally a ballast problem when start time is still long after replacing a lamp. In two-lamp circuits, immediately replace any lamp that burns out. Replace both lamps when the good lamp shows signs of blackening. See Rapid-Start Circuit Testing.

RAPID-START CIRCUIT TESTING

> To test a rapid-start circuit, apply the procedure:
> 1. Turn the power to the circuit OFF by placing the line switch in the OFF position.
> 2. Measure filament or starting voltage. Remove the lamp and connect a voltmeter across the opposing lamp holders to measure starting voltage. Connect the voltmeter across the end of each holder (normally red-red, blue-blue, and yellow-yellow) to measure the filament voltage.
> 3. Turn power to the circuit ON.
>
> Filament voltage should equal 3 V to 9 V, depending on the lamp. Starting voltage is given by the manufacturer and is normally about 2 to 2.5 times the supply voltage. A zero voltage reading indicates an open circuit. A low-voltage reading normally indicates a ballast problem.

Application — Troubleshooting High-Intensity Discharge Lamps

Measuring Line Voltage

The voltage at the input of an HID lamp fixture should be checked to ensure it is at the proper level. Non-starting, non-cycling, and lamp problems result when the voltage is not within a given limit. For constant wattage-rated ballasts, the line voltage should be within ±10% of the ballast (or fixture) nameplate rating. For reactor or high-reactance ballasts, the line voltage should be within ±5% of the ballast (or fixture) nameplate rating. See Measuring Line Voltage.

MEASURING LINE VOLTAGE

To measure the line voltage at an HID fixture, apply the procedure:
1. Open the lamp fixture so that the voltage may be measured directly at the lamp input.
2. Measure the voltage between the two incoming power lines. Ensure that the voltmeter is set to an AC voltage setting high enough to measure the voltage between the two incoming power lines. Check the measured voltage against the lamp's rated voltage. The voltage should be within ±10% for a constant-wattage ballast and ±5% for a reactor or high-reactance ballast.

Warning: Mercury-vapor and metal-halide lamps may cause serious skin burns and eye inflammation when the lamp's bulb is broken. The burns and eye inflammation result from the release of shortwave ultraviolet radiation. The lamp's bulb normally absorbs the shorter ultraviolet waves. The bulb should contain a fusible filament that extinguishes the arc when the bulb breaks. Always turn the power to the circuit OFF before troubleshooting the lamp. Replace any lamp that is cracked or damaged.

Ballast Output Voltage

The voltage output of a ballast should be checked when the incoming voltage is correct. An open circuit voltage measurement is required to see when the ballast is delivering the proper starting voltage to the lamp. See Measuring Ballast Output Voltage.

MEASURING BALLAST OUTPUT VOLTAGE

Warning: Never remove or insert an HID lamp when power is ON. Ensure that power is turned OFF. Measure the voltage to ensure power is OFF.

To measure the ballast output voltage, apply the procedure:
1. Measure the main supply voltage to ensure it is within the proper range.
2. Turn the power to the fixture OFF.
3. Remove the lamp from the fixture.
4. Connect the voltmeter to the lamp socket. Ensure that the voltmeter is set to a voltage setting high enough when connecting a voltmeter to the lamp socket (between the center pin and outside shell).

Caution: Disconnect the starter before measuring the ballast output voltage on a high-pressure sodium lamp. The high starting voltage pulse may damage some voltmeters.

5. Turn the power to the fixture ON.
6. Measure the voltage at the lamp socket. The voltage must be within the set limits. See Recommended Ballast Output Voltage Limits in Appendix.

There is a problem with the fixture when the voltage at the socket is not correct (or no voltage is present), and the incoming supply voltage is correct. The problem may be in the lamp socket, capacitor, ballast, or wiring. The problem is not with the lamp because the lamp is not receiving the correct voltage.

Warning: Metalarc lamps must be installed in fixtures that are enclosed with tempered glass or other manufacturer-approved materials. Metalarc lamps operating at high pressure (50 psi) may burst and shatter the lamp's bulb when they fail. This results in the discharge of hot quartz particles. These particles may be as hot as 1000°C.

Ballast Output Current

A ballast must deliver the proper voltage and current to a lamp. The output current of a ballast should be measured to ensure the ballast is delivering the correct starting current. A short-circuit lamp test may be used to measure ballast output current. In a short-circuit lamp test, the lamp is removed and an AC ammeter is connected across the socket. See Measuring Ballast Output Current.

MEASURING BALLAST OUTPUT CURRENT

To measure ballast output current, apply the procedure:
1. Turn the power to the fixture OFF.
2. Remove the lamp from the fixture.
3. Connect the ammeter to the lamp socket. Ensure that the ammeter is set to a current setting high enough when connecting an in-line ammeter to the lamp socket (between the center pin and outside shell).
4. Turn the power to the fixture ON.
5. Measure the current at the lamp socket. The current must be within the set limits. See Recommended Short-Circuit Current Test Limits in Appendix.

There is a problem with the fixture when the current at the lamp socket is not correct (or no current is present) and the incoming supply voltage is correct. The problem may be in the lamp socket, capacitor, ballast, or wiring. The problem is not with the lamp because the lamp is not receiving the correct voltage.

Normal End of Lamp Life

Low-pressure sodium lamps maintain their light output even at the end of their lamp life. At the end of their lamp life, starting is intermittent and finally impossible. There may be some blackening of the arc tube at the end of the lamp life.

Mercury-vapor and metal-halide lamps have a low light output at the end of their lamp life. They also start having intermittent starting problems. Some blackening of the arc tube occurs.

High-pressure sodium lamps start to cycle ON and OFF at the end of their lamp life. Some blackening of the bulb occurs. Most lamps have a brownish color on the arc tube due to sodium deposits.

Troubleshooting Guides

Troubleshooting guides state a problem, possible cause(s) and corrective action(s) that may be taken. These guides may be used to quickly determine potential problems and possible courses of action. See Troubleshooting Guides.

INCANDESCENT LAMP TROUBLESHOOTING GUIDE

Problem	Possible Cause	Corrective Action
Short lamp life	Voltage higher than lamp rating	Ensure that voltage is equal to or less than lamp's rated voltage. Lamp life is decreased when the voltage is higher than the rated voltage. A 5% higher voltage shortens lamp life by 40%.
	Lamp exposed to rough service conditions or vibration	Replace lamp with one rated for rough service or resistant to vibration
Lamp does not turn ON after new lamp is installed	Fuse blown, poor electrical connection, faulty control switch	Check circuit fuse or CB. Check electrical connections and voltage out of control switch. Replace switch when voltage is present into, but not out of the control switch.

FLUORESCENT LAMP TROUBLESHOOTING GUIDE

Problem	Circuit Type	Possible Cause	Corrective Action
Lamp blinks and has shimmering effect during lighting period	All types	Depletion of emission material on electrodes	Replace with new lamp
New lamp blinks	All types	Loose lamp connection	Reseat lamp in socket securely. Ensure lamp holders are rigidly mounted and properly spaced
		Low voltage applied to circuit	Ensure that voltage is within ±7% of rated voltage
		Cold area or draft hitting lamp	Protect lamp with enclosure. Use a low temperature-rated ballast
	Preheat	Defective or old starter	Replace starter. Lamp life is reduced when starter is not replaced

continued

continued

Lamp does not light or is slow starting	All types	Lamp failure	Replace lamp. Test faulty lamp in another fixture. Check circuit fuse or CB. Check voltage at fixture
		Loss of power to the fixture or low voltage	Troubleshoot fluorescent fixture when voltage is present at the fixture. Replace broken or cracked holder. Check for poor wire connection
	Preheat	Normal end of starter life	Replace starter
	Rapid-start	Failed capacitor in ballast	Replace ballast
		Lamp not seated in holder	Seat lamp properly in holder. In rapid-start circuits, the holder includes a switch that removes power when the lamp is removed due to high voltage present
Bulb ends remain lighted after switch is turned OFF	Preheat	Starter contacts stuck together	Replace starter. Lamp life is reduced when starter is not replaced
Short lamp life	All types	Frequent turning ON and OFF of lamps	Normal operation is based on one start per three hour period of operation time. Short lamp life must be expected when frequent starting cannot be avoided
		Supply voltage excessive or low	Check the supply voltage against the ballast rating. Short lamp life must be expected when supply voltage is not within ±7% of lamp rating
		Low ambient temperature. Low temperature causes a slow start	Protect lamp with enclosure. Use a low temperature-rated ballast
	Instant-start	One lamp burned out and other burning dimly due to series-start ballast circuit	Replace burned-out lamp. Ballast is damaged when lamp is not replaced
		Wrong lamp type. May be using rapid-start or preheat lamp instead of instant-start	Replace lamp with correct type
Light output decreases	All types	Light output decreases over first 100 hours of operation	Rated light output is based on output after 100 hours of lamp operation. Before 100 hours of operation, light output may be as much as 10% higher than normal
		Low circuit voltage	Check the supply voltage against the ballast rating. Short lamp life and low light output must be expected when supply voltage is not within ±7% of lamp rating
		Dirt build-up on lamp and fixture	Clean bulb and fixture

HID LAMP TROUBLESHOOTING GUIDE

Problem	Possible Cause	Corrective Action
Lamp does not start	Normal end of life operating characteristic	Replace with new lamp
	Loose lamp connection	Re-seat lamp in socket securely. Ensure lamp holder is rigidly mounted and properly spaced
	Defective photocell used for automatic turn ON	Replace defective photocell
	Low line voltage applied to circuit	Ensure that voltage is within ±7% of rated voltage
	Cold area or draft hitting lamp	Protect lamp with enclosure. Use a low temperature-rated ballast
	Defective ballast	Replace ballast
Lamp cycles ON and OFF or flickers when ON	Normal end of life operating characteristic	Replace with new lamp
	Poor electrical connection or loose bulb	Check electrical connections and socket contacts
	Line voltage variations	Ensure that voltage is within ±7% of rated voltage. Move lamps to separate circuit when lamps are on same circuit as high-power loads. High-power loads cause a voltage dip when turned ON. This voltage dip may cause the lamp to turn OFF
Short lamp life	Wrong wattage lamp or ballast	HID lamps and ballast must be matched in size. Lamp life is shortened when a large-wattage lamp is used. The same size and type ballast must be installed when replacing a ballast
	Power tap set too low for line voltage	Check ballast taps and ensure they are set for the correct supply voltage
	Defective sodium ballast starter	Replace starter
Low light output	Normal end of life operating characteristic	Replace with new lamp
	Dirty lamp or lamp fixture	Keep lamp fixture clean
	Early blackening of bulb caused by incorrect lamp size or ballast	Ensure that lamp size, lamp type, and ballast match

TROUBLESHOOTING LIGHTING CIRCUITS

Name _____ Date _____ **Activities**

⭘ Activity 17-1. Troubleshooting Incandescent Lamps

1. Using Incandescent Lamp Characteristics and Incandescent Lamp Circuit 1, determine each value.

_____ **A.** The wattage of the lamp based on Meter Reading 1 is _____ W.

_____ **B.** The lumen of the lamp based on Meter Reading 1 is _____ lm.

_____ **C.** The lamp life changes by _____% based on Meter Reading 1.

_____ **D.** The wattage of the lamp based on Meter Reading 2 is _____ W.

_____ **E.** The lumen of the lamp based on Meter Reading 2 is _____ lm.

_____ **F.** The lamp life changes by _____% based on Meter Reading 2.

INCANDESCENT LAMP CIRCUIT 1

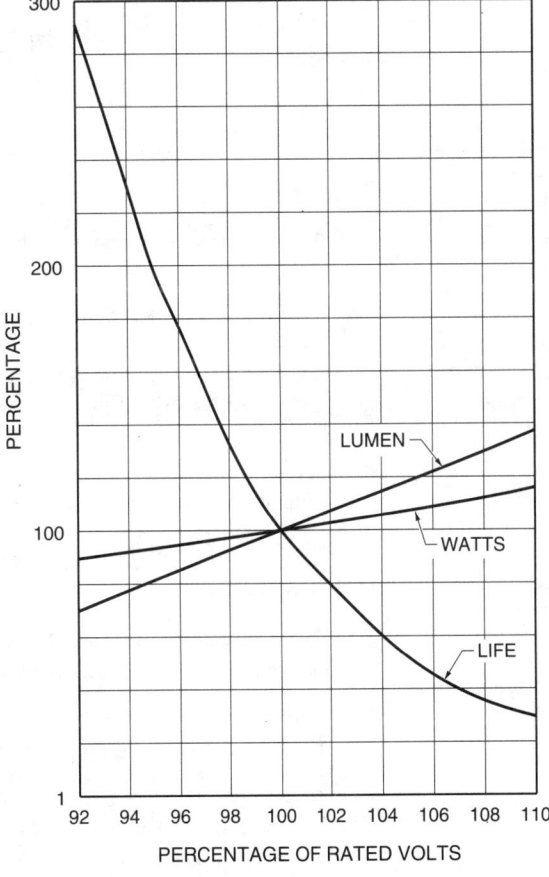

INCANDESCENT LAMP CHARACTERISTICS

324 TROUBLESHOOTING ELECTRICAL/ELECTRONIC SYSTEMS

2. Using Incandescent Lamp Circuit 2 and the meter readings, determine each value. *Note:* Assume all components in the circuit are good for questions A-D.

_____ A. Meter 1 reads _____ V when the switch is in the OFF position.

_____ B. Meter 2 reads _____ V when the switch is in the OFF position.

_____ C. Meter 1 reads _____ V when the switch is in the ON position.

_____ D. Meter 2 reads _____ V when the switch is in the ON position.

_____ E. The problem is a(n) _____ if Meter 1 reads 0 VAC when the switch is in the ON position and 0 VAC when the switch is in the OFF position and the lamp does not light.

_____ F. The problem is a(n) _____ if Meter 2 reads 115 VAC when the switch is in the ON position and 115 VAC when the switch is in the OFF position and the lamp is always ON.

INCANDESCENT LAMP CIRCUIT 2

3. Using Incandescent Lamp Circuit 3 and the meter readings, determine each value. *Note:* Assume all components in the circuit are good for questions A-E.

_____ A. Meter 1 connected in Test Position 1 reads _____ V regardless of the position of the two switches.

_____ B. Meter 1 connected in Test Position 2 reads _____ V when both switches are up.

_____ C. Meter 1 connected in Test Position 2 reads _____ V when both switches are down.

_____ D. Meter 2 connected in Test Position 1 reads _____ V when Switch 1 is in the up position and Switch 2 is in the down position.

_____ E. Meter 2 connected in Test Position 1 reads _____ V when Switch 1 is in the up position and Switch 2 is in the up position.

_____ F. The problem is a(n) _____ if Meter 1 reads 115 VAC in Test Position 1, 0 VAC in Test Position 2 (at all times), and Meter 2 reads 0 VAC in either test position (regardless of either switch position).

Troubleshooting Lighting Circuits

INCANDESCENT LAMP CIRCUIT 3

◯ Activity 17-2. Troubleshooting Fluorescent Lamps

1. Using Fluorescent Lamp Circuit 1 and the meter readings, determine each value. *Note:* Assume all components in the circuit are good for questions A–I.

_____ **A.** Meter 1 reads _____ V before the Starting Switch is pressed.

_____ **B.** Meter 2 reads _____ A before the Starting Switch is pressed.

_____ **C.** Meter 3 reads _____ A before the Starting Switch is pressed.

_____ **D.** Meter 1 reads _____ V when the Starting Switch is held down.

_____ **E.** Meter 2 reads _____ A when the Starting Switch is held down.

326 TROUBLESHOOTING ELECTRICAL/ELECTRONIC SYSTEMS

_____ **F.** Meter 3 reads _____ A when the Starting Switch is held down.

_____ **G.** Meter 1 reads _____ V when the Starting Switch is open and the lamp is ON.

_____ **H.** Meter 2 reads _____ A when the Starting Switch is open and the lamp is ON.

_____ **I.** Meter 3 reads _____ A when the Starting Switch is open and the lamp is ON.

_____ **J.** Meter 1 reads _____ V when the Starting Switch is held down and the filament or ballast is open.

_____ **K.** Meter 2 reads _____ A when the Starting Switch is held down and the filament or ballast is open.

_____ **L.** Meter 3 reads _____ A when the Starting Switch is held down and the filament or ballast is open.

FLUORESCENT LAMP CIRCUIT 1

2. Using Fluorescent Lamp Circuit 2 and the meter readings, determine each value. *Note:* Assume all components in the circuit are good for questions A-C.

_____ **A.** Meter 1 reads _____ V when the lamps are ON.

_____ **B.** Meter 2 reads _____ V when the lamps are ON.

_____ **C.** Meter 3 reads _____ A when the lamps are ON.

_____ **D.** Meter 1 reads _____ V when the line switch is closed, one of the lamps is not properly seated (or removed), and neither lamp is ON.

_____ **E.** Meter 2 reads _____ V when the line switch is closed, one of the lamps is not properly seated (or removed), and neither lamp is ON.

_____ **F.** Meter 3 reads _____ A when the line switch is closed, one of the lamps is not properly seated (or removed), and neither lamp is ON.

FLOURESCENT LAMP CIRCUIT 2

3. Using Fluorescent Lamp Circuit 3 and the meter readings, determine each value. *Note:* Assume all components in the circuit are good for questions A and B.

_____ **A.** Meter 1 reads _____ V when the lamps are ON.

_____ **B.** Meter 2 reads _____ A when the lamps are ON.

_____ **C.** Does Meter 2 reading increase or decrease if the line voltage is increased 5%.

FLUORESCENT LAMP CIRCUIT 3

Activity 17-3. Troubleshooting High-Intensity Discharge Lamps

1. Using the wiring diagram, connect the devices in HID Lamp Circuit 1. Add a wire nut to insulate any wires not used.

HID LAMP CIRCUIT 1

2. Using the wiring diagram, connect the devices in HID Lamp Circuit 2. Add a wire nut to insulate any wires not used.

HID LAMP CIRCUIT 2

Troubleshooting Lighting Circuits 329

3. Using the wiring diagram, connect the devices in HID Lamp Circuit 3. Add a wire nut to insulate any wires not used.

HID LAMP CIRCUIT 3

4. Using the wiring diagram, connect the devices in HID Lamp Circuit 4. Add a wire nut to insulate any wires not used.

HID LAMP CIRCUIT 4

5. Using the wiring diagram, connect the devices in HID Lamp Circuit 5. Add a wire nut to insulate any wires not used.

HID LAMP CIRCUIT 5

SOLENOIDS AND HEATING ELEMENTS

18 Applications

☐ Application — Solenoid-Operated Valves

A *solenoid* is an electromechanical device that converts electrical energy to mechanical energy. Solenoids are used to operate motor starters, contactors, relays, valves, clutches, and many other applications. See Solenoid.

The amount of force a solenoid develops depends on the number of turns in the coil and the amount of applied current. The larger the coil, the greater the force. The current draw is the highest (inrush current) when a solenoid is first energized. The current draw is less (rated current) after the solenoid has moved through its rated travel distance.

SOLENOID

Solenoid-Operated Valves

A *solenoid-operated valve* is a valve that uses electrical force to directly or indirectly change the position of a valve's spool. Solenoid-operated valves are used in most fluid power circuits to control the direction of fluid flow. See Fluid Power Symbols in Appendix. Directional control valves are identified by the ways, number of positions, and types of actuators. See Solenoid-Operated Directional Control Valves.

Ways

A *way* is the flow path through the valve. Most directional control valves are either two-way, three-way, or four-way. The number of ways required depends on the application. Two-way directional control valves have two main ports that allow or stop the flow of fluid. Two-way valves are used as shutoff, check, and quick-exhaust valves. Three-way valves allow or stop fluid flow or exhaust. They are used to control single-acting cylinders, fill-and-drain tanks, and nonreversible fluid motors. Four-way directional control valves have four (or five) main ports that change fluid flow from one port to another. Four-way valves are used to control the direction of double-acting cylinders and reversible-fluid motors.

Positions

A directional control valve is placed in different positions to start, stop, or change the direction of fluid flow. A *position* is the number of positions within the valve that the spool is placed in to direct fluid flow through the valve. Two-way and three-way valves have two positions. Four-way valves may have two or three positions.

Actuators

Directional control valves must have a means to change the valve position. An *actuator* is a device that changes the position of a valve. Directional control valve actuators include pilots, manual levers, mechanical levers, springs, and solenoids. Solenoid-operated, directional control valves are either solenoid-operated/spring return, or double solenoid-operated.

332 TROUBLESHOOTING ELECTRICAL/ELECTRONIC SYSTEMS

Normally-Closed and Normally-Open Valves

A *normally-closed (NC) valve* is a valve that does not allow pressurized fluid to flow out of the valve in the spring-actuated position. A *normally-open (NO) valve* is a valve that allows pressurized fluid to flow out of the valve in the spring-actuated position. Two-way and three-way valves may be either normally open or normally closed. The valve must have a spring position to be normally open or normally closed.

Valve Numbering System

The ways, positions, port size, actuators, coil voltage, and current type must be specified when ordering a new or replacement valve. Valve manufacturers provide a numbering system to simplify valve selection and ordering. The numbering system normally consists of letters and numbers that represent the different valve models. See Valve Selections.

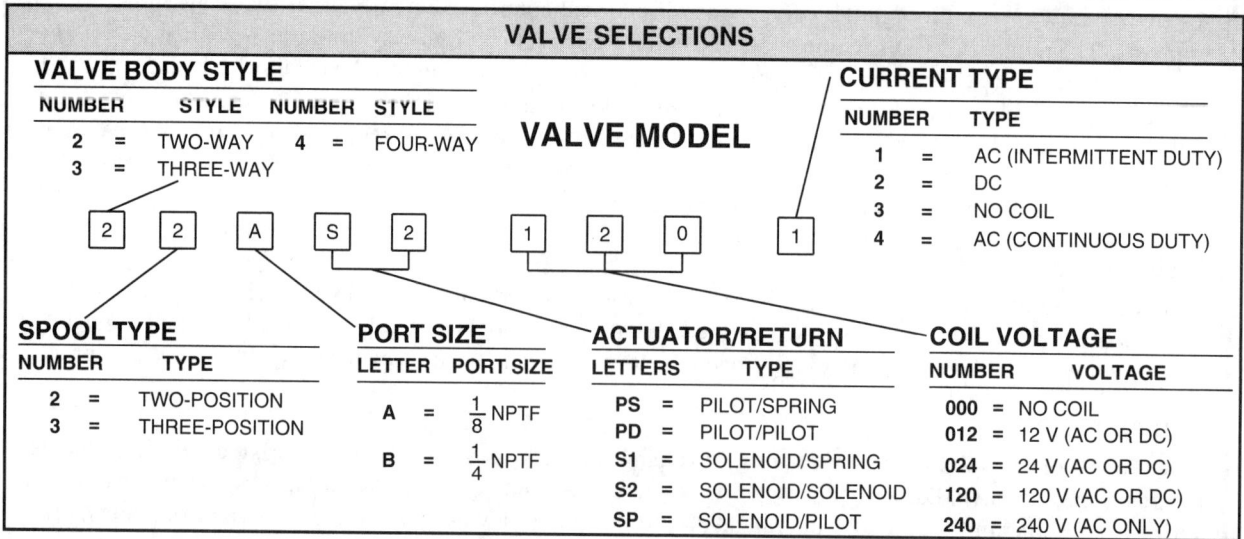

Incorrect Voltage

The voltage applied to a solenoid-operated valve should be ±10% of the rated solenoid value. The voltage is measured directly at the valve when the solenoid is energized. A voltmeter is set on an AC voltage range for AC solenoids. The voltmeter is set on a DC voltage range for DC solenoids. The range setting must be greater than the applied voltage. See Solenoid Voltage Measurement.

A solenoid overheats when the voltage is excessive. The heat destroys the insulation on the coil wire and burns out the solenoid. The solenoid has difficulty moving the spool inside the valve when the voltage is too low. The slow operation causes the solenoid to draw its high inrush current longer. Lasting high inrush current causes excessive heat.

SOLENOID VOLTAGE MEASUREMENT

Incorrect Frequency

Solenoid valves are available with frequency ratings of 50 Hz or 60 Hz. Solenoids with a frequency rating of 50 Hz are used for export applications. Solenoids with a frequency rating of 60 Hz are used for domestic applications. Imported machines may have solenoid valves with a frequency rating of 50 Hz. A solenoid may operate if the frequency is not correct, but may have a higher failure rate and may produce noise.

Double-Acting Solenoid Valve

A *double-acting solenoid valve* is a valve that uses one solenoid to develop a force in one direction and a second solenoid to develop a force in the opposite direction. One solenoid moves the spool to one position and the other solenoid moves the spool to the other position. The proper operation of a double-acting solenoid is to energize one solenoid at a time. The solenoid overheats and burns out when both solenoids are energized simultaneously. See Double-Acting Solenoid Valve.

DOUBLE-ACTING SOLENOID VALVE

An *interlock* is a mechanical or electrical method used to prevent two different circuits from being energized at the same time. An interlock is included in applications that require double-acting solenoids. A voltmeter is connected to both solenoids when testing a circuit in operation. Both meters should never indicate a voltage simultaneously. A control circuit problem exists when both solenoids have power simultaneously. See Solenoid Interlock.

SOLENOID INTERLOCK

Rapid Cycling

A solenoid draws several times its rated current when first connected to power. This high inrush current produces heat. In normal applications, the heat is low and dissipates over time. Rapid cycling does not allow the heat to dissipate quickly. The heat buildup burns the coil insulation and causes solenoid failure. A high-temperature solenoid should be used in applications requiring a solenoid to be cycled more than 10 times per minute. A solenoid may be forced-cooled in some applications.

An AC solenoid heats faster than a DC solenoid. A DC solenoid should be used in fast cycling applications. An AC solenoid with a continuous duty rating should be used if circuit design does not provide for DC. Some manufacturers include a diode in series with the coil of AC solenoids to give the solenoid a continuous duty rating. This rating is normally marked on the valve as CD. The voltage after the diode in an AC-solenoid valve marked CD is pulsating DC, and the voltage before the diode is AC.

Excessive Valve Force

A solenoid moves the valve spool in a solenoid-operated valve. The spool has extra pressure applied to it when undersized valves are used. The solenoid has to apply a greater force to overcome the extra pressure. The greater force produces a higher-than-rated current in the solenoid. The higher-than-rated current is present longer than normal. To prevent this problem, ensure that the valve is correctly sized or slightly oversized.

Manufacturers provide specification curves that illustrate the amount of force a solenoid produces at its rated voltage. Most curves also illustrate the amount of force produced at less than rated voltage and the current draw of the solenoid. The lower the applied voltage, the less the delivered force and current draw. A solenoid must deliver enough force throughout its stroke to operate properly. See Solenoid Specifications.

SOLENOID SPECIFICATIONS

Excessive valve force may be required at times even when the valve is correctly sized. This occurs in cold hydraulic systems. Cold temperatures cause oil to become thick, which may overload a solenoid-operated valve. Use a hydraulic oil that is suitable for low temperature when ambient temperature is low. A higher temperature coil or larger solenoid may also be used in low temperature applications.

Solenoid force should be as close to the load requirements as possible. Excessive force reduces solenoid life. Inadequate force keeps the solenoid from operating.

Transients

In most industrial applications, the power supplying a solenoid comes from the same power lines that supply electric motors and other solenoids. High transient voltages are placed on the power lines as these inductive loads are turned ON and OFF. Transient voltages may damage the insulation on the solenoid coil, nearby contacts, and other loads. The transient voltages may be suppressed by using snubber circuits.

Environmental Conditions

For proper operation, a solenoid must operate within its rating and not be mechanically damaged or damaged by the surrounding atmosphere. A solenoid coil is subject to heat during normal operation. This heat comes from the combination of fluid flowing through the valve, the temperature rise from the coil when energized, and the ambient temperature of the solenoid.

Solenoid-operated valves may be purchased with or without covers. Solenoid-operated valves that do not have covers are intended for mounting inside a dust-tight cabinet. Any valve that is not mounted inside a dust-tight cabinet must have a cover to protect the solenoid coils. The cover protects the solenoid coils from water, oil, dirt, and mechanical contact and should always be kept on the solenoid coils. The cover may be attached to the valve with a small chain to prevent it from being lost. See Solenoid-Operated Valve Cover.

A solenoid coil is damaged when heat increases the temperature of the coil insulation above its rating. The coil is damaged by the breakdown of the insulation. The insulation on the coil helps seal out moisture, dust, oil, corrosive material, and heat. The insulation also provides electrical insulation.

SOLENOID-OPERATED VALVE COVER

SOLENOID INSULATION CLASSES		
Class	Maximum Operating Temperature*	Insulation
A	105	Epoxy or paper
B	130	Epoxy
F	155	Epoxy
H	180	Epoxy

* in °C

Solenoids that have high temperature-rated insulation are used in high-temperature applications. Solenoids use various classes of insulation. See Solenoid Insulation Classes.

High humidity and a frequently changing atmosphere cause condensation. Condensation may cause corrosion that hinders solenoid operation and aid in the deterioration of coil insulation. Condensation in low-temperature areas causes ice buildup. Ice buildup may prevent proper solenoid movement.

A solid molded coil or solenoid designed for oil immersion is used when condensation is a problem. This keeps the moisture away from the coil insulation. A solenoid-operated valve is placed where it cannot be mechanically damaged. The valve is placed in an enclosure when required. Only solenoid valves that include tight covers are used.

Application — Troubleshooting Solenoid-Operated Valves

Four symptoms of a faulty solenoid include: failure to operate when energized, failure to operate when de-energized, noisy operation, and erratic operation. A solenoid that fails to operate when energized does not allow the valve to shift position or electrical contacts to operate. A solenoid that fails to operate when de-energized does not allow the valve to shift to the original position or electrical contacts to change to their normal condition.

A solenoid that operates noisily does not cause improper operation, although the noise may become a problem. A solenoid that operates erratically causes intermittent problems in the system and is a sign of a problem that normally gets worse. See Faulty Solenoid Problems.

FAULTY SOLENOID PROBLEMS

Problem	Possible Causes	Comments
Failure to operate when energized	Complete loss of power to the solenoid	Normally caused by blown fuse or control circuit problem
	Low voltage applied to the solenoid	Voltage should be at least 85% of solenoid's rated value
	Burned out solenoid coil	Normally evident by pungent odor caused by burnt insulation
	Shorted coil	Normally a fuse is blown and continues to blow when changed
	Obstruction of plunger movement	Normally caused by a broken part, misalignment, or the presence of a foreign object
	Excessive pressure on solenoid plunger	Normally caused by excessive system pressure in solenoid-operated valves
Failure to operate spring-return solenoids when de-energized	Faulty control circuit	Normally a problem of the control circuit not disengaging the solenoid's hold or memory circuit.
	Obstruction of plunger movement	Normally caused by a broken part, misalignment, or the presence of a foreign object
	Excessive pressure on solenoid plunger	Normally caused by excessive system pressure in solenoid-operated valves
Failure to operate electrically-operated return solenoids when de-energized	Complete loss of power to solenoid	Normally caused by a blown fuse or control circuit problem
	Low voltage applied to solenoid	Voltage should be at least 85% of solenoid's rated value
	Burned out solenoid coil	Normally evident by pungent odor caused by burnt insulation
	Obstruction of plunger movement	Normally caused by broken part, misalignment, or presence of a foreign object
	Excessive pressure on solenoid plunger	Normally caused by excessive system pressure in solenoid-operated valves
Noisy operation	Solenoid housing vibrates	Normally caused by loose mounting screws
	Plunger pole pieces do not make flush contact	An air gap may be present causing the plunger to vibrate. These symptoms are normally caused by foreign matter
Erratic operation	Low voltage applied to the solenoid	Voltage should be at least 85% of the solenoid's rated voltage
	System pressure may be low or excessive	Solenoid size is inadequate for the application
	Control circuit is not operating properly	Conditions on the solenoid have increased to the point where the solenoid cannot deliver the the required force

Troubleshooting Solenoids

A voltmeter and ohmmeter are required when troubleshooting a solenoid. See Troubleshooting Solenoids.

To troubleshoot a solenoid, apply the procedure:
1. Turn electrical power to the solenoid or solenoid circuit OFF.
2. Measure the voltage at the solenoid to ensure the power is OFF.
3. Remove the solenoid cover and visually inspect the solenoid. Look for a burnt coil, broken parts, or other problems. Replace the coil when burnt. Replace the broken parts when available. Replace the valve, contactor, starter, or solenoid-operated device when the parts are not available. *Note:* Determine the fault before installing a new coil when a solenoid has failed due to a burnt or shorted coil. The new coil burns out if the fault is not corrected. Always observe solenoid operation after a solenoid is replaced.
4. Disconnect the solenoid wires from the electrical circuit when no obvious problem is observed.
5. Check the solenoid continuity. Connect the meter leads to the solenoid wires with all power turned OFF. The meter should indicate a resistance reading of ±15% of the coil's normal reading.

Unknown readings are obtained by testing a good solenoid. A low or zero reading indicates a short or partial short. Replace the solenoid if there is a short. No movement of the needle on an analog meter or infinity resistance on a digital meter indicates the coil is open and defective. Replace the solenoid if the open is not obvious. *Note:* Set the ohmmeter to (R × 100) if the solenoid uses a small gauge wire. The small wire has high resistance.

TROUBLESHOOTING SOLENOIDS

▢ Application — Heating Elements

A *heating element* is a heat-producing device that produces heat from the resistance to the flow of electric current. A heating element generally has a long life with little or no problems. Design, operation, or application problems are present if a heating element has early problems or does not last long. Most problems are caused by using the wrong type of heating element for the application or by using the wrong size heating element.

Heating Element Types

Many types of heating elements are available. The two main types are the open coil and enclosed coil. The open-coil heating element is the least expensive and presents a safety problem when not properly protected. The open-coil heating element is used in hot-air applications such as space heaters, hair dryers, and other small appliances that have a protective enclosure. See Open and Enclosed-Coil Heating Elements.

OPEN AND ENCLOSED-COIL HEATING ELEMENTS

The enclosed-coil heating element is the safest and most commonly used heating element in industrial applications. The *enclosed-coil heating element* has the resistance wire which produces the heat surrounded by a refractory insulating material. The refractory insulation is enclosed in a protective metal tube or cover. The cover protects the resistance wire from physical damage and improves thermal conductivity. The potential of electrical shock is also reduced because the live conductor is not exposed to the outside of the heating element. The most common enclosed-coil heating elements are the tubular, strip, ring, band, and combination types. See Enclosed-Coil Heating Elements.

Heating Element Ratings

Heating elements are rated by the manufacturer as to watt density. *Watt density* is the watts per square inch (W/sq in.) on the outer cover of the heating element. Excessive watt density wastes energy, shortens heating element life, and causes wide heat fluctuations. Low watt density causes long heating times or inadequate heat.

Heating a material to a high temperature quickly is accomplished by increasing the watt density of the heating element. This reduces the life of the heating element. The watt density of a heating element depends on the rate of heat absorption of the heated material. For example, water conducts (absorbs) heat away from a heating element faster than a heavy tar (asphalt). A heating element used to heat water may have a high watt density rating. Water conducts heat away from a heating element about five times faster than a heavy tar so the watt density of a water-heating element (approximately 50 W/in.) may be five times higher than a heavy tar-heating element (approximately 10 W/in.). The heavy tar may be heated faster by using a high-rated heating element, which reduces the life of the heating element.

Outer Cover Material

The type of material heated determines the type of outer cover material of the heating element. A copper jacket is normally used for water and cooking oil heating applications. A steel jacket is normally used for oil and kerosene heating applications. A stainless steel jacket or alloy is normally used for corrosive solutions and high-temperature air heating. See Heating Element Specifications.

Terminal Contamination

Heating elements are connected to supply voltage at terminal points. The electrical resistance of the connection increases when the terminals are contaminated. Moisture, salt, oil, and acid are major causes of contamination.

HEATING ELEMENT SPECIFICATIONS

Material State	Heated Material	Outer Cover Material	Maximum Watt Density*	Material Operating Temperature**
Liquid	Gasoline	Iron/steel	18-22	300
	Oil (low viscosity)	Steel/copper	20-25	Up to 180
	Oil (medium viscosity)	Steel/copper	12-18	Up to 180
	Oil (high viscosity)	Steel/copper	5-7	Up to 180
	Vegetable oil (cooking)	Copper	28-32	400
	Water (washroom)	Copper	70-90	140
	Water (process)	Copper	40-50	212
Air/Gas	Still air (ovens)	Steel/stainless steel	28-32 18-22 8-10 2-3	Up to 700 Up to 1000 Up to 1200 Up to 1500
	Air (moving at 10 fps)	Aluminum steel	30-34 23-26 14-16 2-3	Up to 700 Up to 1000 Up to 1200 Up to 1500
Solid	Asphalt	Iron/steel	9-12 8-10 6-8 5-6	Up to 200 Up to 300 Up to 400 Up to 500
	Molten tin (heated in pot)	Iron/steel	20-22	600
	Steel tubing (heated indirectly)	Iron/steel	45-48 50-52 54-56	500 750 1000

* in W/in.
** in °F

Terminal connections are protected to prevent contamination. Moisture barriers or terminal seals are used to protect the connection. A variety of moisture barriers and terminal seals are available. These include rubber, epoxy, silicon, putty, and ceramic-to-metal hermetic seals. Rubber is used in applications up to 150°F. Epoxy is used in applications up to 300°F. Silicon is used in applications up to 500°F. Putty is used in applications up to 600°F. Ceramic-to-metal hermetic seals are used in applications up to 1000°F. See Terminal Protection.

TERMINAL SEALS	
Type	Max. Temp.*
Rubber	150
Epoxy	300
Silicon	500
Putty	600
Ceramic	1000

* in °F

TERMINAL PROTECTION

Application — Troubleshooting Heating Elements

The problems that occur in heating circuits include no heat, inadequate heat, excessive heat, or the system is giving electrical shocks. Inadequate or excessive heat normally indicates a control circuit problem. No heat normally indicates a blown fuse, a heating element problem, or control circuit problem. Electrical shocks are caused by a high leakage current that occurs when insulation breaks down. To determine the problem in a heating circuit, the circuit is tested for an open or short circuit, the power draw of the heating element is measured, and the insulation is tested.

Open or Short Circuit Test

Heating elements may open or short-circuit due to deterioration, physical damage, or misuse. No heat is produced when a heating element opens or shorts. The fuse (or CB) blows when a heating element shorts. The only sign that a partial short occurs in part of a heating element is a reduction of heat output. See Open or Short Circuit Test.

Warning: Avoid contact with any part of the heating circuit or element that may cause burns.

To test for an open or short circuit, apply the procedure:

1. Turn power to the system OFF when the heating element is suspected of having an open or short.
2. Measure the voltage at the heating element to ensure the power is OFF.
3. Disconnect the heating element from the circuit.

OPEN OR SHORT CIRCUIT TEST

4. Measure the resistance of the heating element. The heating element should have some resistance. The higher the rated heater output, the lower the resistance reading. The lower the rated heater output, the higher the resistance reading. A heating element has a short when the meter measures 0 Ω. A heating element has an open when the meter measures infinity. The heating element should be removed and tested when there is a short or open indication.
5. Reconnect the heating element when there is a resistance reading. Ensure the ohmmeter is removed from the circuit.
6. Turn power to the system ON.
7. Set the temperature control unit to turn ON the heating element.
8. Measure the voltage into the heating element. The problem is located before the heating element when no power is present at the heating element. Turn power to the circuit OFF. The problem may be a blown fuse, faulty control circuit, or open connection. A power draw test is performed if there is correct voltage at the heating element but not the correct heat output.

Power Draw Test

A heating element converts electrical energy (watts) into heat. Most heaters are rated in watts. For example, a portable space heater is rated for 750 W to 1500 W for normal operation on a 15 A or 20 A circuit. The amount of electrical power a heating element draws is an indication of the condition of the heating element.

A wattmeter may be used to measure the electrical power draw of a heating element. A voltmeter and ammeter are normally used to measure the electrical power draw of a heating element because voltage and current are in phase in a heating element. See Power Draw Test.

POWER DRAW TEST

To perform a power draw test on a heating element, apply the procedure:
1. Set the temperature control unit to turn the heating element ON.
2. Measure the voltage into the heating element.
3. Measure the current flowing through the heating element.

Multiply the measured voltage and current to obtain the amount of power draw. The amount of power draw should equal ±10% of the manufacturer's power rating. A reduction in heat output is produced when the power draw is less than 10% of the manufacturer's power rating. An increase in heat output is produced when the power draw is greater than 10% of the manufacturer's power rating. A reduced output normally indicates poor electrical connections or partially open elements. An increased output normally indicates partially shorted elements or high leakage current.

Insulation Breakdown Test

Heating elements absorb moisture over time. This causes current to leak from the current carrying conductors inside the heating element to the outer cover. Current leakage is a function of insulation resistance. Current leakage increases as insulation resistance decreases. Insulation resistance values on dry heating elements should be high. Dry insulation readings should equal several thousand megohms. Over time, the resistance decreases and the leakage current increases. A safety problem is present when the insulation value decreases greatly. See Insulation Breakdown Test.

INSULATION BREAKDOWN TEST

To perform an insulation breakdown test, apply the procedure:
1. Turn power to the system OFF.
2. Measure the voltage at the heating element to ensure the power is OFF.
3. Measure the resistance of the outer cover of the heating element to ground. The resistance reading should be high. The resistance reading should equal 1000 times (or more) the rated voltage. For example, a 240 V heater should have a resistance reading of 240,000 Ω to ground. The heating element should be dried and retested if the reading is low.
4. Remove the megohmmeter and turn power ON when there is a high resistance reading.
5. Set the heating control circuit to turn ON the heating element.
6. Connect an ammeter between each power line terminal and the outer cover. Moisture is detected by an increase in normal electrical leakage from the element cover to ground. Leakage current should be less than 5 mA.

Constant Wattage Heating Element Drying

Any heating element that has an insulation resistance value of less than 250,000 Ω from the outer cover to ground should be dried before energizing. The preferred method of drying a heating element is to remove the element and bake it in an oven. The heating element should be baked at 250°F to 350°F for about 12 hours. Measure the insulation resistance after drying.

The low voltage drying method may be used when it is not practical to place a heating element in an oven. The low voltage method connects the heating element to 25% to 50% of the element's rated voltage. The low voltage is applied for 24 to 48 hours. The lower the resistance reading, the lower the applied voltage, and the longer the time period. Full voltage may be applied when the resistance reading equals 1000 times (or more) the rated voltage.

SOLENOIDS AND HEATING ELEMENTS

18

Name _____ Date _____   **Activities**

◯ Activity 18-1. Solenoid-Operated Valves

1. Using the Glue Dispense Circuit, answer the questions.

_____   **A.** The valve has _____ positions.
_____   **B.** The valve is a _____-way valve.
_____   **C.** The valve is _____-actuated when glue is dispensed.
_____   **D.** The valve is _____-actuated when no glue is dispensed.

GLUE DISPENSE CIRCUIT

2. Using the Device Clamping Circuit, answer the questions.

_____   **A.** The valve has _____ positions.
_____   **B.** The valve is a _____-way valve.
_____   **C.** The valve is _____-actuated when the cylinder is advanced.
_____   **D.** The valve is _____-actuated when the cylinder is retracted.

DEVICE CLAMPING CIRCUIT

344 TROUBLESHOOTING ELECTRICAL/ELECTRONIC SYSTEMS

3. Using the Conveyor Positioning Circuit, answer the questions.

_____ **A.** The valve has _____ positions.

_____ **B.** The valve is a _____-way valve.

_____ **C.** The valve is _____-actuated when the cylinder is advanced.

_____ **D.** The valve is _____-actuated when the cylinder is retracted.

CONVEYOR POSITIONING CIRCUIT

4. Using the Valve Control Circuit, answer the questions.

_____ **A.** The valve has _____ positions.

_____ **B.** The valve is a _____-way valve.

_____ **C.** The valve is _____-actuated when the cylinder is advanced.

_____ **D.** The valve is _____-actuated when the cylinder is retracted.

VALVE CONTROL CIRCUIT

5. In the Basic Glue Dispense Circuit, the glue gun dispenses glue anytime the photoelectric switch detects a box. The glue starts and stops on the edge of each box. This presents a maintenance problem because glue may be sprayed on the conveyor belt. An improved circuit is used to prevent the problem. Using the Improved Glue Dispense Circuit, answer the questions.

_____ A. The glue is dispensed _____ s after the photoelectric switch senses the edge of the box.

_____ B. The glue is dispensed for _____ s.

_____ C. When does the lamp turn ON?

_____ D. Does the Manual Test Button activate the glue gun even when the selector switch is in the OFF position?

BASIC GLUE DISPENSE CIRCUIT

IMPROVED GLUE DISPENSE CIRCUIT

6. In the Basic Device Clamping Circuit, the cylinder is extended anytime the Foot Switch is pressed. This may present a safety problem. Using the Improved Device Clamping Circuit, answer the questions.

_____ A. Does the system pressure have to be above or below 100 psi before the cylinder advances?

_____ B. How is the cylinder advanced after the pressure switch is closed?

346 TROUBLESHOOTING ELECTRICAL/ELECTRONIC SYSTEMS

BASIC DEVICE CLAMPING CIRCUIT

IMPROVED DEVICE CLAMPING CIRCUIT

7. In the Basic Valve Control Circuit, the Product Flow Valve Cylinder is moved anytime a pushbutton is pressed. This may present a maintenance problem. Using the Basic and Improved Valve Control Circuit, answer the questions.

    _____ **A.** In the basic circuit, can both solenoids be energized simultaneously?

    _____ **B.** In the improved circuit, can both solenoids be energized simultaneously?

    _____ **C.** In either circuit, can the cylinder be placed in a position other than fully-extended or fully-retracted?

BASIC VALVE CONTROL CIRCUIT

IMPROVED VALVE CONTROL CIRCUIT

8. In the Basic Conveyor Positioning Circuit, the conveyor arm is moved anytime the selector switch is moved to a different position. This may present a safety problem. Using the Improved Conveyor Positioning Circuit, answer the questions.

_____ A. When does the alarm sound?

_____ B. The alarm sounds for _____ s.

_____ C. Does the gate change position as soon as the selector switch is moved to a different position?

_____ D. Is the gate moved before, during, or after the sounding of the alarm?

BASIC CONVEYOR POSITIONING CIRCUIT

IMPROVED CONVEYOR POSITIONING CIRCUIT

Activity 18-2. Troubleshooting Solenoid-Operated Valves

1. In the Glue Dispense Circuit, the timer is rated at .050 A, the solenoid at .60 A, and the lamp at .10 A. Using the Glue Dispense Circuit, answer the questions. *Note:* Assume all components in the circuit are good.

_____ A. Meter 1 reads _____ A when the Glue Gun Selector Switch is in the ON position and no box is in front of the photoelectric switch.

_____ B. Meter 2 in Test Position 1 reads _____ V when the Glue Gun Selector Switch is in the ON position and no box is in front of the photoelectric switch.

_____ C. Meter 2 in Test Position 2 reads _____ V when the Glue Gun Selector Switch is in the ON position and no box is in front of the photoelectric switch.

_____ **D.** Meter 2 in Test Position 3 reads _____ V when the Glue Gun Selector Switch is in the ON position and no box is in front of the photoelectric switch.

_____ **E.** Meter 3 reads _____ V when the Glue Gun Selector Switch is in the ON position and no box is in front of the photoelectric switch.

_____ **F.** Meter 4 reads _____ A when the Glue Gun Selector Switch is in the ON position and no box is in front of the photoelectric switch.

_____ **G.** Meter 1 reads _____ A when the Glue Gun Selector Switch is in the ON position, a box is in front of the photoelectric switch, and the glue gun is dispensing glue.

_____ **H.** Meter 2 in Test Position 1 reads _____ V when the Glue Gun Selector Switch is in the ON position, a box is in front of the photoelectric switch, and the glue gun is dispensing glue.

_____ **I.** Meter 2 in Test Position 2 reads _____ V when the Glue Gun Selector Switch is in the ON position, a box is in front of the photoelectric switch, and the glue gun is dispensing glue.

_____ **J.** Meter 2 in Test Position 3 reads _____ V when the Glue Gun Selector Switch is in the ON position, a box is in front of the photoelectric switch, and the glue gun is dispensing glue.

_____ **K.** Meter 4 reads _____ V when the Glue Gun Selector Switch is in the ON position, a box is in front of the photoelectric switch, and the glue gun is dispensing glue.

_____ **L.** Meter 4 reads _____ A when the Glue Gun Selector Switch is in the ON position, a box is in front of the photoelectric switch, and the glue gun is dispensing glue.

GLUE DISPENSE CIRCUIT

2. In the Device Clamping Circuit, the cylinder cannot be advanced unless the system pressure is at least 100 psi. This prevents the cylinder from advancing and not having enough pressure to hold the part in place. The problem may be in the pushbutton, foot switch, pressure switch, solenoid, electrical system, or fluid power system if the cylinder does not advance. To aid in troubleshooting a circuit, a push-to-test pushbutton is added to a circuit. Redraw the circuit adding a push-to-test pushbutton that overrides Pushbutton 1 and Pushbutton 2, the foot switch, and pressure switch. Add an alarm that sounds anytime the system pressure drops below 100 psi. *Note:* No additional components are required other than the push-to-test pushbutton and alarm. The components must be rearranged for the alarm to sound anytime pressure drops below 100 psi.

DEVICE CLAMPING CIRCUIT

3. In the Valve Control Circuit, the valve is opened or closed anytime a pushbutton is pressed. Redraw the circuit adding a selector switch that allows the valve to operate only when placed in the "valve operational" position. When the selector switch is placed in the OFF position, the valve cannot be moved. Add a red indicator lamp to the circuit to indicate the valve is operational and a green indicator lamp to indicate the valve is not operational.

VALVE CONTROL CIRCUIT

Activity 18-3. Heating Elements

Power is consumed when an electric heating element produces heat. To calculate the amount of power required to heat air, apply the formula:

$$kW = \frac{cfm \times TR}{3000}$$

where
kW = power required to heat air (in kilowatts)
cfm = cubic feet per minute air flow
(measured at 70°F inlet air and 14 psi)
TR = temperature rise (in °F)
3000 = constant

1. Complete Heating Air Power Requirements* by determining the power required to heat the air.

Air Flow Rate (in cfm)	HEATING AIR POWER REQUIREMENTS* Temperature Rise (°F)							
	50	100	150	200	250	300	350	400
100	1.67	3.33	5	6.67	8.33	10	11.67	13.33
200	3.33	6.67	10	13.33	16.67	20	23.33	26.67
300	5	10	15	20	25	30	35	40
400	6.67	13.33	20	26.67	33.33	40	46.67	53.33
500	8.33	16.67	25	33.33	41.67	50	58.33	66.67
600	10	20	30	40	50	60	70	80
700	11.67	23.33	35	46.67	58.33	70	81.67	93.33
800	13.33	26.67	40	53.33	66.67	80	93.33	106.67

* in kilowatts

The power required to heat liquid in a tank depends on the amount of tank insulation. Uninsulated tanks require additional power due to heat loss. To calculate the amount of power required to heat water in a tank that is adequately insulated, apply the formula:

$$kW = \frac{gal. \times TR}{325 \times ht}$$

where
kW = power required to heat water (in kilowatts)
gal. = amount or water (in gallons)
TR = temperature rise (in °F)
325 = constant
ht = heat-up time (in hours)

2. Complete Heating Water Power Requirements* by determining the amount of power required to heat the water.
   Note: Assume 1 hr heat-up time.

HEATING WATER POWER REQUIREMENTS*					
Liquid Amount (in gal.)	Temperature Rise (°F)				
	20	40	80	100	140
10					
20					
50					
100					
150					
200					
400					
500					

* in kilowatts

To calculate the amount of current required to produce a given amount of power, apply the formulas:

For 1ϕ:

$$I = \frac{P}{E}$$

where

I = current (in amperes)
P = power (in watts)
E = voltage (in volts)

For 3ϕ:

$$I = I = \frac{P}{E \times \sqrt{3}}$$

PRESSURE VESSEL

HEATING ELEMENT ASSEMBLY

THREE-PHASE VOLTAGE VALUES
For 208 V × 1.732, use 360
For 230 V × 1.732, use 398
For 240 V × 1.732, use 416
For 440 V × 1.732, use 762
For 460 V × 1.732, use 797
For 480 V × 1.732, use 831

3. Complete Power Heating Requirements* by determining the amount of current required to produce the given amount of power.

POWER HEATING REQUIREMENTS*						
kW	1ϕ			3ϕ		
	120 V	240 V	440 V	208 V	240 V	440 V
1						
5						
10						
25						
50						
100						

* in amperes

Activity 18-4. Troubleshooting Heating Elements

Manufacturers specify a rated voltage and current for each heating element type. Heating elements are often used at voltages other than the rated voltage. When the voltage is not equal to the rated voltage, the amount of current draw from the power lines does not equal the rated current. A higher-than-rated voltage produces a higher-than-rated current and a higher-than-rated power output. A lower-than-rated voltage produces a lower-than-rated current and a lower-than-rated power output.

To calculate power, apply the formulas:

For 1ϕ:
$P = I \times E$

For 3ϕ:
$P = I \times E \times \sqrt{3}$

1. Using the meter readings, determine the power.

   _____ **A.** Power equals _____ W.

   _____ **B.** Power equals _____ W.

354 TROUBLESHOOTING ELECTRICAL/ELECTRONIC SYSTEMS

_____ C. Power equals _____ W.

_____ D. Power equals _____ W.

_____ E. Power equals _____ W.

MECHANICAL AND SOLID-STATE SWITCHES — Applications (19)

▢ Application — Troubleshooting Mechanical Switches

Mechanical and solid-state switches are used to switch ON and OFF the flow of electricity in a circuit. A *mechanical switch* is any switch that uses silver contacts to start and stop the flow of current in a circuit. Mechanical switches may include normally-open, normally-closed, or combination switching contacts. Mechanical switches may be manually, mechanically, or automatically operated.

Manually-operated switches are used when a circuit is controlled by a person. Manually-operated switches include pushbutton, foot switches, toggle switches, and keyboards.

Mechanically-operated switches are used when a circuit is controlled by the movement of an object. A limit switch is a mechanically-operated switch.

Automatically-operated switches are used when a circuit is controlled by the change in a given condition. Automatically-operated switches include flow, level, pressure, temperature, humidity, and gas-operated switches. See Mechanical Switches.

MECHANICAL SWITCHES

A suspected fault with a mechanical switch is tested using a voltmeter. The voltmeter is used to test the voltage into and out of the switch. See Mechanical Switch Testing.

To test a mechanical switch, apply the procedure:
1. Measure the voltage into the switch. Connect the voltmeter between the neutral and hot conductor feeding the switch. The voltmeter lead may be connected to ground instead of neutral if the neutral conductor is not available in the same box in which the switch is located. The problem is located upstream from the switch when there is no voltage present or the voltage is not at the correct level. The problem may be a blown fuse or open circuit. Voltage must be reestablished to the switch before the switch may be tested.
2. Measure the voltage out of the switch. There should be a voltage reading when the switch contacts are closed. There should not be a voltage reading when the switch contacts are open. The switch has an open and must be replaced if there is no voltage reading in either switch position. The switch has a short and must be replaced if there is a voltage reading in both switch positions.

Warning: Always ensure power is OFF before changing a control switch. Use a voltmeter to ensure the power is OFF.

356 TROUBLESHOOTING ELECTRICAL/ELECTRONIC SYSTEMS

MECHANICAL SWITCH TESTING

▪ Application — Troubleshooting Solid-State Switches

A solid-state switch has no moving parts (contacts). A solid-state switch uses a triac, SCR, current sink (NPN) transistor, or current source (PNP) transistor output to perform the switching function. The triac output is used for switching AC loads. The SCR output is used for switching high-power DC loads. The current sink and current source outputs are used for switching low-power DC loads. Solid-state switches include normally-open, normally-closed, or combination switching outputs. See Solid-State Switches.

Two-Wire Solid-State Switches

A *two-wire solid-state switch* has two connecting terminals or wires (exclusive of ground). A two-wire switch is connected in series with the controlled load. A two-wire solid-state switch is also called a load-powered switch because it draws operating current through the load. The operating current flows through the load when the switch is not conducting (load OFF). This operating current is inadequate to energize most loads. Operating current is also called residual current or leakage current by some manufacturers.

OUTPUT SWITCHING DEVICES	
Device	Use
TRIAC	SWITCH AC LOADS
SCR	SWITCH HIGH-POWER DC LOADS
NPN TRANSISTOR / PNP TRANSISTOR	SWITCH LOW-POWER DC LOADS

SOLID-STATE SWITCHES

Operating current may be measured with an ammeter when the load is OFF. See Operating Current and Load Current.

The current in a circuit is a combination of the operating current and load current when a switch is conducting (load ON). A solid-state switching device must be rated high enough to carry the current of the load. Load current is measured with an ammeter when the load is ON.

OPERATING CURRENT

LOAD CURRENT

The current draw of a load must be sufficient to keep the solid-state switch operating when the switch is conducting (load ON). *Minimum holding current* is the minimum current that ensures proper operation of a solid-state switch. Minimum holding current values range from 2 mA to 20 mA.

Operating current and minimum holding current values are usually not a problem when a solid-state switch controls low-impedance loads, such as motor starters, relays, and solenoids. Operating current and minimum holding current values may be a problem when a switch controls high-impedance loads, such as programmable controllers and other solid-state devices. The operating current may be high enough to affect the load when the switch is not conducting. For example, a programmable controller may see the operating current as an input signal.

A load resistor must be added to the circuit to correct this problem. A load resistor is connected in parallel with the load. The load resistor acts as an additional load which increases the total current in the circuit.

Load resistors range in value from 4.5 kΩ to 7 kΩ. A 5 kΩ, 5 W resistor is used in most applications. See Load Resistor.

Two-wire solid-state switches connected in series affect the operation of the load because of the voltage drop across the switches. A two-wire switch drops about 3 V to 8 V. The total voltage drop across the switches equals the sum of the voltage drop across each switch. No more than three solid-state switches should be connected in series. See Two-Wire Switches.

Two-wire solid-state switches connected in parallel affect the operation of the load because each switch has its operating current flowing through the load. The load may turn ON if the current through the load becomes excessive. The total operating current equals the sum of the operating current of each switch. No more than three solid-state switches should be connected in parallel.

LOAD RESISTOR

TWO-WIRE SWITCHES

A suspected fault in a two-wire solid-state switch may be tested using a voltmeter. The voltmeter is used to test the voltage into and out of the switch. See Two-Wire Solid-State Switch Testing.

To test a two-wire solid-state switch, apply the procedure:
1. Measure the supply voltage into the switch. The problem is located upstream from the switch when there is no voltage present or the voltage is not at the correct level. The problem may be a blown fuse or open circuit. Voltage must be reestablished to the switch before the switch may be tested.
2. Measure the voltage out of the switch. The voltage should equal the supply voltage minus the rated voltage drop (3 V to 8 V) of the switch when the switch is conducting (load ON). Replace the switch if the voltage output is not correct.

Warning: Always ensure power is OFF before changing a control switch. Use a voltmeter to ensure the power is OFF.

TWO-WIRE SOLID-STATE SWITCH TESTING

Three-Wire Solid-State Switches

A *three-wire solid-state switch* has three connecting terminals or wires (exclusive of ground). A three-wire solid-state switch is connected in series with the control load and also connected to the power lines. A three-wire solid-state switch draws operating current directly from the power lines. The operating current does not flow through the load.

A three-wire solid-state switch is also called a line-powered switch because it draws operating current from the power lines. The two types of three-wire solid-state switches are the current source (PNP) switch and the current sink (NPN) switch. See PNP Switch and NPN Switch.

PNP SWITCH

NPN SWITCH

360 TROUBLESHOOTING ELECTRICAL/ELECTRONIC SYSTEMS

Three-wire solid-state switches connected in series affect the operation of the load because each switch downstream from the last switch must carry the load current and the operating current of each switch. An ammeter may be used to measure operating and load current values. The measured values must not exceed the manufacturer's maximum rating. See Three-Wire Switches.

THREE-WIRE SWITCHES

Three-wire solid-state switches connected in parallel affect the operation of the load because the non-conducting switch may be damaged due to reverse polarity. A blocking diode should be added to each switch output to prevent reverse polarity on the switch.

A suspected fault with a three-wire solid-state switch may be tested using a voltmeter. The voltmeter is used to test the voltage into and out of the switch. See Three-Wire Solid-State Switch Testing.

THREE-WIRE SOLID-STATE SWITCH TESTING

To test a three-wire solid-state switch, apply the procedure:
1. Measure the voltage into the switch. The problem is located upstream from the switch when there is no voltage present or the voltage is not at the correct level. The problem may be a blown fuse or open circuit. Voltage must be reestablished to the switch before the switch may be tested.
2. Measure the voltage out of the switch. The voltage should be equal to the supply voltage when the switch is conducting (load ON). Replace the switch if the voltage out of the switch is not correct.

Warning: Always ensure power is OFF before changing a control switch. Use a voltmeter to ensure the power is OFF.

Application — Protecting Switch Contacts

Switches are rated for the amount of current they may switch. The switch rating is usually specified for switching a resistive load. Resistive loads are the least destructive loads to switch. Most loads that are switched are inductive loads, such as solenoids, relays, and motors. Inductive loads are the most destructive loads to switch.

A large induced voltage appears across the switch contacts when inductive loads are turned OFF. The induced voltage causes arcing at the switch contacts. Arcing may cause the contacts to burn, stick, or weld together. Contact protection should be added when frequently switching inductive loads to prevent or reduce arcing. See Contact Protection.

A diode is added in parallel with the load to protect contacts that switch DC. The diode does not conduct when the load is ON. The diode conducts when the switch is open, providing a path for the induced voltage in the load to dissipate.

A resistor and capacitor are connected across the switch contacts to protect contacts that switch AC. The capacitor acts as a high impedance (resistor) load at 60 Hz, but becomes a short circuit at the high frequencies produced by the induced voltage of the load. This allows the induced voltage to dissipate across the resistor when the load is switched OFF.

DC-SWITCHED CONTACTS

AC-SWITCHED CONTACTS

CONTACT PROTECTION

Application — Proximity Sensor Installation

Proximity sensors have a sensing head that produces a radiated sensing field. The sensing field detects the target of the sensor. The sensing field must be kept clear of interference for proper operation. *Interference* is any object other than the object to be detected that is sensed by the sensor. Interference may come from objects close to the sensor or from other sensors. General clearances are required for most proximity sensors.

Flush-Mounted Inductive and Capacitive Proximity Sensors

A distance equal to or greater than twice the diameter of the sensors is required between sensors when flush-mounting inductive and capacitive proximity sensors. The diameter of the largest sensor is used for installation when two sensors of different diameters are used. For example, at least 16 mm is required between sensors if two 8 mm inductive proximity sensors are flush-mounted. See Flush-Mounted Proximity Sensor.

Non-Flush-Mounted Inductive and Capacitive Proximity Sensors

A distance of three times the diameter of the sensor is required within or next to a material that may be detected when using non-flush-mounting inductive and capacitive proximity sensors. For example, at least 48 mm is required between sensors if two 16 mm capacitive proximity sensors are non-flush-mounted. See Non-Flush-Mounted Proximity Sensor.

Three times the diameter of the largest sensor is required when inductive and capacitive proximity sensors are installed next to each other. Spacing is measured from center to center of the sensors.

Six times the rated sensing distance is required for proper operation when inductive and capacitive proximity sensors are mounted opposite each other. Six times the rated sensing distance is required because the sensing field causes false readings on the other.

FLUSH-MOUNTED PROXIMITY SENSOR

NON-FLUSH-MOUNTED PROXIMITY SENSOR

Mounting Photoelectric Sensors

A photoelectric sensor transmits a light beam. The light beam detects the presence (or absence) of an object. Only part of the light beam is effective when detecting the object. The *effective light beam* is the area of light that travels directly from the transmitter to the receiver. The object is not detected if the object does not completely block the effective light beam.

The receiver is positioned to receive as much light as possible from the transmitter when mounting photoelectric sensors. Greater operating distances are allowed and more power is available for the system to see through dirt in the air and on the transmitter and receiver lenses because more light is available at the receiver. The transmitter is mounted on the clean side of the detection zone because light scattered by dirt on the receiver lens affects the system less than light scattered by dirt on the transmitter lens. See Correct Sensor Mounting and Incorrect Sensor Mounting.

Mechanical and Solid-State Switches

CORRECT SENSOR MOUNTING

INCORRECT SENSOR MOUNTING

Application — Protecting Pressure Switches

A *pressure switch* is a switch that is designed to activate its contacts at a preset pressure. A pressure switch is rated according to its operating pressure range. A pressure switch may be damaged if its maximum pressure limit is exceeded.

Protection for a pressure switch should be added in any system in which a higher pressure than the maximum limit is possible. A pressure relief valve is installed to protect the pressure switch. A pressure relief valve should be set just below the pressure switch's maximum limit. The valve opens when the system pressure increases to the setting of the relief valve.

Caution: The output of the relief valve must be connected to a proper drain (or return line) if the product under pressure is a gas or a fluid. See Pressure Switch Protection.

PRESSURE SWITCH PROTECTION

Application — Flow Switches

A *flow switch* is a switch designed to detect the movement (flow) of a product through a pipe or duct. Most flow switches use a paddle to detect the movement of the product. The paddle is designed to detect the product movement with the least possible pressure drop across the switch. A flow switch must be installed correctly to ensure it does not interfere with the movement of the product. Most flow switches are designed to operate in the vertical position. Allow at least three pipe diameters (ID) on each side of the flow switch when mounting a flow switch. See Flow Switch Mounting.

For example, an application that moves a product through a 1½″ diameter pipe. The minimum horizontal distance of straight pipe required on each side of the flow switch is 4½″ (1½ × 3 = 4½).

FLOW SWITCH MOUNTING

Application — Level Control Problems

One-Level Registration

A *level switch* is a proximity sensor used to detect the presence of a liquid or solid in a tank. In level control applications, the distance between the maintained maximum and minimum levels must be determined. The distance may be small or large. Maintenance problems develop if the distance is not correct for the application.

One-level registration uses one sensor to determine the product level between the ON and OFF points of an operation. A small distance is maintained when using one-level registration. In one-level registration, the discharge pump motor starts when the sensor comes in contact with the product. The discharge pump motor stops when the sensor is no longer in contact with the product. See One-Level Registration.

Two-Level Registration

Two-level registration uses two sensors to determine the product level between the ON and OFF points of an operation. The distance may vary to almost any length when using two-level registration. In two-level registration, the discharge pump motor starts when the product comes in contact with the top sensor. The pump motor stays ON even when the top sensor is no longer in contact with the product. The pump motor stops when the bottom sensor is no longer in contact with the product. The discharge pump motor does not restart until the product comes in contact with the top sensor. See Two-Level Registration.

The smaller the distance to be maintained, the more times the pump motor must cycle ON and OFF. The motor heats faster when started frequently because motors draw high inrush current when starting. Consideration must be given to the speed of the product flowing from the tank. The faster the level in the tank drops, the faster the motor must cycle. A motor cycles faster if the product in the tank is turbulent. A pump does not have to cycle as often if two-level registration is used.

The type of product must be taken into consideration in determining the distance and time between the maximum and minimum level. For example, problems may develop with two-level registration if a common house paint is maintained in a fill tank. Problems develop because the paint accumulates as it is allowed to dry on the side of the tank. This may clog or impede the discharge or fill action.

ONE-LEVEL REGISTRATION

TWO-LEVEL REGISTRATION

MECHANICAL AND SOLID-STATE SWITCHES

Name _____ Date _____ **Activities**

Activity 19-1. Troubleshooting Mechanical Switches

1. Voltmeters are added to Light Circuit 1. Determine the meter readings for the given conditions.

LIGHT CIRCUIT 1

_____ **A.** Voltmeter 1 in Test Position 1 reads _____ V when the switch is in the OFF position and the circuit is operating properly.

_____ **B.** Voltmeter 1 in Test Position 2 reads _____ V when the switch is in the OFF position and the circuit is operating properly.

_____ **C.** Voltmeter 1 in Test Position 1 reads _____ V when the switch is in the ON position and the circuit is operating properly.

_____ **D.** Voltmeter 1 in Test Position 2 reads _____ V when the switch is in the ON position and the circuit is operating properly.

365

366 TROUBLESHOOTING ELECTRICAL/ELECTRONIC SYSTEMS

_____ E. Voltmeter 1 in Test Position 1 reads _____ V when the switch is in the ON position and the switch has an open.

_____ F. Voltmeter 1 in Test Position 2 reads _____ V when the switch is in the ON position and the switch has an open.

_____ G. Voltmeter 1 in Test Position 1 reads _____ V when the switch is in the OFF position and the switch has a short.

_____ H. Voltmeter 1 in Test Position 2 reads _____ V when the switch is in the OFF position and the switch has a short.

_____ I. Voltmeter 2 reads _____ V when the switch is in the ON position and the circuit is operating properly.

_____ J. Voltmeter 3 reads _____ V when the switch is in the ON position and the circuit is operating properly.

2. Voltmeters are added to Light Circuit 2. Determine the meter readings for the given conditions.

LIGHT CIRCUIT 2

_____ A. Voltmeter 1 reads _____ V when the Lght Switch is in the OFF position and the circuit is operating properly.

_____ B. Voltmeter 1 reads _____ V when the Light Switch is in the ON position and the circuit is operating properly.

_____ C. Voltmeter 2 reads _____ V when the Receptacle Switch is in the OFF position and the circuit is operating properly.

_____ D. Voltmeter 2 reads _____ V when the Receptacle Switch is in the ON position and the circuit is operating properly.

_____ E. Voltmeter 1 reads _____ V when the Light Switch is in the OFF position and one lamp is burned out (open).

_____ F. Voltmeter 1 reads _____ V when the Light Switch is in the OFF position and both lamps are burned out (open).

_____ G. Voltmeter 1 reads _____ V when the Light Switch is in the ON position and one lamp is burned out (open).

_____ H. Voltmeter 1 reads _____ V when the Light Switch is in the ON position and both lamps are burned out (open).

◯ Activity 19-2. Troubleshooting Solid-State Switches

1. Connect Voltmeter 1 to measure the voltage into the switch, Voltmeter 2 to measure the voltage out of the switch, and add a load resistor to the circuit.

_____ 2. A(n) _____-powered switch is controlling the load in Test Circuit 1.

_____ 3. A(n) _____ is used as the switching device in Test Circuit 1.

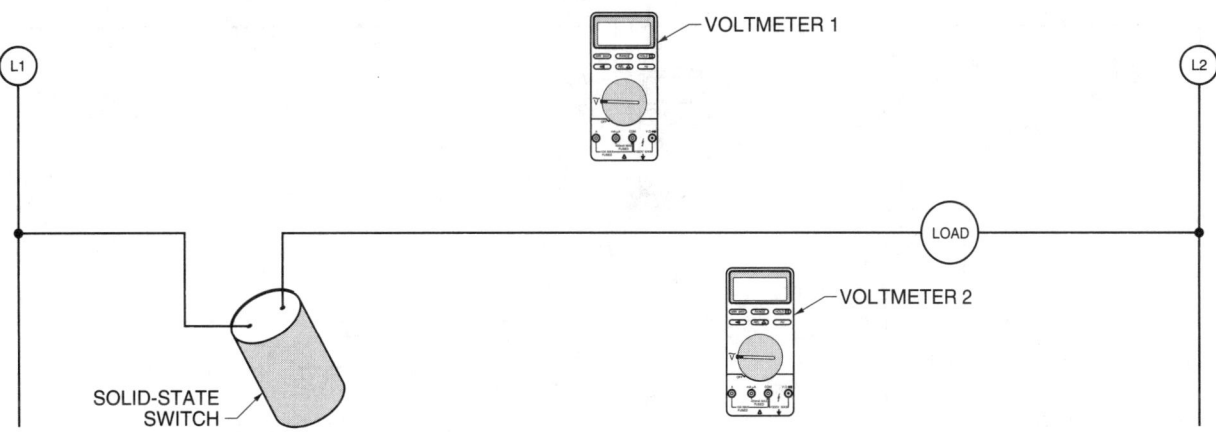

TEST CIRCUIT 1

4. Connect Voltmeter 1 to measure the voltage into the switch and Voltmeter 2 to measure the voltage out of the switch.

_____ **5.** A(n) _____-powered switch is controlling the load in Test Circuit 2.

_____ **6.** A(n) _____ is used as the switching device in Test Circuit 2.

TEST CIRCUIT 2

7. Connect Voltmeter 1 to measure the voltage into the switch and Voltmeter 2 to measure the voltage out of the switch.

_____ **8.** A(n) _____-powered switch is controlling the load in Test Circuit 3.

_____ **9.** A(n) _____ is used as the switching device in Test Circuit 3.

TEST CIRCUIT 3

Activity 19-3. Protecting Switch Contacts

1. Add a diode to the Automobile Starter Motor Circuit so that the induced voltage of the starter relay is dissipated to protect the electronic computer module circuits.

AUTOMOBILE STARTER MOTOR CIRCUIT

2. Add contact protection to the Air Conditioning Circuit to protect the thermostat contacts from the induced voltage of the evaporator blower relay and compressor motor relay. Add contact protection to protect the evaporator blower relay contacts and the compressor relay contacts.

AIR CONDITIONING CIRCUIT

Activity 19-4. Proximity Sensor Installation

_____ 1. The minimum distance required between the sensors for proper operation is _____ mm.

_____ 2. The minimum distance required between the sensor and the surrounding material for proper operation is _____ mm.

Activity 19-5. Protecting Pressure Switches

1. Add protection to the Fluid Power Circuit to protect the pressure switch from excessive fluid pressure. Add protection to the pressure switch contacts to protect them from induced voltage from the solenoid.

FLUID POWER CIRCUIT

Activity 19-6. Flow Switches

Determine the minimum length of straight pipe required on each side of the flow switch.

_____ 1. A ¼" diameter pipe requires ____" of straight pipe on either side of the flow switch.

_____ 2. A ½" diameter pipe requires ____" of straight pipe on either side of the flow switch.

_____ 3. A ¾" diameter pipe requires ____" of straight pipe on either side of the flow switch.

_____ 4. A 1" diameter pipe requires ____" of straight pipe on either side of the flow switch.

_____ 5. A 1¼" diameter pipe requires ____" of straight pipe on either side of the flow switch.

_____ 6. A 2" diameter pipe requires ____" of straight pipe on either side of the flow switch.

_____ 7. A 3" diameter pipe requires ____" of straight pipe on either side of the flow switch.

Activity 19-7. Level Control Problems

A level switch monitors the level of hot water in an automatic coffee brewer. The level switch sends a signal to the solenoid valve which opens to allow water to refill the reservoir.

_____ 1. The automatic coffee brewer uses _____-level registration.

372 TROUBLESHOOTING ELECTRICAL/ELECTRONIC SYSTEMS

Two level switches maintain the product level in a tank. An additional level switch is used to sound an overfill alarm if the product rises too high.

_____ 2. The tank uses _____-level registration.

A high-level switch and low-level switch control the product level in a tank. The high-level switch sounds an alarm if the product rises too high. The low-level switch sounds an alarm if the product falls too low.

_____ 3. The tank uses _____-level registration circuits.

DIODES, TRANSISTORS, AND THYRISTORS

20 Applications

Application — Diodes

A *diode* is an electronic device that allows current to flow in only one direction. A diode has an anode and a cathode. The *anode* is the positive lead. The *cathode* is the negative lead. See Diode.

A diode allows current to flow only when a positive voltage is applied to the anode (forward bias). The diode does not allow current to flow when the anode and cathode have the same polarity or if a positive voltage is applied to the cathode (reverse bias).

A diode is rated for the maximum forward current it can safely conduct and the maximum reverse voltage that can be applied. The maximum forward current rating applies to the amount of current the diode can withstand while conducting electrons in the forward direction. Exceeding the maximum forward current rating causes the diode to overheat and may cause the diode to fail. The maximum reverse voltage rating applies to the amount of voltage that may be applied in the reverse direction without allowing the diode to conduct (voltage breakover) in the opposite direction. See Diode Current and Voltage.

373

374 TROUBLESHOOTING ELECTRICAL/ELECTRONIC SYSTEMS

Application — Transistors

A *transistor* is a three-element device made of semiconductor material. A transistor is used in circuits as either a switch or an amplifier. Most transistors are used as current control devices. The three elements of a transistor are the emitter (E), base (B), and collector (C). The lead with the arrow is the emitter. A transistor is an NPN transistor if the arrow points out on the emitter. A transistor is a PNP transistor if the arrow points in on the emitter. See Transistor Symbols.

An *NPN transistor* has a thin layer of P-type material placed between two pieces of N-type material. N-type material carries current in the form of electrons because the material is manufactured to have extra electrons. P-type material carries current in the form of positive charges (holes) because the material is manufactured to have a deficiency of electrons. A *PNP transistor* has a thin layer of N-type material placed between two pieces of P-type material.

TRANSISTOR SYMBOLS

In a transistor, a very small current applied to one lead controls a large current flowing through the other two leads. A transistor may be used to start or stop current flow (switch) or to increase current flow (amplify). Any transistor may be used as a switch or an amplifier. Some transistors are designed to operate better as a switch while others are designed to operate better as an amplifier.

Transistor as a Switch

A transistor may be used as a solid-state switch in DC circuits. A transistor does not allow current to flow between the emitter and collector until a current is applied to the base of the transistor. Current flows through the emitter and collector when current is applied to the base of the transistor. A very low voltage may be used in the base circuit. A high voltage may be used in the emitter/collector circuit. For example, a 1.5 VDC base voltage can switch a 60 VDC emitter/collector circuit. See Transistor Circuit Current Flow.

Diodes, Transistors, and Thyristors 375

TRANSISTOR CIRCUIT CURRENT FLOW

Current flows in both the emitter/base circuit and the emitter/collector circuit when the transistor is switched ON. The emitter/base current is very low (about 5%) compared to the emitter/collector current. A transistor allows a small control current to control a high load current.

A transistor acts as a relay when used as a switch. A relay allows a small coil voltage and current to control a high load voltage and current. For example, a pushbutton on a beverage gun may be used to apply a low current to the base of a transistor. See Transistor Switch Application.

TRANSISTER SWITCH APPLICATION

Current flows through the emitter/collector circuit when current is applied to the base of the transistor. The beverage-dispense solenoid is energized when current flows through the emitter/collector circuit. The solenoid is energized as long as current flows through the base circuit (pushbutton pressed).

A transistor used as a switch allows a very small input signal to control a high output. In the beverage gun application, the output current may be 100 times higher (or more) than the input current. In this application, the transistor has a current gain of 100. *Current gain* of a transistor (beta) is a measure of the ratio of the collector current to the base current. See Transistor Gain.

$$\text{CURRENT GAIN} = \frac{I_{COLLECTOR}}{I_{BASE}}$$

$$\text{CURRENT GAIN} = \frac{200 \text{ mA } (.2 \text{ A})}{2 \text{ mA } (.002 \text{ A})}$$

CURRENT GAIN = 100

TRANSISTOR GAIN

Transistor as an Amplifier

The current in the emitter/collector circuit is switched completely ON or OFF when the transistor is used as a switch. This is accomplished by raising the base current high enough to allow full current flow through the emitter/collector circuit. The current flow through the emitter/collector circuit may be limited to any value less than full current flow. The amount of current flow is limited by limiting the base current to a value less than full current.

The transistor acts as a control valve allowing some of the current to flow through the emitter/collector circuit when the base current is limited. The amount of current flowing through the emitter/collector circuit is proportionately higher than the amount of current flowing through the emitter/base circuit.

Transistors amplify weak signals in radios, TVs, and industrial control equipment. For example, an NPN transistor may be used to amplify a signal from a microphone to a speaker. See Transistor Amplification Application.

TRANSISTOR AMPLIFICATION APPLICATION

The microphone converts sound waves to electric current pulses. The electric current pulses from the microphone are fed to the base of the transistor. The transistor blocks the current flow through the speaker when there are no pulses (no sound at the microphone). A much larger but proportional quantity of current flows through the speaker when there is current at the base. A transistor is used because the current produced by the microphone is too small to drive the speaker.

Application — Thyristors

A *thyristor* is a solid-state switching device that switches current ON by a quick pulse of control current. A thyristor does not require control current to remain ON once the current is switched ON. The most common thyristors are the silicon controlled rectifier (SCR) and the triac.

An *SCR* is a thyristor that is triggered into conduction in only one direction. A *triac* is a thyristor that is triggered into conduction in either direction. Thyristors have two terminals for load current and one terminal for control current. The three terminals of an SCR are the anode (A), cathode (K), and gate (G). The three terminals of a triac are the main terminal 1 (MT_1), main terminal 2 (MT_2), and gate (G). See Thyristors.

THYRISTORS

A thyristor is generally used to control large amounts of current. SCRs and triacs are used for most solid-state industrial switching circuits. SCRs and triacs control motor speed, heat output, brightness of lights, and other applications requiring a controlled solid-state switching.

SCR DC Voltage Control

An SCR is commonly used to control the voltage in a DC motor control circuit. Current flows between the anode and cathode when control current is applied to the gate. The gate control circuit turns the SCR ON once during each full cycle of the AC sine wave. The point at which the gate control circuit triggers the SCR determines the point on the AC voltage sine wave at which current starts to flow. This point also determines the amount of output voltage. See SCR DC Voltage Control.

378 TROUBLESHOOTING ELECTRICAL/ELECTRONIC SYSTEMS

SCR DC VOLTAGE CONTROL

The gate control circuit may start current flowing at any point between points 2 and 4, varying the voltage to the motor from zero to full half-wave. Only full half-wave is possible because the SCR allows current to flow in only one direction.

Triac AC Voltage Control

A triac is commonly used to control the voltage in an AC heating control circuit. Current flows between MT_1 and MT_2 when control current is applied to the gate. The triac is triggered for each half-wave of the AC sine wave. The point at which the control circuit triggers the triac determines the point on the wave at which current begins to flow. This point also determines the amount of output voltage applied to the heating element. The higher the output voltage, the hotter the heating element. See Triac AC Voltage Control.

TRIAC AC VOLTAGE CONTROL

The gate control circuit may start current flowing at any point between points 2 and 4, varying the voltage to the heating element. When triggered at the beginning of each half-wave (point 4), there is full voltage output and the heating element is the hottest.

Thyristor Triggering Methods

A thyristor is triggered into conduction by applying a pulse of control current to the gate. Once turned ON, the thyristor remains ON as long as there is a minimum level of holding current flowing through the circuit. A thyristor is turned OFF by reducing the current flowing through the circuit below the holding current. See Thyristor Triggering.

THYRISTOR TRIGGERING

The correct method of thyristor turn ON is applying a proper signal to the gate of the thyristor. The incorrect methods of thyristor turn ON are voltage breakover turn ON, static turn ON, and thermal turn ON.

Gate Turn ON

Gate turn ON is a method of thyristor turn ON that occurs when the proper signal is applied to the gate at the correct time. Gate turn ON is the only correct way to turn ON a thyristor. For an SCR, the gate signal must be positive with respect to the cathode polarity for the thyristor to turn ON. See Gate Turn ON.

GATE TURN ON

A triac is triggered into conduction in either direction by a gate current. The simplest triac-triggering circuit uses two fixed resistors, a variable resistor and capacitor. The variable resistor determines the triggering point and the fixed resistor limits current flow. The capacitor is used to trigger the gate once it has charged to a value high enough to deliver sufficient gate current.

A diac is usually added to improve the gate triggering circuit. A *diac* is a two terminal bidirectional device used primarily to control the triggering of triacs. The advantage of using a diac is that it delivers a sharp pulse of gate current rather than a sinusoidal gate current. See Triac Gate Triggering Methods.

TRIAC GATE TRIGGERING METHODS

Voltage Breakover Turn ON

Voltage breakover turn ON is a method of thyristor turn ON that occurs when the voltage across the thyristor terminals exceeds the maximum voltage rating of the device. This causes localized heating in the thyristor and may damage the component. A varistor connected across the thyristor terminals prevents voltage breakover turn ON. See Voltage Breakover Turn ON.

VOLTAGE BREAKOVER TURN ON

Static Turn ON

Static turn ON is a method of thyristor turn ON that occurs when a fast-rising voltage is applied across the anode and cathode terminals of an SCR or the main terminals of a triac. Manufacturers refer to this turn ON point as the dv/dt rating. This rating defines the amount of voltage over a given time period that turns ON the device. A typical rating is 250 V/1s. Static turn ON does not damage the thyristor provided the surge current is limited. A snubber circuit is added across the thyristor terminals if static turn ON is a problem. See Static Turn ON.

STATIC TURN ON

Thermal Turn ON

All solid-state components are heat sensitive. *Thermal turn ON* is a method of thyristor turn ON that occurs when the heat level exceeds the limit of a thyristor (normally 110°C). Once started, the component is usually damaged or destroyed. Thermal turn ON is eliminated by using the correct heat sink.

DIODES, TRANSISTORS, AND THYRISTORS

20

Name _____ Date _____ **Activities**

◐ Activity 20-1. Diodes

An *alternator* is a device that converts mechanical energy (spinning motion) to electrical energy. An alternator generates AC in its stator. Diodes convert the AC to DC. Draw the current path through the diodes for the given polarity of the transformers or windings. Determine the negative and positive polarity terminals of the alternator output.

1.

2.

3.

◯ Activity 20-2. Transistors

Transistors may be connected in series or parallel to develop circuit logic.

1. Complete Truth Table 1 by determining whether the load is ON or OFF for each input switch combination. Identify the equivalent logic function (AND, OR, etc.) of the transistor arrangement.

   _____ A. The equivalent logic function of Circuit 1 is _____.

CIRCUIT 1

TRUTH TABLE 1			
Input			Output
Pushbutton	Temperature Switch	Level Switch	Load
Open	Open	Open	
Open	Open	Closed	
Open	Closed	Open	
Closed	Open	Open	
Closed	Closed	Open	
Open	Closed	Closed	
Closed	Closed	Closed	

2. Complete Truth Table 2 by determining whether the load is ON or OFF for each input switch combination. Identify the equivalent logic function (AND, OR, etc.) of the transistor arrangement.

   _____ A. The equivalent logic function of Circuit 2 is _____.

CIRCUIT 2

TRUTH TABLE 2		
Input		Output
Pushbutton	Limit Switch	Load
Closed	Open	
Open	Closed	
Open	Open	
Closed	Closed	

3. Diodes and transistors may be used to control the direction of rotation of a DC motor. Determine the positive and negative power bus. Draw the current path from the positive power bus to the negative power bus when Transistor 1 and Transistor 2 are switched ON. Mark the positive and negative side of the motor when Transistor 1 and Transistor 2 are switched ON.

_____ **A.** Power bus _____ is the negative power bus.

_____ **B.** Power bus _____ is the positive power bus.

4. Determine the positive and negative power bus. Draw the current path from the positive power bus to the negative power bus when Transistor 3 and Transistor 4 are switched ON. Mark the positive and negative side of the motor when Transistor 3 and Transistor 4 are switched ON.

_____ **A.** Power bus _____ is the negative power bus.

_____ **B.** Power bus _____ is the positive power bus.

Activity 20-3. Thyristors

Phase control is the most common use of thyristor power control. A thyristor blocks current flow until it is turned ON by the control circuit. Voltage is applied to the load when the thyristor is turned ON. *Conduction angle* is the time during which the thyristor is ON. The conduction angle determines the peak voltage (V_{max}), rms voltage (V_{rms}), and average voltage (V_{avg}) applied to the load. Using Voltage/Conduction Angle, determine the peak voltage, rms voltage, and average voltage.

1. The conduction angle is 180°.

    _____ A. V_{max} equals _____ V.
    _____ B. V_{rms} equals _____ V.
    _____ C. V_{avg} equals _____ V.

2. The conduction angle is 120°.

    _____ A. V_{max} equals _____ V.
    _____ B. V_{rms} equals _____ V.
    _____ C. V_{avg} equals _____ V.

3. The conduction angle is 90°.

    _____ A. V_{max} equals _____ V.
    _____ B. V_{rms} equals _____ V.
    _____ C. V_{avg} equals _____ V.

4. The conduction angle is 20°.

    _____ A. V_{max} equals _____ V.
    _____ B. V_{rms} equals _____ V.
    _____ C. V_{avg} equals _____ V.

NOTE: INPUT VOLTAGE = 115 V

VOLTAGE/CONDUCTION ANGLE

5. The conduction angle is 180°.

_____ **A.** V_{max} equals _____ V.

_____ **B.** V_{rms} equals _____ V.

_____ **C.** V_{avg} equals _____ V.

6. The conduction angle is 160°.

_____ **A.** V_{max} equals _____ V.

_____ **B.** V_{rms} equals _____ V.

_____ **C.** V_{avg} equals _____ V.

7. The conduction angle is 90°.

_____ **A.** V_{max} equals _____ V.

_____ **B.** V_{rms} equals _____ V.

_____ **C.** V_{avg} equals _____ V.

8. The conduction angle is 60°.

_____ **A.** V_{max} equals _____ V.

_____ **B.** V_{rms} equals _____ V.

_____ **C.** V_{avg} equals _____ V.

9. The conduction angle is 20°.

_____ **A.** V_{max} equals _____ V.

_____ **B.** V_{rms} equals _____ V.

_____ **C.** V_{avg} equals _____ V.

10. The conduction angle is 0°.

_____ **A.** V_{max} equals _____ V.

_____ **B.** V_{rms} equals _____ V.

_____ **C.** V_{avg} equals _____ V.

VOLTAGE/CONDUCTION ANGLE

NOTE: INPUT VOLTAGE = 230 V

386 TROUBLESHOOTING ELECTRICAL/ELECTRONIC SYSTEMS

11. Complete the R/C Triac Gate Triggering Circuit by adding a capacitor and conductors to the circuit. Connect the ammeter to measure the load current. Connect the voltmeter to measure the load voltage.

R/C TRIAC GATE TRIGGERING CIRCUIT

12. Complete the SCR Gate Triggering Circuit by adding an ON Pushbutton, OFF Pushbutton, and SCR to the circuit. Connect the ammeter to measure the load current. Connect the voltmeter to measure the load voltage.

SCR GATE TRIGGERING CIRCUIT

21

TESTING DIODES, TRANSISTORS, AND THYRISTORS — Applications

☐ Application — Diode Testing

Ohmmeter Diode Test

Forward bias is the condition of a diode while it conducts current. The anode in a forward-biased diode has a positive polarity compared to the cathode. *Reverse bias* is the condition of a diode when it acts as an insulator. The cathode of a reverse-biased diode has a positive polarity compared to the anode. A diode acts as a closed switch when it is forward-biased and as an open switch when it is reverse-biased. An ohmmeter is used to conduct a basic diode check. See Ohmmeter Diode Test.

OHMMETER DIODE TEST

To use an ohmmeter to test a diode, apply the procedure:
1. Ensure that all power in the circuit is OFF. Test for voltage using a voltmeter to ensure power is OFF.
2. Set the meter on the resistance setting.
3. Connect the meter leads to the diode. Record the meter reading.
4. Reverse the meter leads. Record the meter reading.

The diode is forward-biased when the positive (red) probe is on the anode and the negative (black) probe is on the cathode. The forward-biased resistance of a good diode should be between 1000 Ω and 10 MΩ. The resistance reading is high when the diode is forward-biased because the current from the meter's voltage source flows through the diode and results in a resistance measurement.

The diode is reverse-biased when the positive (red) lead is on the cathode and the negative (black) lead is on the anode. The reverse-biased resistance of a good diode should equal infinity. An infinite resistance is displayed as an overload (OL) on a digital meter. The diode is bad if the meter readings are the same in both directions.

> **Multimeter Diode Test**
>
> Testing a diode using an ohmmeter does not always indicate whether a diode is good or bad. Testing a diode that is connected in a circuit with an ohmmeter may give false readings because other components may be connected in parallel with the diode under test. The best way to test a diode is to measure the voltage drop across the diode when it is forward-biased.
>
> A good diode has a voltage drop across it when it is forward-biased and conducting current. The voltage drop is between .5 V and .8 V for the most commonly used silicon diodes. Some diodes are made of germanium and have a voltage drop between .2 V and .3 V.
>
> A multimeter set on the diode test position is used to test the voltage drop across a diode. In this position, the meter produces a small voltage between the test leads. The meter displays the voltage drop when the leads are connected across a diode. See Multimeter Diode Test.

MULTIMETER DIODE TEST

To test a diode using the diode test position on a multimeter, apply the procedure:

1. Ensure that all power in the circuit is OFF. Test for voltage using a voltmeter to ensure power is OFF.
2. Set the meter on the diode test position.
3. Connect the meter leads to the diode. Record the meter reading.
4. Reverse the meter leads. Record the meter reading.

The meter displays a voltage drop between .5 V and .8 V (silicon diode) or .2 V and .3 V (germanium diode) when a good diode is forward-biased. The meter displays an OL when a good diode is reverse-biased. The OL reading indicates the diode is acting like an open switch. An open (bad) diode does not allow current to flow through it in either direction. The meter displays an OL reading in both directions when the diode is open. A shorted diode gives the same voltage drop reading in both directions. This reading is normally about .4 V.

Application — Transistor Testing

A transistor becomes defective from excessive current or temperature. A transistor normally fails due to an open or shorted junction. The two junctions of a transistor may be tested with an ohmmeter. See Transistor Testing.

To test an NPN transistor for an open or shorted junction, apply the procedure:
1. Connect a multimeter to the emitter and base of the transistor. Measure the resistance.
2. Reverse the meter leads and measure the resistance. The emitter/base junction is good when the resistance is high in one direction and low in the opposite direction.

Note: The ratio of high to low resistance should be greater than 100 : 1. Typical resistance values are 1 kΩ (with the positive lead of the meter on the base) and 100 kΩ (with the positive lead of the meter on the emitter). The junction is shorted when both readings are low. The junction is open when both readings are high.

3. Connect the meter to the collector and base of the transistor. Measure the resistance.
4. Reverse the meter leads and measure the resistance. The collector/base junction is good when the resistance is high in one direction and low in the opposite direction.

Note: The ratio of high to low resistance should be greater than 100 : 1. Typical resistance values are 1 kΩ (with the positive lead of the meter on the base) and 100 kΩ (with the positive lead of the meter on the collector).

5. Connect the meter to the collector and emitter of the transistor. Measure the resistance.
6. Reverse the meter leads and measure the resistance. The collector/emitter junction is good when the resistance reading is high in both directions.

The same test used for an NPN transistor is used for testing a PNP transistor. The difference is that the meter test leads must be reversed to obtain the same results.

TRANSISTOR TESTING

Clamping Diode

A clamping diode is used when transistors control inductive loads. A *clamping diode* is a diode that prevents voltage in one part of the circuit from exceeding the voltage in another part. A clamping diode protects a transistor from high-voltage surges. See Clamping Diode.

All coils have inductance. *Inductance* is the property of a circuit that causes it to oppose a change in current due to energy stored in a magnetic field. The collapsing magnetic field of a coil produces a high-voltage spike when switched OFF. This spike may damage or destroy a transistor. A diode prevents the spike from passing through the transistor because the diode conducts the excessive voltage away from the transistor.

CLAMPING DIODE

Application — Thyristor Testing

SCR Testing

High-power SCRs should be tested using a test circuit. Low-power and some high-power SCRs may be tested using a multimeter set on the resistance scale. See SCR Testing.

To test an SCR using a multimeter, apply the procedure:
1. Set the multimeter on the R × 100 scale.
2. Connect the negative lead of the meter to the cathode.
3. Connect the positive lead of the meter to the anode. The meter should read infinity.
4. Short circuit the gate to the anode using a jumper wire. The meter should read almost 0 Ω. Remove the jumper wire. The low-resistance reading should remain.
5. Reverse the meter leads so that the positive lead is on the cathode and the negative lead is on the anode. The meter should read almost infinity.
6. Short circuit the gate to the anode with a jumper wire. The resistance on the meter should remain high.

SCR TESTING

Triac Testing

Triacs should be tested under operating conditions using an oscilloscope. A multimeter may be used to make a rough test with the triac out of the circuit. See Triac Testing.

To test a triac using a multimeter, apply the procedure:
1. Set the multimeter on the R × 100 scale.
2. Connect the negative meter lead to main terminal 1.
3. Connect the positive meter lead to main terminal 2. The meter should read infinity.
4. Short circuit the gate to main terminal 2 using a jumper wire. The meter should read almost 0 Ω. The zero reading should remain when the lead is removed.
5. Reverse the meter leads so that the positive lead is on main terminal 1 and the negative lead is on main terminal 2. The meter should read infinity.
6. Short circuit the gate of the triac to main terminal 2 using a jumper wire. The meter should read almost 0 Ω. The zero reading should remain after the lead is removed.

TRIAC TESTING

Application — Electronic Heat Problems

All solid-state switching components produce heat. Heat affects the operation, life, and reliability of solid-state switching components. Excessive heat destroys all solid-state components. Heat is produced by current passing through a solid-state component because of the friction of the electrons. Heat must be kept to a safe level. This is done by limiting the amount of current passing through the component or removing the heat. Many solid-state components control large amounts of power. This produces large amounts of excess heat that must be dissipated by heat sinks for proper operation of the solid-state components.

Heat Sinks

A *heat sink* is a device that conducts and dissipates heat away from a component. A heat sink is mounted on solid-state components to assure good heat conduction from the solid-state component to the heat sink. The heat sink must fit the solid-state component with no air space between for efficient heat conduction. Thermal grease or pads are used to ensure that no gap exists between the solid-state component and the heat sink. Solid-state components must be evenly spaced to assure uniform heat transfer when more than one is used on the same heat sink. Even spacing prevents hot spots that may damage or warp the mounting board.

Mounting Surface

The surface of the heat sink and solid-state component must be flat and compatible. Mounting holes should be only as large as required by the fasteners. The mounting surface must be free from all foreign material, film, and oxidation. The mounting surface may be polished with #000 steel wool. Clean with an alcohol based cleaner after polishing.

Thermal Compounds

Thermal grease or a thermal pad should be applied after cleaning. Thermal compounds fill air gaps between the metal surfaces. The thermal grease should cover the entire surface area between the solid-state component and the heat sink. Remove all excessive thermal grease after mounting. See Mounting Heat Sinks.

Natural convection occurs as warm air rises. The natural rising of warm air causes air to flow over a heated component. To take advantage of this natural airflow, always provide for both top and bottom vents in cabinets.

Allow the moving air to flow over each component with as little obstruction as possible. Use vertical fins and place high heat-producing components near the top. See Airflow Over Heat Sinks.

MOUNTING HEAT SINKS

AIRFLOW OVER HEAT SINKS

To assure good airflow with forced convection, stagger the solid-state components so they have about the same amount of air flowing over them. The placement of a fan should be near the bottom, with the top vented. Add a filter to the intake of the fan to keep the fan moving as much air as possible. Change or clean the filter and fan as needed.

21
TESTING DIODES, TRANSISTORS, AND THYRISTORS

Name _____ Date _____ **Activities**

◔ Activity 21-1. Diode Testing

1. Using the meter readings, determine the anode and cathode lead of the diode. *Note:* Assume each diode is good.

_____ **A.** Lead 1 is the _____. _____ **C.** Lead 3 is the _____.
_____ **B.** Lead 2 is the _____. _____ **D.** Lead 4 is the _____.

[Meter reading: 1500 Ω — LEAD 1, LEAD 2 on DIODE]

[Meter reading: O.L. — LEAD 3, LEAD 4 on DIODE]

2. Using the meter readings, determine if the diode is forward-biased or reverse-biased.

_____ **A.** Diode 1 is _____-biased. _____ **B.** Diode 2 is _____-biased.

[Meter reading: 2200 Ω — DIODE 1]

[Meter reading: O.L. — DIODE 2]

393

3. Using the meter readings, determine if the diode is good, open, or shorted.

A. Diode 1 is _____.

B. Diode 2 is _____.

C. Diode 3 is _____.

D. Diode 4 is _____.

Activity 21-2. Transistor Testing

1. Using the meter readings, determine if the transistor junction is good, open, or shorted.

_____ **A.** Junction 1 is _____.

_____ **B.** Junction 2 is _____.

_____ **C.** Junction 3 is _____.

_____ **D.** Junction 4 is _____.

396 TROUBLESHOOTING ELECTRICAL/ELECTRONIC SYSTEMS

E. Junction 5 is _____.

F. Junction 6 is _____.

G. Junction 7 is _____.

H. Junction 8 is _____.

2. Add a clamping diode to each circuit. *Note:* Connect to correct polarity.

A.

B.

Activity 21-3. Thyristor Testing

Using the meter readings, determine if the SCR is good or bad.

_____ 1. The SCR is _____.

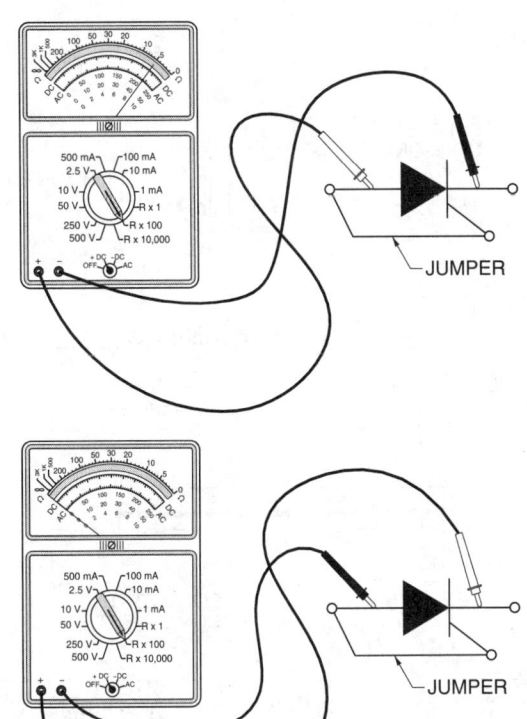

Using the meter readings, determine if the triac is good or bad.

_____ 2. The triac is _____.

Activity 21-4. Electronic Heat Problems

Determine if the mounting of the heat sink and components is correct or incorrect.

_____ 1. The mounting is _____.

_____ 2. The mounting is _____.

_____ 3. The mounting is _____.

_____ 4. The mounting is _____.

Determine if the airflow over the heat sink and components is good or poor.

_____ 5. The airflow is _____.

_____ 6. The airflow is _____.

_____ 7. The airflow is _____.

_____ 8. The airflow is _____.

PROGRAMMABLE CONTROLLERS

22 Applications

Application — Programmable Controllers

A *programmable controller* is a solid-state control device that is programmed and reprogrammed to automatically control machines and industrial processes. A programmable controller contains a power supply, programming terminal, processor, and input and output modules. The *power supply* is the device that provides all the voltages required for the internal operation of the programmable controller. See Programmable Controller.

PROGRAMMABLE CONTROLLER

The *programming terminal* is a keyboard-operated device that is used to insert the control program into memory, make changes to the stored program, and display the program circuit and data. The programming terminal allows the circuit to be monitored before or after the program is placed into operation.

The *processor (CPU)* is part of the programmable controller that performs logic functions, program storage, and supervisory functions. The processor scans and monitors the status of all the input devices and output devices in response to the programmed instructions in memory. The processor energizes or de-energizes output devices as a result of the pre-programmed instructions.

An *input module* is the part of a programmable controller that receives information from pushbuttons, temperature switches, pressure switches, photo-electric and proximity switches, and other input devices. An *output module* is the part of a programmable controller that delivers the output voltage required to control alarms, lights, solenoids, motor starters, and other output devices.

Processor Modes

The processor of a programmable controller has different modes. The different modes allow the programmable controller to be taken on-line (system running), or off-line (system on stand-by). Processor modes include the program, run, and test modes. The program mode is used for developing the logic of the control circuit. In the program mode, the circuit is monitored and the program is edited, changed, saved, and transferred.

The run mode is used to execute the program. In the run mode, the circuit may be monitored and the inputs and outputs forced. Program changes cannot be made in the run mode. The test mode is used to check the program without energizing output circuits or devices. In the test mode, the circuit is monitored and inputs and outputs are forced (without actually energizing the load connected to the output). See Processor Modes.

Warning: A programmable controller is switched from the program mode to the run mode by placing the controller in the run mode. The machine or process is started when the controller is placed in the run mode. Extreme care must be taken to assure that no damage to personnel or equipment occurs when switching the controller to the run mode. Only qualified personnel should change processor modes and key-operated switches should always be used in any dangerous application.

PROCESSOR MODES

▢ Application — Status Indicators and Error Codes

A programmable controller is designed to withstand a more severe physical environment than a standard personal computer. The programmable controller's program is designed to keep production operating with as little downtime as possible. Manufacturers include status indicators and error messages (or codes) to help a troubleshooter find a problem that occurs.

Status Indicators

A *status indicator* is a light that shows the condition of the components in the programmable controller system. Each major part of the system normally has status indicators. The general location of a problem may be determined by observing the status indicators. Status indicators are a valuable troubleshooting aid when trying to determine a problem in the system. The two types of status indicators are the input/output status indicators, and the operating and fault status indicators. See Status Indicators.

Input/Output Status Indicators

An input/output (I/O) status indicator shows the status of the input and output devices. The status indicators on the input module are energized when an electrical signal is received at an input terminal. This occurs when an input contact is closed or signal is present. The status indicators on the output module are energized when an output device is energized. Each input device and output device has its own status indicator.

Operating and Fault Status Indicators

The processor of a programmable controller is programmed to look for potential problems (self diagnostics). The processor performs error checks as part of its normal operation and sends status information to the appropriate status indicators. Typical diagnostic checks include monitoring the input power, the processor's operating mode, CPU faults, forced inputs or outputs, and a low battery condition.

STATUS INDICATORS

Operating and fault status indicators include the power status, PC run, CPU fault, forced I/O, and battery low indicators. The power status indicator turns ON to indicate that the processor is energized and power is being applied. This status indicator should normally be ON. The PC run indicator turns ON when the processor is in the run mode. Care must be taken when the run indicator is ON because the controller activates the loads as programmed. This indicator is OFF when the processor is placed in the program mode. The CPU fault indicator turns ON when the processor has detected an error in the controller. The processor automatically turns OFF all loads and stops operation when this indicator is ON. An error message is normally displayed on the monitor screen indicating the problem.

The forced I/O indicator turns ON when one or more input or output device has been forced ON or OFF. All force commands must be removed from the program before normal operation is resumed. A battery is used to provide back-up power for the processor memory in case of an external power failure. The battery-low indicator turns ON when the battery should be replaced or if the battery is not charging.

The condition of the status indicators must be checked when troubleshooting a problem in a system using a programmable controller. Potential problems may be determined based on the condition of the status indicators. See Status Indicator Condition.

STATUS INDICATOR CONDITION

Status Indicator	Problem	Possible Cause	Corrective Action
☐ POWER ☐ PC RUN ☐ CPU FAULT ☐ FORCED I/O ☐ BATTERY LOW	No or low system power	Blown fuse, tripped CB, or open circuit	Test line voltage at power supply. Line voltage must be within 10% of the controller's rated voltage. Check for proper power supply jumper connections when voltage is correct. Replace the power supply module when the module has power coming into it but is not delivering the correct power
■ POWER ☐ PC RUN ☐ CPU FAULT ☐ FORCED I/O ☐ BATTERY LOW	Programmable controller not in run mode	Improper mode selection for system operation	Place in run mode. Ensure that all personnel are clear before placing the system in run mode
■ POWER ☐ PC RUN ■ CPU FAULT ☐ FORCED I/O ☐ BATTERY LOW	Fault in controller	Faulty memory module, memory loss, or memory error, normally caused by a high-voltage surge, short circuit, or improper grounding	Turn power OFF and re-start system. Remove power and replace the memory module when fault indicator is still ON. Load backup program on new memory module and reboot system
■ POWER ☐ PC RUN ☐ CPU FAULT ☐ FORCED I/O ▨ BATTERY LOW	Fault in controller due to inadequate or no power	Loss of memory when power was OFF and battery charge was inadequate to maintain memory	Replace battery and reload program
■ POWER ■ PC RUN ☐ CPU FAULT ■ FORCED I/O ☐ BATTERY LOW	System does not operate as programmed	Input or output device(s) in forced condition	Monitor program and determine forced input and output device(s). Disable forced input or output device(s) and test system
■ POWER ■ PC RUN ☐ CPU FAULT ☐ FORCED I/O ☐ BATTERY LOW	System does not operate	Defective input device, input module, output device, output module, or program	Monitor program and check condition of status indicators on the input and output modules. Reload program when there is a program error

Error Codes

The manufacturer normally provides a programmable controller which includes a self-troubleshooting program as part of the normal program. The monitor screen displays the problem when a pre-programmed error is detected in the system. The problem may be directly displayed as a message on the monitor screen or may be displayed as an error code number. Simple problems such as exceeding the counter's limit are normally displayed directly on the monitor screen.

Problems that require service are typically displayed by a coded number. The coded number must be cross-referenced to a listing in the service manual. Error code numbers and possible causes vary for each manufacturer. The corrective action taken for the problem is normally the same. See Error Codes.

ERROR CODES				
Error Code	Problem	Possible Cause	Corrective Action
01	Memory error occuring in run mode	Voltage surge, improper grounding, high noise interference on lines, inadequate power supply	Add surge suppression to incoming power lines and all inductive output lines. Check power supply voltage for correct level. Reload program with backup program and reboot system
02	Processor does not meet required software level	Software is not compatible with the hardware system. Normally occurs when updated software is loaded into an older system	Consult the software specifications or manufacturer to determine the required hardware level. Upgrade system hardware to meet the software requirements
03	Power failure of an expansion I/O module	Power was removed from module, inadequate power supply, or power has dipped below the minimum specification of the module	Measure voltage at the module power supply and correct any problems. Reboot the system to return to normal operation when message still appears after power is restored
04	Program is trying to address an I/O module in an empty slot on rack	Program is set to the wrong rack and/or slot number	Set the program to a new address number or insert the correct I/O module in the slot number being addressed
05	Module has been detected as being inserted under power	Power was not disabled before a module was inserted into a slot	Reboot the system to return to normal operation when message still appears after power is restored. Never insert a module while power is applied to the unit. Always turn power OFF before inserting or removing a module

Application — Force and Disable

The force command opens or closes an input device or turns ON or OFF an output device. The force command is designed for use when system troubleshooting. Forcing an input or output device allows checking the circuit using software. See Force and Disable.

An input device may be forced to test the circuit operation. Forcing an input device may also be used when service is required on a defective input device. The defective input device may be forced ON until the device may be serviced if the input device is not critical to production. The force command is removed after the device is fixed.

An output device turns ON regardless of the programmed circuit's logic when the force ON command is used. The output device remains ON until the force OFF command is used. Extra care must be taken when using the force command because the force command overrides all safety features designed for the program.

The disable command prevents an output device from operating. The disable command is the opposite of the force command. The disable command is used to prevent one or all of the output devices from operating. Ensure that all force and disable commands are removed before returning a system to normal operation.

FORCE AND DISABLE

Application — Startup Procedure

A *startup procedure* is a logical, systematic procedure that isolates and eliminates potential problems that may cause personal injury and equipment damage. Common startup problems include equipment malfunctions, wiring mistakes, and programming errors.

A startup procedure should always be used when a system is first installed and after software or hardware changes are made. A startup procedure may be used any time troubleshooting is required. A programmable controller startup procedure includes:
- Inspecting the system
- Disabling motion producing output devices
- Verifying the program
- Troubleshooting input modules and devices
- Troubleshooting output modules and devices
- Testing the system

Inspecting the System

An inspection of all parts of a system is required before any power is applied to the system. Inspection is performed to ensure that each module is in the correct location, securely mounted, correctly wired, and properly programmed. See System Inspection.

SYSTEM INSPECTION

To inspect a system, apply the procedure:
1. Ensure that the controller and all associated hardware is securely mounted in the proper location. Loose connections cause vibration which may cause plug-in circuit boards to loosen, open circuits, and arcing. Verify each module slot position.
2. Check all wire connections from the main disconnect to the controller input devices, output devices, expansion and communication cables, and jumper links. A *jumper link* is a dual-voltage unit used to set the power rating of the controller. A jumper link must be in the correct location.

Check to ensure that all wire connections are tight and that proper wiring procedures are followed. Proper wiring procedures include physically separating the main power lines from the input lines, input lines from the output lines, power lines from communication lines, and AC lines from DC lines. Separating different line types reduces the possibility of introducing electromagnetic interference on the lines. Electromagnetic interference may produce false signals and cause program errors.

3. Ensure that the controller is grounded. Proper grounding is an important safety precaution in any electrical installation. Proper grounding is especially important in a programmable controller application because improper grounding may lead to interference being induced into the controller. The induced interference may cause false turn ON of output devices, which may cause personnel or equipment damage. Refer to equipment manufacturer grounding recommendations. See Grounding.

To prevent problems, the grounding path must be:
- Permanent
- Continuous and uninterrupted
- Of minimum resistance
- Of sufficient size to carry any potential fault current

4. Ensure that incoming voltage is correct according to the programmable controller's specifications and wiring. Excessive voltage damages the controller. Inadequate voltage may cause malfunctions. One very damaging problem occurs when 230 V is applied to a controller set for 115 V. This normally occurs when the jumper links are not placed in the correct position.

Spare parts should be available because many programmable controller problems are solved by the substitution of PC boards and modules. Most manufacturers provide a list of recommended spare parts to stock in-house. This list normally includes a single replacement module for each CPU and power supply, and spare I/O modules equal to 10% of the total number used in the system. Output modules fail more often than input modules because they are connected to high current loads.

Disabling Motion Producing Output Devices

All motion producing output devices should be disabled to prevent damage to personnel and equipment. These include motors, solenoids, and valves. Output devices such as lights generally do not have to be disabled. The output devices may be enabled after the circuit input devices and programs are checked. Disabling the output devices does not prevent the system and outputs from being checked. Disabling output devices reduces the possibility of injury or damage. See Disabling Output Devices.

GROUNDING

DISABLING OUTPUT DEVICES

Motors are disabled by turning OFF and locking out the disconnect switch feeding the main power to the motor. By locking out power at the disconnect, the motor starter is energized by the programmable controller but the motor is not capable of turning ON. Heating elements and solenoids that are on their own circuit are disabled by removing the main fuse or by placing the circuit breaker in the OFF position.

An output device may be disabled by disconnecting the wires directly at the output device. Ensure that the voltage is OFF by checking it with a voltmeter before disconnecting any wires. An output device may also be disabled by disconnecting the wires at the output module of the programmable controller. This is only recommended when there is no other practical way of disabling the output device. Ensure that the wire and terminal screws are clearly marked to prevent the wire from being re-connected to the wrong terminal.

Verifying the Program

Most programmable controllers include a test mode in addition to a program and run mode. In the test mode, the processor executes the program without actually turning ON the output devices. System operation is simulated without energizing the loads.

The program is tested because the program may have changed unexpectedly due to higher-than-rated temperatures, electromagnetic interference, high transient voltages, or unauthorized tampering. Check the program for changes any time a system does not seem to be operating properly and the input and output devices appear to be working.

Ensure that there is a backup program available. The program in the controller can always be re-loaded using the backup program if there is a problem. See Saving Programs.

SAVING PROGRAMS

Troubleshooting Input Modules

Signals and information are sent to a programmable controller using input devices such as pushbuttons, limit switches, level switches, and pressure switches. The input devices are connected to the input module of the programmable controller. The controller does not receive the proper information if the input device or input module is not operating correctly. See Troubleshooting Input Modules.

To troubleshoot an input module, apply the procedure:
1. Measure the supply voltage at the input module to ensure that there is power supplied to the input device(s). Test the main power supply of the controller when there is no power.
2. Measure the voltage from the control switch. Connect the meter directly to the same terminal screw to which the input device is connected.

The voltmeter should read the supply voltage when the control switch is closed. The voltmeter should read the full supply voltage when the control device uses mechanical contacts. The voltmeter should read nearly the full supply voltage when the control device is solid-state. Full supply voltage is not read because .5 V to 6 V is dropped across the solid-state control device. The voltmeter should read zero or little voltage when the control switch is open.

3. Monitor the status indicators on the input module. The status indicators should illuminate when the voltmeter indicates the presence of supply voltage.
4. Monitor the input device symbol on the programming terminal monitor. The symbol should be highlighted when the voltmeter indicates the presence of supply voltage.

Replace the control device if the control device does not deliver the proper voltage. Replace the input module if the control device delivers the correct voltage, but the status indicator does not illuminate.

TROUBLESHOOTING INPUT MODULES

Programmable Controllers

Troubleshooting Input Devices

Input devices such as pushbuttons, limit switches, pressure switches, and temperature switches are connected to the input module(s) of a programmable controller. Input devices send information and data concerning circuit and process conditions to the controller. The processor receives the information from the input devices and executes the program. All input devices must operate correctly for the circuit to operate properly. See Troubleshooting Input Devices.

TROUBLESHOOTING INPUT DEVICES

To troubleshoot an input device, apply the procedure:
1. Place the controller in the test or program mode. This step prevents the output devices from turning ON. Output devices are turned ON when the controller is placed in the run mode.
2. Monitor the input devices using the input status indicators (located on each input module), the programming terminal monitor, or the data file. A *data file* is a group of data values (inputs, timers, counters, and outputs) that is displayed as a group and whose status may be monitored.
3. Manually operate each input device starting with the first input. Never reach into a machine when manually operating an input device. Always use a wood stick or other non-conductive device.

The input status indicator located on the input module should illuminate and the input symbol should be highlighted in the control circuit on the monitor screen when a normally-open input device is closed. The bit status on the programming terminal monitor screen should be set to 1 indicating a high or presence of voltage.

The input status indicator located on the input module should turn OFF and the input symbol should no longer be highlighted in the control circuit on the monitor screen when a normally-closed input device is open. The bit status on the programming terminal monitor screen should be set to 0 indicating a low or absence of voltage.

Select the next input device and test it when the status indicator and associated bit status match. Continue testing each input device until all inputs have been tested. Troubleshoot the input device and output device when the status indicator and associated bit status do not match.

Troubleshooting Output Modules

A programmable controller turns ON and OFF the output devices (loads) in the circuit according to the program. The output devices are connected to the output module of the programmable controller. No work is produced in the circuit when the output devices or output module are not operating correctly. The problem may lie in the output module, output device, or controller when an output device does not operate. See Troubleshooting Output Modules.

TROUBLESHOOTING OUTPUT MODULES

To troubleshoot an output module, apply the procedure:
1. Measure the supply voltage at the output module to ensure that there is power supplied to the output devices. Test the main power supply of the controller when there is no power.
2. Measure the voltage delivered from the output module. Connect the meter directly to the same terminal screw to which the output device is connected.

The voltmeter should read the supply voltage when the program energizes the output device. The voltmeter should read full supply voltage when the output module uses mechanical contacts. The voltmeter should read almost full supply voltage when the output module uses a solid-state switch. Full voltage is not read because .5 V to 6 V is dropped across the solid-state switch. The voltmeter should read zero or little voltage when the program de-energizes the output device.

3. Monitor the status indicators on the output module. The status indicators should be energized when the voltmeter indicates the presence of supply voltage.
4. Monitor the output device symbol on the programming terminal monitor. The output device symbol should be highlighted when the voltmeter indicates the presence of supply voltage. Replace the output module when the output module does not deliver the proper voltage. Troubleshoot the output device when the output module does deliver the correct voltage, but the output device does not operate.

Troubleshooting Output Devices

Output devices such as motor starters, solenoids, contactors, and lights are connected to the output modules of a programmable controller. An output device performs the work required for the application. The processor energizes and de-energizes the output devices according to the program. All output devices must operate correctly for the circuit to operate properly. See Troubleshooting Output Devices.

TROUBLESHOOTING OUTPUT DEVICES

412 TROUBLESHOOTING ELECTRICAL/ELECTRONIC SYSTEMS

> To troubleshoot an output device, apply the procedure:
> 1. Place the controller in the test or program mode. Placing the controller in the test or program mode prevents the output devices from turning ON. Output devices turn ON when the controller is placed in the run mode.
> 2. Monitor the output devices using the output status indicators (located on each output module), the programming terminal monitor, or the data file.
> 3. Activate the input device that controls the first output device. Check the program displayed on the monitor screen to determine which input device activates which output device. Never reach into a machine to activate an input device.
>
> Select the next output device and test it when the status indicator and associated bit status match. Continue testing each output device until all output devices have been tested. Troubleshoot the input device and output device when the status indicator and associated bit status do not match.
>
> **Testing the System**
>
> The complete system may be tested after all input devices, output devices, and programs have been tested. Always have an individual ready to operate the emergency stop pushbutton when testing the system. The emergency stop pushbutton de-energizes the master control relay and removes power from the output devices. An emergency stop pushbutton that operates a master stop relay is required on all programmable controller installations. See Emergency Stop Circuit.

EMERGENCY STOP CIRCUIT

All personnel must remain clear of the controller and equipment when power is applied to test a system. There may be an intermittent problem and any sudden, unexpected machine motion could result in injury. See Testing the System.

To test the system, apply the procedure:
1. Apply power to the first output device. Leave the other output devices de-energized.
2. Place the controller in the run mode.
3. Cycle the control circuit until the first output device is activated.
4. Place the controller in the program or test mode when the first output device operates properly.
5. Apply power to the second output device with the controller in the program or test mode. Troubleshoot the output device when the device does not operate. Check the program for programming errors when the output device does not operate properly.
6. Place the controller in the run mode.
7. Cycle the control circuit until the second output device is activated.
8. Place the controller in the program or test mode when the second output device operates properly.
9. Apply power to the next output device with the controller in the program or test mode. Test the next output device and all other output devices using the same method as the first and second output devices.

TESTING THE SYSTEM

Some manufacturers provide flow charts to aid in troubleshooting a programmable controller circuit. See Output Troubleshooting Flow Chart.

The manufacturer's service manual and flow chart may be used when a problem is not easily found. Using a flow chart aids in giving direction to the troubleshooting process.

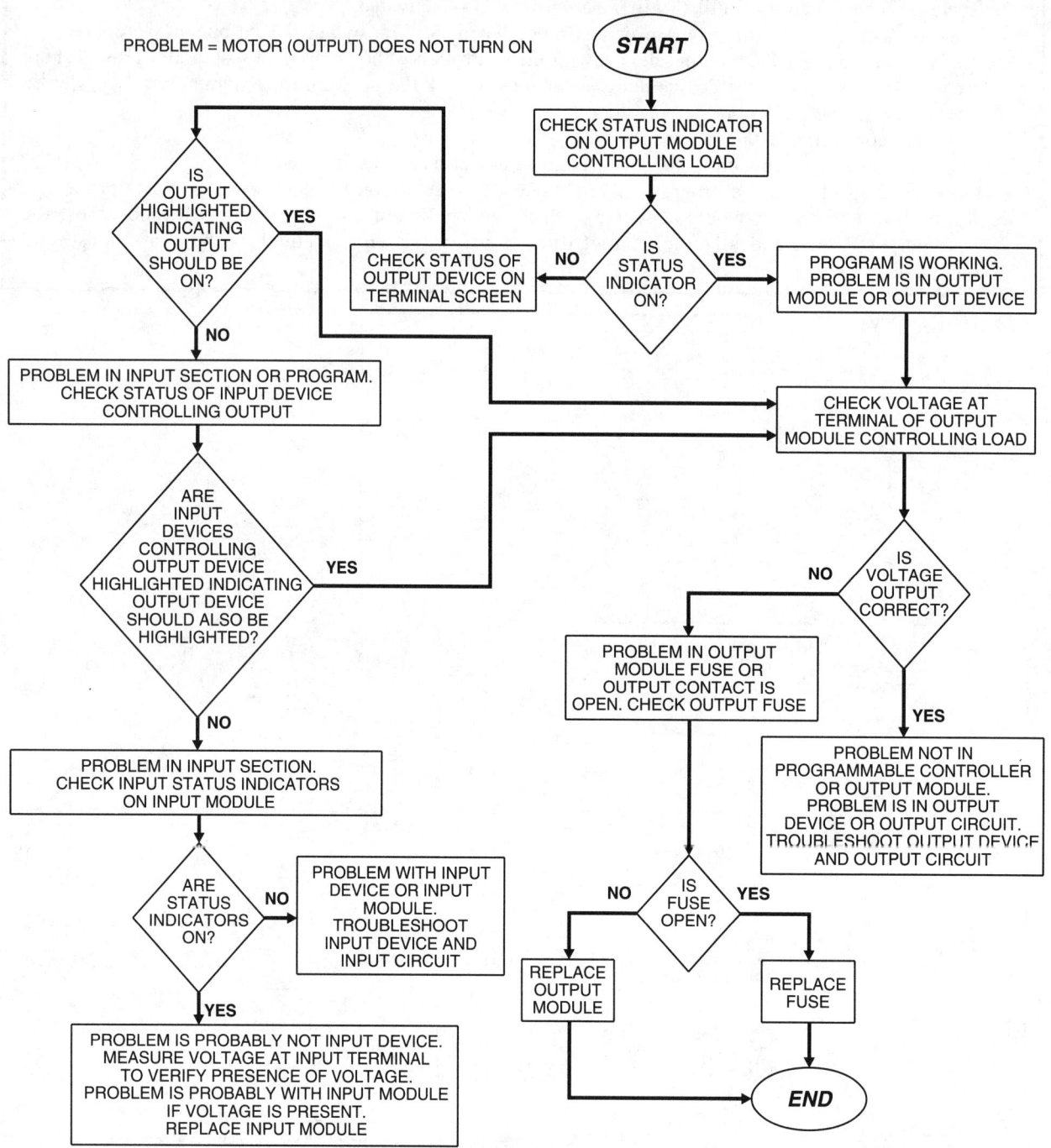

OUTPUT TROUBLESHOOTING FLOW CHART

Application — PC Preventive Maintenance

Programmable controllers are designed to operate for a long time with little maintenance. Most problems that occur are with the input devices and output devices connected to the controller. Preventive maintenance steps should be taken to keep the programmable controller working properly.

Enclosures

The PC boards used in the controller must be protected from dirt, oil, moisture, and other contaminants. To protect the PC boards, the controller must be installed in an enclosure suitable for the environment. The enclosure provides the main protection from atmospheric conditions. A NEMA 12 enclosure is used for most indoor general purpose areas. See Enclosure Types in Appendix.

A NEMA 12 enclosure is designed to be dust tight. A metal enclosure is used with programmable controllers because a metal enclosure helps minimize the effect of electromagnetic interference. Electromagnetic interference may cause false operations in the system. Ensure metal enclosures are properly grounded. Grounding provides a safe path for unwanted signals.

All programmable controllers produce heat that must be dissipated away from the controller. Proper spacing of components within the enclosure is required. Components are spaced so that the heat from one component does not cause problems in another. Use an enclosure that is large enough to allow room around each component to help dissipate the heat. The temperature inside the enclosure must not exceed the maximum operating temperature of the controller (normally 60°C). A fan is required in higher ambient temperatures.

A fan used to circulate air through the cabinet containing a programmable controller should pull the outside air into the cabinet. A filter must be used on the intake side of the fan to remove dirt from the air before it enters the fan and cabinet. This allows clean air to circulate over the components and out the vent holes. As air passes through the filter, the filter becomes dirty. The filter should be cleaned or replaced every three months (sooner in dirty environments).

Electrical Noise Suppression

Electrical noise is unwanted signals that are present on a power line. Electrical noise enters through input devices, output devices, and power supply lines. Unwanted noise pickup may be reduced by placing the controller away from noise-generating equipment such as motors, motor starters, welders, and drives.

Noise suppression should be included in every programmable controller installation because it is impossible to eliminate noise in an industrial environment. Certain sensitive input devices (analog, digital, and thermocouple) require a shielded cable to reduce electrical noise.

A shielded cable uses an outer conductive jacket (shield) to surround the two inner signal-carrying conductors. The shield blocks electromagnetic interference. The shield must be properly grounded to be effective. Proper grounding includes grounding the shield at only one point. A shield grounded at two points tends to conduct current between the two grounds. See Input Shielding.

INPUT SHIELDING

416 TROUBLESHOOTING ELECTRICAL/ELECTRONIC SYSTEMS

A high-voltage spike is produced when inductive loads (motors, solenoids, coils) are turned OFF. These spikes may cause problems in a programmable controller system. High-voltage spikes should be suppressed to prevent problems. A snubber circuit is used to suppress a voltage spike. Typical snubber circuits use an RC (resistor/capacitor), MOV (metal oxide varistor), or a diode depending on the load. See Snubber Circuits.

SNUBBER CIRCUITS

General Preventive Maintenance

General preventive maintenance steps are taken to prevent problems when servicing a programmable controller. Programmable controller general preventive maintenance steps include:
- Changing or cleaning filters. Programmable controllers that use a fan to help cool the circuits generally include a filter as part of the fan unit. The airflow is reduced or stopped as the filter gets dirty.
- Keeping the programmable controller modules clean, especially the fins on the power supply and I/O modules. The fins are designed to dissipate heat produced by the solid-state electronic devices. Wipe the fins with a clean cloth when dirty.
- Checking to ensure that the modules are placed tightly in their sockets. Modules are designed to be inserted and removed. The modules may loosen over time when vibration is present. A loose module often causes an intermittent problem which is the hardest problem to troubleshoot.
- Keeping manufacturer service bulletins up-to-date and available.

PROGRAMMABLE CONTROLLERS

22

Name _____   Date _____

Activities

Activity 22-1. Programmable Controllers

1. Using the Conveyor Circuit Line Diagram, complete the Conveyor Circuit table by identifying the type (input or output), name (master stop, pressure switch, etc.), and function (start Conveyor 1 motor, etc.) of the circuit's inputs and outputs.

CONVEYOR CIRCUIT LINE DIAGRAM

CONVEYOR CIRCUIT			
Device	**Type**	**Name**	**Function**
A	Input	Master stop	Stop any or all motors that are running
B			
C			
D			
E			
F			
G			
H			
I			
J			
K			
L			
M			
N			
O			

418 TROUBLESHOOTING ELECTRICAL/ELECTRONIC SYSTEMS

2. Connect the inputs in the Conveyor Circuit to the input modules so they are listed A, B, C, etc. Mark any extras as spares.

3. Connect the outputs in the Conveyor Circuit to the output module so they are listed A, B, C, etc. Mark any extras as spares.

4. Using the Tank Circuit Line Diagram, complete the Tank Circuit table by identifying the type (input or output), name (run pushbutton, temperature switch, etc.), and function (start pump motor, etc.) of the circuit's inputs and outputs.

TANK CIRCUIT LINE DIAGRAM

	TANK CIRCUIT		
Device	Type	Name	Function
A	Input	Selector switch (automatic position)	Place pump circuit in automatic condition
B			
C			
D			
E			
F			
G			
H			
I			
J			
K			
L			
M			
N			
O			
P			
Q			

420 TROUBLESHOOTING ELECTRICAL/ELECTRONIC SYSTEMS

5. Connect the inputs in the Tank Circuit to the input modules so they are listed A, B, C, etc. Mark any extras as spares.

6. Connect the outputs in the Tank Circuit to the output module so they are listed A, B, C, etc. Mark any extras as spares.

Programmable Controllers 421

○ Activity 22-2. Status Indicators and Error Codes

The condition of the I/O status indicators is used when troubleshooting a control circuit. A troubleshooter must identify which status indicator(s) should be ON and which status indicator(s) should be OFF to interpret a problem. Using Pump Unloader Circuit Line Diagram, and Pump Unloader Circuit Input and Output Connections, determine the condition (ON or OFF) for each status indicator.

PUMP UNLOADER CIRCUIT LINE DIAGRAM

PUMP UNLOADER CIRCUIT INPUT AND OUTPUT CONNECTIONS

1. Circuit Condition 1 has normal (no problem) status indication when the programmable controller is in the run mode. The Selector Switch is in the run position and the Stop Pushbutton has been pressed and released.

Ⓐ _____   Ⓕ _____   Ⓚ _____   Ⓠ _____
Ⓑ _____   Ⓖ _____   Ⓝ _____   Ⓡ _____
Ⓒ _____   Ⓗ _____   Ⓛ _____   Ⓢ _____
Ⓓ _____   Ⓘ _____   Ⓞ _____   Ⓣ _____
Ⓔ _____   Ⓙ _____   Ⓜ _____   Ⓤ _____
                          Ⓟ _____

422 TROUBLESHOOTING ELECTRICAL/ELECTRONIC SYSTEMS

2. Circuit Condition 2 has normal (no problem) status indication when the programmable controller is in the run mode. The Selector Switch is in the run position and the Start Pushbutton has been pressed and released.

(A) _____ (F) _____ (K) _____ (Q) _____
(B) _____ (G) _____ (N) _____ (R) _____
(C) _____ (H) _____ (L) _____ (S) _____
(D) _____ (I) _____ (O) _____ (T) _____
(E) _____ (J) _____ (M) _____ (U) _____
                           (P) _____

STATUS: POWER, PC RUN, CPU FAULT, FORCED I/O, BATTERY LOW
INPUT / OUTPUT

3. Circuit Condition 3 has normal (no problem) status indication when the programmable controller is in the run mode. The Selector Switch is in the jog position and the Jog Pushbutton is held down.

(A) _____ (F) _____ (K) _____ (Q) _____
(B) _____ (G) _____ (N) _____ (R) _____
(C) _____ (H) _____ (L) _____ (S) _____
(D) _____ (I) _____ (O) _____ (T) _____
(E) _____ (J) _____ (M) _____ (U) _____
                           (P) _____

Using the status indicators, Pump Unloader Circuit Line Diagram, and Pump Unloader Circuit Input and Output Connections, determine the problem of each circuit condition.

4. The Start (or Jog) Pushbutton does not start the pump motor. The Selector Switch is in the run position.

_____ **A.** The problem is a(n) _____.

STATUS: ■ POWER, ■ PC RUN, ☐ CPU FAULT, ☐ FORCED I/O, ☐ BATTERY LOW

5. The Start (or Jog) Pushbutton does not start the pump motor. The Selector Switch is in the run position.

_____ **A.** The problem is _____.

STATUS: ■ POWER, ☐ PC RUN, ☐ CPU FAULT, ☐ FORCED I/O, ☐ BATTERY LOW

Activity 22-3. Force and Disable

Forcing an input or output device allows troubleshooting a circuit using software. Using the Tank Circuit Line Diagram and Tank Circuit Input and Output Connections, determine the inputs and outputs that are forced when troubleshooting each problem.

TANK CIRCUIT LINE DIAGRAM

TANK CIRCUIT INPUT AND OUTPUT DIAGRAMS

1. The operator reports that the Discharge Motor does not start when the Start Pushbutton is pressed.

   _____ A. _____ is forced to determine if the Start Pushbutton is operating.

   _____ B. The force _____ (ON, OFF, open, or closed) instruction is used to determine if the Start Pushbutton is operating.

424 TROUBLESHOOTING ELECTRICAL/ELECTRONIC SYSTEMS

_____ C. _____ is forced to determine if the Overload Contact has tripped.

_____ D. The force _____ (ON, OFF, open, or closed) instruction is used to determine if the Overload Contact has tripped.

_____ E. Input _____ is forced to bypass the pushbutton station.

_____ F. The force _____ (ON, OFF, open, or closed) instruction is used to bypass the pushbutton station.

_____ G. _____ is forced to override all inputs in the circuit and test if the starter is operating.

_____ H. The force _____ (ON, OFF, open, or closed) instruction is used to override all inputs in the circuit and test if the starter is operating.

2. The Tank Circuit has been in operation for several years and a routine inspection is scheduled. The routine schedule requires a test of the circuit's visual and audible indicators.

_____ A. Output _____ is forced to determine if the High Temperature Alarm is operating.

_____ B. The force _____ (ON, OFF, open, or closed) instruction is used to determine if the High Temperature Alarm is operating.

_____ C. Output _____ is forced to determine if the Pump Motor Indicator Light is operating.

_____ D. The force _____ (ON, OFF, open, or closed) instruction is used to determine if the Pump Motor Indicator Light is operating.

_____ E. Output _____ is forced to determine if the Discharge Motor Indicator Light is operating.

_____ F. The force _____ (ON, OFF, open, or closed) instruction is used to determine if the Discharge Motor Indicator Light is operating.

3. The High Temperature Alarm has sounded. The Cooling Motor is not running.

_____ A. Output _____ is forced to shut the alarm OFF during the troubleshooting call.

_____ B. The force _____ (ON, OFF, open, or closed) instruction is used to force the output device to shut the alarm OFF during the troubleshooting call.

_____ C. Input _____ is forced to shut the alarm OFF during the troubleshooting call.

_____ D. The force _____ (ON, OFF, open, or closed) instruction is used to force the input to shut the alarm OFF during the troubleshooting call.

Activity 22-4. Startup Procedure

1. Connect the programmable controller to 115 V.

2. Connect the programmable controller to 230 V.

3. Determine the meter readings if the Tank Circuit is operating properly. The Selector Switch is in the auto position, FS1 is closed, the Start Pushbutton has been pressed and released, and TS1 is closed.

TANK CIRCUIT

_____ **A.** Meter 1 reading equals _____ V.

_____ **B.** Meter 2 reading equals _____ V.

_____ **C.** Meter 3 reading equals _____ V.

426 TROUBLESHOOTING ELECTRICAL/ELECTRONIC SYSTEMS

_____ D. Meter 4 reading equals _____ V.

_____ E. Meter 5 reading equals _____ V.

_____ F. Meter 6 reading equals _____ V.

_____ G. Meter 7 reading equals _____ V.

_____ H. Meter 8 reading equals _____ V.

_____ I. Meter 9 reading equals _____ V.

_____ J. Meter 10 reading equals _____ V.

⬤ Activity 22-5. PC Preventive Maintenance

1. Add voltage spike suppression on each inductive load.

PREVENTIVE MAINTENANCE

Applications — Chapter 23

Application — General Motor Maintenance

Preventive maintenance is maintenance performed to keep machines, assembly lines, production operations, and plant operations running with little or no downtime. Preventive maintenance is performed on equipment that is not operating.

Preventive maintenance helps prevent motor problems from occurring. A motor that is well-maintained and used in an application for which it is suited can have a long service life. Maintenance may be performed as needed or as part of a routine scheduled maintenance program.

Water-Soaked Motors

Motors may become water-soaked from flooding, broken water pipes, or lack of proper protection. Normally, water alone does not damage a motor if it is serviced shortly after soaking. Servicing includes a thorough cleaning, drying, and testing.

Motor Cleaning

A water-soaked motor must be cleaned as soon as possible. Cleaning normally requires a cleaning solvent to remove grease, oil, and contaminations. Warm water is normally enough to clean a motor that has been soaked in relatively clean water.

A motor must be thoroughly dried after the motor has been cleaned. Applying power to wet or damp motor windings damages the motor far more than water alone. A motor that has been cleaned may need relubrication because cleaning may remove some of the bearing lubrication. See Motor Cleaning.

MOTOR CLEANING

To clean a motor, apply the procedure:
1. Turn power to the motor circuit OFF. Ensure that all power is OFF.
2. Disconnect the motor from the circuit.
3. Apply water and an approved cleaning solvent to the motor. Water pressure should not exceed 25 psi. Water temperature should not exceed 200°F. Allow the cleaning solvent sufficient time to work per manufacturer's recommendations.
4. Clean the motor with a stiff brush and water. Rinse off all the cleaning solvent.

Motor Drying

A motor must be thoroughly dried before any power is applied. Motors may be slow-dried or force-dried. Slow-drying a motor is better than force-drying. A motor is slow-dried by placing it in a warm room or in the sun for two to four days.

Motors are force-dried by heating or by applying low voltage. To force-dry a motor by heating, apply 150°F to 200°F heat for approximately 18 hours to 30 hours depending on the motor type and temperature. To force-dry a motor by applying low voltage, apply approximately 10% of the motor's rated voltage for two to four hours to dry the coils.

Motor Testing

Always check the resistance of a motor before reapplying power. The motor insulation is checked to ensure it has the proper resistance value. A megohmmeter or ohmmeter is used to test the resistance of the insulation.

Although a motor may operate satisfactorily with a reading of less than 1 MΩ, a resistance of 1 MΩ or more is recommended as the minimum resistance. Additional drying or running the motor for a short time increases the resistance.

Check for signs of spitting or smoking when restarting the motor. Spitting indicates moisture in the motor. Smoking indicates an insulation breakdown. In either case, turn the power OFF and completely dry and retest the motor.

Motor Bearing Lubrication

Motors are lubricated at the factory to provide long operation under normal service conditions without relubrication. Excessive and frequent lubrication may damage a motor. The time period between lubrications depends on the motor's service conditions, ambient temperature, and environment. See Motor Lubrication.

Always follow lubrication instructions provided with the motor. These instructions are normally listed on

ADD ELECTRIC MOTOR OIL TO SLEEVE BEARINGS AFTER THREE YEARS OF NORMAL OPERATION

LUBRICATE BALL BEARINGS AFTER FIVE YEARS OF NORMAL OPERATION

MOTOR LUBRICATION

the nameplate or terminal box cover. There may be separate lubrication instructions furnished with the motor. Relubricate sleeve bearings and ball bearings per schedule if lubrication instructions are not available.

For normal operation, add electric motor oil (or SAE #10 or #20 non-detergent oil) to sleeve bearings after three years. For heavy-duty operation, relubricate once a year. For light operation, relubricate every four years.

Ball bearings are designed for many years of operation without relubrication. The schedule for relubricating ball bearings varies from one year to 10 years, depending on the motor's service conditions, ambient temperature, and environment. When relubricating, use standard, long-life ball-bearing grease.

Motor Mounting/Remounting

Motors other than dripproof types may be mounted in any position or at any angle. Dripproof motors must be mounted in the normal horizontal position to meet the requirements of the enclosure. Motors with sleeve bearings must be mounted with the oil cap up.

Mount a motor securely to the mounting base of equipment or to a rigid, flat surface. Metal surfaces are preferred. For direct-coupled installations, align the motor shaft and coupling carefully. Place shims under the motor base as required to assure proper alignment. Use flexible couplings whenever possible. In belt applications, align pulleys and adjust belt tension so that approximately $\frac{1}{64}''$ of belt deflection occurs per inch of span. Position sleeve-bearing motors so that the belt is away from the oil in the sleeve bearing (located approximately under the oiler of the motor). See Motor Mounting.

MOTOR MOUNTING

Dirt and Corrosion

Remove dirt and corrosion by brushing, vacuuming, or blowing. After removing obvious dirt, feel for air being discharged from cooling ports of the motor. The internal air passages may be clogged if the air flow is weak.

As time passes, small amounts of corrosion are normal. Large amounts of corrosion may indicate a problem. Check for signs of corrosion. After removing corrosion, use corrosion-resistant, high-temperature paint to repaint the motor.

Heat, Noise, and Vibration

Excessive heat or vibration is determined by touching the motor frame and bearings lightly. Listen closely for any abnormal noise. Problems with heat are normally due to improper ventilation, excessive motor cycling, or overloads. Vibration and noise are caused by misalignment and loose parts. Tighten all mountings connecting the motor and motor load. Realign the motor if excessive vibration remains.

Scheduled Maintenance

A maintenance schedule is essential to the satisfactory operation of electric motors. The frequency of scheduled maintenance on motors varies. Motor maintenance checklists for scheduled semiannual and annual maintenance provide a record of maintenance performed. A motor repair record provides a record of motor repairs. See Annual Motor Maintenance Checklist and Motor Repair and Service Record in Appendix.

Application — Recording Meters

A *recording meter* is a meter that provides a written record of readings over a time period. Recording meters are used to record quantities such as voltage, current, power, speed, frequency, temperature, humidity, etc.

Recording meters record either analog traces or actual numerical values onto paper. The two basic types of recording meters are the vertical chart recorder and the circular chart recorder. A *vertical chart recorder* is a graphic pen and ink recorder that has measured values drawn on a continuous roll of graph paper. Vertical chart recorders are normally used in permanent installations and are designed to record over long time periods (days or weeks). A *circular chart recorder* is a graphic pen and ink recorder that has measured values drawn on a rotating circular chart. Circular chart recorders are normally used for temporary measurements and normally record over a 24-hour time period. See Recording Meters.

RECORDING METERS

Recording meters are used to record either application or circuit information. Application information, such as temperature and humidity levels, may be used to determine potential process problems. For example, a permanent record of temperature and humidity may be recorded in a food storage area.

Circuit information, such as voltage and power levels, may be used to determine potential control problems. For example, a permanent record of voltage spikes in a computer room may be used to determine intermittent problems.

Application — Lockout/Tagout

Per OSHA standards, equipment must be locked out and tagged out before any preventive maintenance or servicing is performed. *Lockout* is the process of removing the source of power and installing a lock which prevents the power from being turned ON. *Tagout* is the process of placing a danger tag on the source of power which indicates that the equipment may not be operated until the danger tag is removed. See Lockout/Tagout.

A danger tag has the same importance and purpose as a lock and is used by itself only when a lock does not fit the disconnect device. The danger tag shall be attached at the disconnect device with a tag tie or equivalent and shall have space for the worker's name, craft, and other required information. A danger tag must withstand the elements and expected atmosphere for as long as the tag remains in place.

A lockout/tagout is used when:
- Servicing electrical equipment that does not require power to be ON to perform the service.
- Removing or bypassing a machine guard or other safety device.
- The possibility exists of being injured or caught in moving machinery.
- Clearing jammed equipment.
- The danger exists of being injured if equipment power is turned ON.

LOCKOUT/TAGOUT

Lockouts and tagouts do not by themselves remove power from a circuit. An approved procedure must be followed when applying a lockout/tagout. Lockouts and tagouts are attached only after the equipment is turned OFF and tested to ensure that power is OFF. The lockout/tagout procedure is required for the safety of workers due to modern equipment hazards. OSHA provides a standard procedure for equipment lockout/tagout. OSHA's procedure is:
1. Prepare for machinery shutdown
2. Machinery or equipment shutdown
3. Machinery or equipment isolation
4. Lockout or tagout application
5. Release of stored energy
6. Verification of isolation

Warning: Personnel should consult OSHA Standard 29CFR1910.147 for industry standards on lockout/tagout.

A lockout/tagout shall not be removed by any other person than the person that installed it, except in an emergency. In an emergency, the lockout/tagout may be removed only by authorized personnel. The authorized personnel shall follow approved procedures. A list of company rules and procedures are given to any person that may use a lockout/tagout.

432 TROUBLESHOOTING ELECTRICAL/ELECTRONIC SYSTEMS

Always remember:
- Use a lockout and tagout when possible.
- Use a tagout when a lockout is impractical. A tagout is used alone only when a lock does not fit the disconnect device.
- Use a multiple lockout when individual employee lockout of equipment is impractical.
- Notify all employees affected before using a lockout/tagout.
- Remove all power sources including primary and secondary.
- Measure for voltage using a voltmeter to ensure that power is OFF.

Lockout Devices

Lockout devices are lightweight enclosures that allow the lock out of standard control devices. Lockout devices are available in various shapes and sizes that allow for the lock out of ball valves, gate valves, and electrical plugs. Lockout devices resist chemicals, cracking, abrasion, and temperature changes, and are available in colors to match ANSI pipe colors. Lockout devices are sized to fit standard industry control device sizes. See Lockout Devices.

ANSI PIPE COLORS	
Color	**Designation**
Red	Fire protection materials
Yellow	Hazardous gas or liquid
Green	Non-hazardous liquid
Blue	Non-hazardous gas

Locks used to lock out a device may be color-coded and individually keyed. The locks are rust-resistant and are available with various size shackles.

Danger tags provide additional lockout and warning information. Various danger tags are available. Danger tags may include warnings such as "Do Not Start," "Do Not Operate," or may provide space to enter worker, date, and lockout reason information. Tag ties must be strong enough to prevent accidental removal and must be self-locking and nonreusable. See Danger Tags.

Lockout/tagout kits are also available. A lockout/tagout kit contains items required to comply with the OSHA lockout/tagout standard. Lockout/tagout kits contain reusable danger tags, tag ties, multiple lockouts, locks, magnetic signs, and information on lockout/tagout procedures. See Lockout/Tagout Kit.

DANGER TAGS

Application — Enclosure Selection

An *enclosure* is a housing used to protect wires and equipment and prevent personnel injury by accidental contact with a live circuit. Enclosures are used to protect electrical equipment such as programmable controllers, motor starters, fuses/breakers, and control circuits. An enclosure provides the main protection from atmospheric conditions. Using the proper enclosure helps prevent problems caused by contamination, moisture, and physical damage. Enclosures are categorized by the protection they provide. An enclosure is selected based on the location of the equipment and the NEC® requirements. See Enclosures. See Enclosure Types in Appendix.

The NEC® classifies hazardous locations according to the properties and quantities of the hazardous material that may be present. Hazardous locations are divided into three classes, two divisions, and seven groups.

Class refers to the generic hazardous material present. Class I applies to locations where flammable gases or vapors may be present in the air in quantities sufficient to produce an explosive or ignitable mixture. Class II applies to locations where

LOCKOUT/TAGOUT KIT

ENCLOSURES

Type	Service Conditions	Sealing Method	Cost
1	No unusual		Base
4	Windblown dust and rain, splashing water, hose-directed water, and ice on enclosure		12 x Base
4X	Corrosion, windblown dust and rain, splashing water, hose-directed water, and ice on enclosure		12 x Base
7	Withstand and contain an internal explosion of specified gases, contain an explosion sufficiently so an explosive gas-air mixture in the atmosphere is not ignited		48 x Base
9	Dust		48 x Base
12	Dust, falling dirt, and dripping noncorrosive liquids		5 x Base

combustible dusts may be present in sufficient quantity to cause an explosion. Class III applies to locations where the hazardous material consists of easily ignitable fibers or flyings that are not normally in suspension in the air in large enough quantities to produce an ignitable mixture.

Division applies to the probability that a hazardous material is present. Division 1 applies to locations where ignitable mixtures exist under normal operating conditions found in the process, operation, or during periodic maintenance. Division 2 applies to locations where ignitable mixtures exist only in abnormal situations. Abnormal situations occur as a result of accidents or when equipment fails.

Air mixtures of gases, vapors, and dusts are grouped according to their similar characteristics. The NEC® classifies gases and vapors in Groups A, B, C, and D for Class I locations and combustible dusts in Groups E, F, and G for Class II locations. See Hazardous Locations. For example, a Type 7 enclosure is required for an indoor application where gasoline is stored (Class I, Group D).

HAZARDOUS LOCATIONS

Class	Group	Material
I	A	Acetylene
	B	Hydrogen, butadiene, ethylene oxide, propylene oxide
	C	Carbon monoxide, ether, ethylene, hydrogen sulfide, morpholine, cyclopropane
	D	Gasoline, benzene, butane, propane, alcohol, acetone, ammonia, vinyl chloride
II	E	Metal dusts
	F	Carbon black, coke dust, coal
	G	Grain dust, flour, starch, sugar, plastics
III	No groups	Wood chips, cotton, flax, and nylon

Application — Conveyor Belt Tracking

Conveyor Terminology

The two drive methods used to move a product along a conveyor belt are the direct drive and roller drive methods. The *direct drive method* has the product riding directly on the driven belt and is used for light loads. The *roller drive method* has the belt-driving rollers on which the product rides. The roller drive method is used for heavy loads. See Direct Drive and Roller Drive Methods.

The right side of a conveyor is the right side when facing the forward direction of material travel. The two ends of a conveyor are identified by the relationship to the forward direction of material travel. The *tail end* of a conveyor is the end where material is fed. The *head end* of a conveyor is the end from which material is discharged. A drive unit is connected to the tail-end pulley or head-end pulley.

DIRECT DRIVE　　　　ROLLER DRIVE

Conveyor Belt Tension

To prevent slippage, a conveyor belt must have the correct tension. Proper belt tension is accomplished by adjusting the take-up pulley. A *take-up pulley* is a pulley that is used for correcting belt tension and is not connected to a drive.

Conveyor Belt Tracking

Proper belt tracking depends on the alignment of the pulleys and rollers. A pulley is a revolving cylinder connected to a drive. A roller is a revolving cylinder not connected to a drive. As a conveyor belt moves, the belt should remain in the center of the pulleys and rollers. When a belt drifts to one side, the belt edge wears or folds up on the conveyor guard.

A conveyor belt drifts toward the side where pulley and roller centers are not parallel and are the closest to each other. Conveyor belt tracking is adjusted by moving the pulley or snub roller. A *snub roller* is a roller not connected to a drive that is used to correct belt alignment. When aligning a belt, all adjustments should be slight with time allowed for the belt to react to the adjustment.

Conveyor Adjustment — Head-End Pulley

If the belt drifts to the right on the head-end pulley during forward material travel, adjust the right side of the head-end snub roller in the forward direction of material travel and/or the left side of the head-end snub roller in the reverse direction of material travel. See Head-End Pulley.

Conveyor Adjustment — Tail-End Pulley

If the belt drifts to the right on the tail-end pulley during forward material travel, adjust the right side of the tail-end snub roller in the reverse direction of material travel and/or the left side of the tail-end snub roller in the forward direction of material travel. See Tail-End Pulley.

Conveyor Adjustment — Center Drive

If the belt drifts to the right side of the center drive and take-up pulleys, adjust the right side of the snub roller in the reverse direction of material travel and/or the left side of the snub roller in the forward direction of material travel. See Center Drive Conveyor.

Conveyor Adjustment — Reverse-Running

If the belt drifts to the right side of the center drive and take-up pulleys, adjust the right side of the snub roller in the forward direction of material travel and/or the left side of the snub roller in the reverse direction. See Reverse-Running Center Drive Conveyor.

PREVENTIVE MAINTENANCE

23

Name _____ Date _____ **Activities**

⬤ Activity 23-1. General Motor Maintenance

Determine if the belt tension is loose, tight, or correct.

_____ 1. The belt tension is _____. _____ 2. The belt tension is _____.

_____ 3. The belt tension is _____. _____ 4. The belt tension is _____.

437

Activity 23-2. Recording Meters

Using the circular chart, answer the questions.

_____ 1. The maximum recorded humidity is _____ %.

_____ 2. The maximum humidity was recorded at _____.

_____ 3. The minimum recorded humidity is _____ %.

_____ 4. The minimum humidity was recorded at _____.

Using the vertical chart, answer the questions.

_____ 5. The maximum recorded temperature is _____ °F.

_____ 6. The maximum temperature was recorded at _____.

_____ 7. The minimum recorded temperature is _____ °F.

_____ 8. The minimum temperature was recorded at _____.

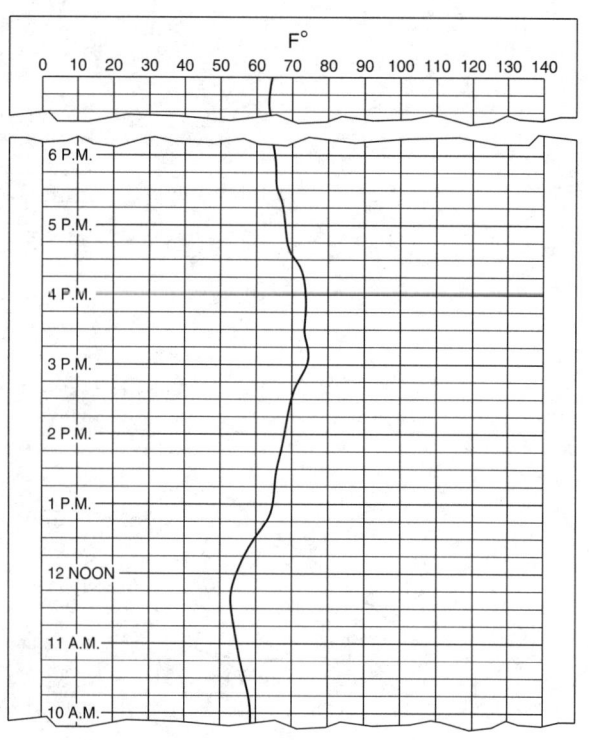

Preventive Maintenance **439**

◯ Activity 23-3. Lockout/Tagout

Determine if a lockout/tagout is required for each service call.

_____ 1. The service call requires the pump to be disconnected from the motor.

_____ 2. The service call requires checking the output of an HID ballast.

_____ 3. The service call requires checking belt tension.

_____ 4. The service call requires checking for hot spots that may be caused by loose connections.

◯ Activity 23-4. Enclosure Selection

Using Enclosure Types in Appendix, determine the enclosure type required for each location.

_____ 1. Required enclosure type is _____.

_____ 2. Required enclosure type is _____.

440 TROUBLESHOOTING ELECTRICAL/ELECTRONIC SYSTEMS

_____ **3.** Required enclosure type is _____.

_____ **4.** Required enclosure type is _____.

⬤ Activity 23-5. Conveyor Belt Tracking

Determine if the roller is adjusted properly or improperly to correct the belt tracking.

_____ **1.** The roller is adjusted _____ to correct the belt tracking.

_____ **2.** The roller is adjusted _____ to correct the belt tracking.

_____ **3.** The roller is adjusted _____ to correct the belt tracking.

_____ **4.** The roller is adjusted _____ to correct the belt tracking.

GLOSSARY

Terms in this glossary are defined as they relate to electrical/electronic systems.

A

abbreviation: A letter or combination of letters that represent a word.

AC sine wave: A symmetrical waveform that contains 360 electrical degrees. See *waveform*.

actuator: A device that changes the valve position. See *device* and *valve*.

adjustable motor base: A mounting base that allows a motor to be easily moved over a short distance.

adjustable motor base

alignment chart: A chart that shows the relationship among three variables. See *chart*.

alternating current (AC): Current that reverses its direction of flow at regular intervals. See *current*.

alternating current (AC) motor: A motor that uses AC to produce rotation.

alternation: Half of a cycle. See *cycle*.

alternator: A device that converts mechanical energy (spinning motion) to electrical energy.

ambient temperature: The temperature of the air surrounding a device. See *device*.

amplitude: The height of a waveform. See *waveform*.

analog display: An electromechanical device that indicates readings by the mechanical motion of a pointer.

analog display

angular misalignment: Misalignment that occurs when two shafts are not parallel.

anode: The positive lead of a diode. See *diode*.

apparent power: The product of voltage and current in a circuit calculated without considering the phase shift that may be present between total voltage and total current in a circuit. See *voltage, current,* and *phase shift*.

application drawing: A drawing that shows the use of a particular piece of equipment or product in an application. See *drawing*.

arcing: The discharge of an electric current across a gap such as when an electric switch is opened.

arc suppressor: A device that dissipates the energy present across opening contacts. See *device* and *contact*.

arc tube: The light-producing element of an HID lamp. See *high-intensity discharge (HID) lamp*.

armature: The rotating part of a DC motor that is mounted on the motor shaft. See *DC motor*.

average value: The mathematical mean of all instantaneous voltage values in a sine wave. See *sine wave*.

B

ballast: A transformer or solid-state circuit that limits current flow and supplies the starting voltage for a fluorescent lamp. See *transformer*.

bar graph: On a digital meter, a graph composed of segments that function as an analog pointer. In drawings and diagrams, a graph that compares statistical data with vertical or horizontal bars. See *graph*.

battery: A DC voltage source that converts chemical energy to electrical energy.

bench testing: Testing performed when equipment under test is brought to a designated service area.

block diagram: A diagram that shows the relationship between individual sections (blocks) of circuits or components. See *diagram* and *component*.

board: A group of electronic components placed on a printed circuit (PC) board that performs a set task.

board

break-in period: The period of time just after the installation of a new piece of equipment in which defects resulting from defective parts, poor manufacturing quality, contamination, or environmental stress appear.

breaks: The number of separate contacts a switch uses to open or close each individual circuit. See *contact*.

brown-out: The deliberate reduction of the voltage level by a power company to conserve power during peak usage times.

brushes: The sliding contacts that provide contact between the external power source and the commutator in a DC motor. See *commutator* and *direct current (DC) motor*.

441

bulb: A glass envelope enclosing the light source of an electric lamp.

C

candela (cd): The unit of luminous intensity produced by a light source in a given direction.

capacitance: The property of an electric device that permits the storage of electrically separated charges when potential differences exist between two conductors. See *device* and *conductor*.

capacitive circuit: A circuit in which current leads voltage. See *current* and *voltage*.

capacitive leakage current: Leakage current that flows through conductor insulation due to a capacitive effect. See *leakage current*.

capacitor: An electronic device used to store an electric charge. See *device*.

capacitor motor: A 1ɸ motor with a capacitor connected in series with the stator windings to produce phase displacement in the starting winding. See *capacitor*, *series*, and *stator*.

capacitor-run motor: A motor that leaves the starting winding and capacitor in the circuit at all times. See *capacitor*.

capacitor start-and-run motor: A motor that uses two capacitors. See *capacitor*.

capacitor-start motor: A motor that has the capacitor connected in series with the starting winding. See *capacitor* and *series*.

cathode: The negative lead of a diode. See *diode*.

cell: A unit that produces electricity at a fixed voltage and current level. See *voltage* and *current*.

centrifugal force: The force that moves rotating bodies away from the center of rotation.

centrifugal switch: In motors, a switch that opens to disconnect the starting winding when the rotor reaches a certain preset speed and reconnects the starting winding when the speed falls below a preset value. See *rotor*.

centrifugal switch

chart: A pictorial representation of data.

circuit analysis method: A method of troubleshooting that uses a logical sequence to determine the reason for device failure. See *troubleshooting* and *device*.

circuit breaker (CB): A reusable overcurrent protective device that opens a circuit automatically at a predetermined overcurrent. See *overcurrent*.

circular chart recorder: A graphic pen and ink recorder that has measures drawn on a rotating circular chart.

clamping diode: A diode that prevents voltage in one part of the circuit from exceeding the voltage in another part. See *diode* and *voltage*.

clamp-on ammeter: A meter that measures current in a circuit by measuring the strength of the magnetic field around a conductor. See *current* and *conductor*.

clamp-on ammeter

code: A regulation or minimum requirement.

coil: Winding consisting of insulated conductors arranged to produce magnetic flux.

color rendering: The appearance of a color when illuminated by a light source. See *illumination*.

common noise: Noise produced between the ground and hot, or the ground and neutral lines.

commutator: A series of copper segments connected to the armature. See *armature*.

complimentary metal-oxide semiconductor ICs (CMOS ICs): A group of ICs that employ MOS transistors. See *transistor*.

component: An individual device used in a board or module. See *device*, *board*, and *module*.

conduction angle: The time during which a thyristor is ON.

conductive leakage current: The small amount of current that normally flows through the insulation of a conductor.

conduction angle

conductor: Any material through which current flows easily. See *current*.

constant-wattage autotransformer ballast: A high-reactance autotransformer ballast with a capacitor added to the circuit. See *high-reactance autotransformer ballast* and *capacitor*.

contact: A conducting part that operates with another conducting part to make or break a circuit.

contact life: The number of times a relay's contacts switch the load controlled by the relay before malfunctioning. See *relay* and *load*.

contactor: A control device that uses a small control current to energize or de-energize the load connected to it. See *device* and *load*.

contact protection circuit: A circuit that protects contacts by providing a nondestructive path for generated voltage as a switch is opened. See *contact* and *voltage*.

contact tachometer: A device that measures the rotational speed of an object through direct contact of the tachometer tip with the object to be measured.

contact thermometer: An instrument that measures temperature at a single point.

continuity checker: An instrument that indicates an open or closed circuit in a circuit in which all power is OFF.

conveyor: A mechanical system that extends from a receiving point to a discharge point and conveys, transports, or transfers material between the two points.

contact thermometer

current: The amount of electrons flowing through an electrical circuit. See *electron*.

current gain: A measure of the ratio of the collector current to the base current in a transistor. See *current* and *transistor*.

cycle: One complete wave of alternating voltage or current. See *voltage* and *current*.

D

data file: A group of data values (inputs, timers, counters, and outputs) that is displayed as a group and whose status may be monitored.

delta configuration: A transformer connection that has each transformer coil connected end-to-end to form a closed loop. See *transformer* and *coil*.

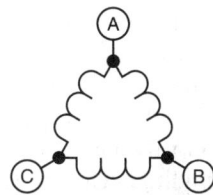

delta configuration

delta-connected 3ϕ motor: A motor that has each phase wired end-to-end to form a completely closed loop circuit. See *phase*.

device: Equipment or mechanism designed to serve a special purpose or perform a special function.

diac: A two-terminal bidirectional device used primarily to control the triggering of triacs. See *device* and *triac*.

diagram: A drawing that represents an object or area. See *drawing*.

dielectric absorption test: A test that checks the absorption characteristics of humid or contaminated insulation. See *insulator*.

dielectric material: A medium in which an electric field is maintained with little or no outside energy supply.

digital display: An electronic device that displays readings as numerical values.

digital logic probe: A special DC voltmeter that detects the presence or absence of a signal.

digital display

diode: An electronic device that allows current to flow in only one direction. See *device*.

direct current (DC): Current that flows in one direction. See *current*.

direct current (DC) compound motor: A motor with a field connected in both series and shunt with the armature. See *field*, *series*, and *armature*.

direct current (DC) motor: A motor that uses direct current connected to a field and armature to produce rotation. See *direct current*, *field*, and *armature*.

direct current (DC) permanent-magnet motor: A motor that uses magnets for the field winding. See *field*.

direct current (DC) series motor: A motor with a field connected in series with the armature. See *field*, *series*, and *armature*.

direct current (DC) shunt motor: A motor with a field connected in shunt (parallel) with the armature. See *field*, *parallel*, and *armature*.

direct drive method: A conveyor drive method in which the product rides directly on the driven belt and is used for light loads. See *conveyor*.

direct light: Light that travels directly from a lamp to the surface being illuminated.

double-acting solenoid valve: A valve that uses one solenoid to develop a force in one direction and a second solenoid to develop a force in the opposite direction. See *valve*.

direct light

drawing: A representation of an object by means of lines.

dry bulb temperature: The temperature of the air taken with a dry temperature probe.

dual-voltage motor: A motor that operates at more than one voltage level. See *voltage*.

E

effective light beam: In a proximity sensor, the area of light that travels directly from the transmitter to the receiver. See *proximity sensor*.

effective value: The value of a sine wave that produces the same amount of heat in a pure resistive circuit as a DC of the same value. See *root-mean-square value* and *sine wave*.

electric discharge lamp: A lamp that produces light by an arc discharged between two electrodes.

electrical balance: Transformer balance that occurs when loads on a transformer are placed so that each coil of the transformer carries the same amount of current. See *load, transformer, coil,* and *current*.

electrical noise: Unwanted signals that are present on a power line.

electrolyte: A conducting medium in which the current flow occurs by ion migration.

electromagnetic overload relay: An electromechanical relay operated by the current flow in a circuit. See *electromechanical relay* and *current*.

electromechanical: Electrically operated producing mechanical motion.

electromechanical relay: A device that allows the connection of two different components, voltage levels, voltage types (AC/DC), or systems. See *relay*.

electromechanical relay

electron: An atomic particle containing a negative electrical charge.

enclosed-coil heating element: A heating element that has the resistance wire which produces the heat surrounded by a refractory insulating material. See *heating element*.

enclosure: A housing used to protect wires and equipment and prevent personnel injury by accidental contact with a live circuit.

erratic relay operation: The proper operation of a relay at times, and the improper operation of the same relay at other times. See *relay*.

exact relay replacement method: A method of troubleshooting that replaces a bad relay with a relay of the same type and size. See *relay*.

F

field: The stationary windings, or magnets of a DC motor. See *DC motor*.

filament: A conductor that has a resistance high enough to cause the conductor to heat. See *conductor* and *resistance*.

floating input: A digital input that is not high or low at all times.

flow chart: A diagram that shows a logical sequence of steps for a given set of conditions. See *diagram*.

flow switch: A switch designed to detect the movement (flow) of a product through a pipe or duct.

SYMBOL	ITEM
○╱○ ▽	FLOW SWITCH

flow switch

fluorescent lamp: An electric discharge lamp in which ionization of mercury-vapor transforms ultraviolet energy generated by the discharge into light.

footcandle (fc): The amount of light produced by a lamp (lumen) divided by the area that is illuminated. See *illumination*.

forward bias: The condition of a diode while it conducts current. See *diode* and *current*.

frequency (f): The number of cycles per second (cps) in an AC sine wave. See *AC sine wave*.

function: A mathematical, graphical, or tabular statement of the influence one device or component has on another device or component.

fuse: An overcurrent protection device with a fusible link that melts and opens the circuit on an overcurrent condition. See *overcurrent*.

G

gate turn ON: A method of thyristor turn ON that occurs when the proper signal is applied to the gate at the correct time. See *thyristor*.

ghost voltage: A voltage that appears on a meter that is not connected to a circuit. See *voltage*.

graph: A pictorial representation of data.

grounded circuit: A circuit in which current leaves its normal path and travels to earth ground. See *current*.

grounding: The connection of all exposed non-current-carrying metal parts to the earth.

H

half-waving: The failure of a relay to turn OFF when the current in a circuit reaches zero but the voltage in a circuit does not reach zero at the same time. See *relay, phase shift,* and *inductive circuit*.

hardware: The physical components in a system. See *component* and *system*.

hard-wired unit: Equipment or unit that has conductors connected to terminal screws. See *conductor*.

harmonic analyzer: A meter that detects and measures harmonic distortion. See *harmonic distortion*.

harmonic distortion: The voltage and/or current in a power line that is a multiple of the fundamental line frequency. See *frequency*.

head end: The end of a conveyor from which material is discharged. See *conveyor*.

heating element: A heat-producing device that produces heat from the resistance to the flow of electric current. See *device* and *resistance*.

heat sink: A device that conducts and dissipates heat away from a component. See *component*.

high-intensity discharge (HID) lamp: A lamp that produces light from an arc tube. See *arc tube*.

high-pressure sodium lamp: An HID lamp that produces light when current flows through vaporized sodium under pressure and temperature. See *high-intensity discharge (HID) lamp*.

heat sink

high-reactance autotransformer ballast: A device that uses two coils (primary and secondary) to regulate both voltage and current. See *device, coil, voltage,* and *current*.

holding current: The minimum current that ensures proper operation of a solid-state switch. See *current*.

I

illumination: The effect of light falling on a surface.

improper phase sequence: The changing of the sequence of any two phases (phase reversal) in a 3φ motor control circuit. See *phase*.

incandescent lamp: An electric lamp that produces light by the flow of current through a tungsten filament inside a gas-filled, sealed glass bulb. See *filament,* and *bulb*.

indirect light: Light that is reflected from one or more objects to the surface being illuminated. See *illumination*.

inductance: The property of a circuit that causes it to oppose a

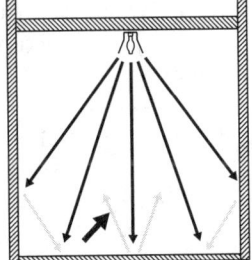

indirect light

change in current due to energy stored in a magnetic field. See *current*.

inductive circuit: A circuit in which current lags voltage. See *current* and *voltage*.

inductive reactance: The opposition to the flow of alternating current in a circuit due to inductance. See *alternating current* and *inductance*.

infrared meter: A meter that measures heat energy by measuring the infrared energy that a material emits.

initial lumen: The amount of light produced when a lamp is new.

in-line ammeter: A meter that measures current in a circuit by inserting the meter in series with the component under test. See *series* and *component*.

in-phase: The state when voltage and current reach their maximum amplitude and zero level simultaneously. See *amplitude*.

input module: The part of a programmable controller that receives information from pushbuttons,

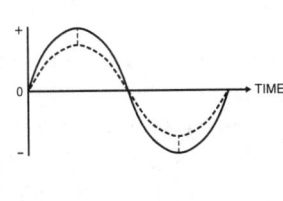

in-phase

temperature switches, pressure switches, photo-electric and proximity switches, and other input devices. See *programmable controller*.

instant-start circuit: A fluorescent lamp-starting circuit that provides sufficient voltage to strike an arc instantly. See *fluorescent lamp*.

insulation spot test: A test that checks motor insulation over the life of the motor.

insulation step voltage test: A test that creates electrical stress on internal insulation cracks to reveal aging or damage not found during other insulation tests.

insulator: Material that current cannot flow through easily. See *current*.

intensity: The level of brightness.

interconnection diagram: A diagram that shows the external connections between unit assemblies or equipment. See *diagram*.

interface: An electronic device that allows different levels or types of components to be used together in the same circuit. See *component*.

interference: In proximity sensors, any object other than the object to be detected that is sensed by the sensor. See *proximity sensor*.

interlock: A mechanical or electrical method used to prevent two different circuits from being energized at the same time.

interlock

isolated grounded receptacle: A receptacle that minimizes electrical noise by providing a separate grounding path. See *electrical noise* and *grounding*.

isometric drawing: A pictorial drawing with a 120° included axis. See *pictorial drawing*.

J

jumper link: A dual-voltage unit used to set the power rating of a programmable controller. See *programmable controller*.

K

L

leakage current: Current that flows through insulation. See *current*.

let-through voltage: The maximum voltage surge allowed to pass through a surge suppressor. See *voltage surge* and *surge suppressor*.

level switch: A proximity sensor used to detect the presence of a liquid or solid in a tank. See *proximity sensor*.

SYMBOL	ITEM
⌒	LEVEL SWITCH

level switch

light meter: A portable instrument that measures the amount of light.

linear scale: A scale divided into equally spaced segments.

linear speed: Distance traveled per unit of time in a straight line.

line (one line) diagram: A diagram that shows with single lines and graphic symbols the logic of an electrical circuit or system of circuits and components. See *diagram* and *component*.

line graph: A graph that shows two related variables plotted on coordinate paper.

load: A device that converts electrical energy to mechanical energy, heat, light, or sound. See *device*.

locked rotor torque: The torque that a motor produces when the rotor is stationary and full power is applied to the motor. See *rotor*.

lockout: The process of removing the source of power and installing a lock which prevents the power from being turned ON.

lockout devices: Lightweight enclosures that allow the lockout of standard control devices.

low-pressure sodium lamp: An HID lamp that operates at a low vapor pressure and uses sodium as the vapor. See *high-intensity discharge (HID) lamp*.

lumen (lm): The unit used to measure the total amount of light produced by a light source.

lumen per watt (lm/W): The number of lumen produced per watt of electrical power.

M

malfunction: The failure of a system, equipment, or part to operate properly.

mean lumen: The average light produced after a lamp has operated about 40% of its rated life.

mechanical life: The number of times a relay's mechanical parts operate before malfunctioning.

mechanical switch: Any switch that uses silver contacts to start and stop the flow of current in a circuit. See *contact* and *current*.

megohmmeter: A device that detects insulation deterioration by measuring high resistance values under high test voltage conditions.

megohmmeter

mercury-vapor lamp: An HID lamp that produces light by an electric discharge through mercury-vapor. See *high-intensity discharge (HID) lamp*.

metal-halide: An element (normally sodium and scandium iodide) which is added to mercury in small amounts.

metal-halide lamp: An HID lamp that produces light by an electric discharge through mercury-vapor and metal-halide in the arc tube. See *high-intensity discharge (HID) lamp*, *metal-halide*, and *arc tube*.

meter loading effect: The additional resistance a meter adds to the total resistance in a circuit. See *resistance*.

minimum holding current: The minimum current that ensures proper operation of a solid-state switch.

module: A group of electronic or electrical components housed in an enclosure that performs a set task. See *component* and *enclosure*.

motor damage: Any damage that occurs to a properly manufactured motor.

motor defect: An imperfection created during the manufacture of a motor.

motor starter: A contactor with overload pro-

motor starter

tection added. See *contactor* and *overload*.

multimeter: A meter that is capable of measuring two or more electrical quantities.

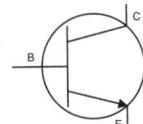
multimeter

N

neon light: A light that is filled with neon gas and uses two electrodes to ignite the gas.

neutron: An atomic particle containing a neutral electrical charge.

nonlinear scale: A scale that is divided into unequally spaced segments.

normally-closed (NC) valve: A valve that does not allow pressurized fluid to flow out of the valve in the spring-actuated position. See *valve* and *position*.

normally-open (NO) valve: A valve that allows pressurized fluid to flow out of the valve in the spring-actuated position. See *valve* and *position*.

NPN transistor: A transistor that has a thin layer of P-type material placed between two pieces of N-type material. See *transistor*.

NPN transistor

O

off-site troubleshooting: Troubleshooting at a location other than where hardware is installed. See *troubleshooting* and *hardware*.

off-state leakage current: The amount of current that leaks through a solid-state switch when the switch is turned OFF. See *current*.

ohmmeter: An instrument that measures resistance. See *resistance*.

Ohm's law: The relationship between voltage, current, and resistance in an electrical circuit. See *voltage*, *current*, and *resistance*.

one-level registration: Determining the product level between the ON and OFF points of an operation using one sensor.

one-time fuse: A fuse that cannot be reused after it has opened. See *fuse*.

on-site troubleshooting: Troubleshooting at the location where hardware is installed. See *troubleshooting* and *hardware*.

open circuit: A circuit that does not provide a path for current to flow. See *current*.

operational diagram: A diagram that shows the operation of individual components or circuits. See *diagram* and *component*.

oscilloscope: An instrument that displays an instantaneous voltage. See *voltage*.

oscilloscope

output module: The part of a programmable controller that delivers the output voltage required to control alarms, lights, solenoids, motor starters, and other output devices. See *programmable controller*.

overcurrent: Any current above the normal current level. See *current*.

overcycling: The process of turning a motor ON and OFF repeatedly.

overload: The application of excessive load to a motor. See *load*.

overload relay: A time-delay device that allows temporary overloads without disconnecting the load. See *device*, *overload*, and *load*.

P

parallel: Two or more components connected creating more than one path for current flow. See *component* and *current*.

parallel circuit: A circuit that contains two or more loads and has more than one path through which current flows.

parallel misalignment: Misalignment that occurs when two shafts are parallel but not on the same line.

peak value: The maximum value of either the positive or negative alternation of a sine wave. See *alternation* and *sine wave*.

peak-to-peak value: The value measured from the maximum positive alternation to the maximum negative alternation of a sine wave. See *alternation* and *sine wave*.

period: The time required to produce one complete cycle of a waveform. See *cycle* and *waveform*.

phase: The fractional part of a period through which the time variable of a periodic quantity has moved. See *period*.

phase shift: The state when voltage and current in a circuit do not reach their maximum amplitude and zero level simultaneously. See *voltage*, *current*, and *amplitude*.

phase unbalance: The unbalance that occurs when power lines are out of phase. See *phase*.

photo tachometer: A device that measures the speed of an object without direct contact with the object.

photocell: A voltage source that converts light energy to electrical energy.

pie chart: A chart that compares parts of a whole in relation to their total use.

pictorial drawing: A drawing that shows the length, height, and depth of an object in one view.

PNP transistor: A transistor that has a thin layer of N-type material placed between two pieces of P-type material. See *transistor*.

PNP transistor

polarity: The positive (+) or negative (−) state of an object.

poles: The number of completely isolated circuits that can pass through a switch at one time.

position: In valves, the number of positions within the valve that the spool is placed in to direct fluid-flow through the valve. See *valve, position,* and *valve spool*.

power: The rate of doing work or using energy.

power circuit: The part of an electrical circuit that connects the loads to the main power lines. See *load*.

power factor (PF): The ratio of true power used in an AC circuit to apparent power delivered to the circuit. See *true power* and *apparent power*.

power supply: The device that provides all the voltages required for the internal operation of a device or system. See *device* and *system*.

preheat circuit: A fluorescent lamp-starting circuit that heats the cathode before an arc is created.

pressure switch: A switch that is designed to activate its contacts at a preset pressure.

SYMBOL	ITEM
○╱○	PRESSURE SWITCH

pressure switch

preventive maintenance: Maintenance performed to keep machines, assembly lines, production operations, and plant operations running with little or no downtime.

primary division: On meter displays, a division with a listed value.

process: A sequence of operations that accomplishes desired results.

primary division

processor (CPU): Part of the programmable controller that performs logic functions, program storage, and supervisory functions. See *programmable controller*.

programmable controller: A solid-state control device that is programmed and reprogrammed to control machines and industrial processes automatically.

programming terminal: A keyboard-operated device that is used to insert the control program into memory, make changes to the stored program, and display the program circuit and data. See *device*.

proton: An atomic particle containing a positive electrical charge.

proximity: The quality or state of being close.

proximity sensor: A device that reacts to the proximity of an actuating means without physical contact.

psychrometric chart: A graph that is used to find various properties of air. See *graph*.

pulley: 1. A shaft-mounted wheel used to transmit power by means of a belt, chain, etc. **2.** In conveyor systems, a revolving cylinder connected to a drive. See *conveyor*.

pull-up resistor: A resistor that has one side connected to the power supply at all times and the other side connected to one side of an input switch.

pure DC power: Power obtained from a battery or DC generator. See *battery*.

Q

R

rapid-start circuit: A fluorescent lamp-starting circuit that brings the lamp to full brightness in about two seconds. See *fluorescent lamp*.

RC circuit: A circuit in which resistance and capacitance are used to help filter the power. See *resistance* and *capacitance*.

reactor ballast: A ballast that connects a coil (reactor) in series with the power line leading to the lamp. See *ballast, coil,* and *series*.

recording meter: A meter that provides a written record of readings over a time period.

relay: An interface that controls one electrical circuit by opening and closing contacts in another circuit. See *interface* and *contact*.

recording meter

renewable fuse: A fuse that is designed so that the fusible link can be replaced.

resistance: Opposition to the flow of electrons.

resistive circuit: A circuit that contains only resistive components, such as heating elements and incandescent lamps. See *component, heating element,* and *incandescent lamp.*

reverse bias: The condition of a diode when it acts as an insulator. See *diode* and *insulator.*

roller: A revolving cylinder not connected to a drive.

roller drive method: A conveyor drive method in which the product rides on the belt-driving rollers. See *conveyor.*

root-mean-square value (rms): The value that produces the same amount of heat in a pure resistive circuit as a DC of the same value. See *effective value.*

rotational speed: Distance traveled per unit of time in a circular direction.

rotor: The rotating part of an AC motor. See *alternating current (AC) motor.*

rough service lamp: A lamp made of a specially processed tungsten filament that has extra supports to resist breakage. See *tungsten* and *filament.*

S

schematic (elementary) diagram: A diagram that shows with graphic symbols the electrical connections and functions of a specific circuit arrangement. See *diagram.*

scope: A device that gives a visual display of voltages. See *device.*

scopemeter: A combination oscilloscope and digital multimeter. See *oscilloscope.*

SCR: A thyristor that is triggered into conduction in only one direction. See *thyristor.*

secondary division: On meter displays, a division that divides primary divisions in halves, thirds, fourths, fifths, etc.

sectional view drawing: A drawing that shows the internal features of an object. See *drawing.*

secondary division

series: Two or more components connected creating one path for current flow. See *component.*

series circuit: A circuit that contains two or more loads and one path through which current flows. See *load.*

series/parallel: A combination of series- and parallel-connected components. See *series, parallel,* and *component.*

shaded-pole motor: An AC motor that uses a shaded stator pole for starting. See *alternating current (AC) motor* and *stator.*

short circuit: A circuit in which current takes a shortcut around the normal path of current flow.

sine wave: A wave that can be expressed as the sine of a linear function. See *function.*

single phasing: The operation of a motor designed to operate on three phases operating on two phases because one phase is lost. See *phase.*

single-voltage motor: A motor that operates at only one voltage level. See *voltage.*

slip: The difference between the synchronous speed and actual speed of a motor. See *synchronous speed.*

snub roller: In conveyor systems, a roller not connected to a drive that is used to correct belt alignment. See *conveyor.*

software: The computer programs and procedures that operate hardware. See *hardware.*

solenoid: An electromechanical device that converts electrical energy to mechanical energy. See *device.*

solenoid

solenoid-operated valve: A valve that uses electrical force to directly or indirectly change the position of a valve spool. See *valve, position,* and *valve spool.*

solid-state relay: An electronic switching device that has no moving parts. See *device.*

split-phase motor: An AC motor that can run on one or more phases. See *alternating current (AC) motor* and *phase.*

standard: An accepted reference or practice.

standard ambient temperature: In transformers, the average temperature of the air that cools a transformer over a 24-hour period. See *transformer.*

startup procedure: A logical, systematic procedure that isolates and eliminates potential problems that may cause personal injury and equipment damage.

static turn ON: A method of thyristor turn ON that occurs when a fast rising voltage is applied across the anode and cathode terminals of an SCR or the main terminals of a triac. See *thyristor, anode, cathode, SCR,* and *triac.*

stator: The stationary part of an AC motor to which power lines are connected. See *alternating current (AC) motor.*

status indicator: A light that shows the condition of the components in a programmable controller system. See *programmable controller.*

status indicator

strobe tachometer: A device that uses a flashing light to measure the speed of a moving object. See *device.*

subdivision: On meter displays, a division that divides secondary divisions in halves, thirds, fourths, fifths, etc. See *secondary division.*

subdivision

substitution: The replacement of malfunctioning equipment or units. See *unit.*

surface leakage current: Current that flows from areas on conductors where insulation is removed to allow electrical connections. See *current* and *conductor.*

sulfidation: The formation of film on a contact surface. See *contact.*

surge suppressor: A protection device that limits voltage surges that may be present on power lines. See *device* and *voltage surge.*

sweep: On oscilloscopes, the movement of the displayed trace across a scope screen. See *oscilloscope, trace,* and *scope.*

symbol: A graphic element that represents a quantity or unit.

synchronous speed: The theoretical speed of a motor based on the motor's number of poles and the line frequency. See *frequency.*

system: A combination of components, units, or modules that are connected to perform work or meet a specific need. See *component, unit,* and *module.*

system analysis: The breakdown of system requirements and components performed when designing, implementing, maintaining, or troubleshooting a system. See *system, component,* and *troubleshooting.*

systems analyst: An individual who troubleshoots at the system level. See *troubleshooting* and *system.*

T

tabular chart: A chart that presents data in columnar form. See *chart.*

tachometer: A device that measures the speed of a moving object. See *device.*

tagout: The process of placing a danger tag on the source of power which indicates that the equipment may not be operated until the danger tag is removed.

tail end: In conveyor systems, the end of a conveyor where material is fed. See *conveyor.*

take-up pulley: In conveyor systems, a pulley that is used for correcting belt tension and is not connected to a drive.

taps: Connecting points that are provided along a transformer coil. See *transformer* and *coil.*

temperature: A measurement of the intensity of heat.

temperature rise: An increase in temperature above ambient temperature. See *ambient temperature.*

terminal: A point of connection in an electrical circuit.

terminal diagram: A diagram that relates the internal circuit of a piece of equipment to its terminal physical configuration. See *diagram.*

test light: A light that is connected to two test leads and gives a visual indication when voltage is present in a circuit. See *voltage.*

thermal overload relay: An electromechanical relay that operates by heat developed in the relay. See *relay.*

test light

thermal resistance (R$_{TH}$): The ability of a device to impede the flow of heat. See *device.*

thermal switch: A switch that operates its contacts when a preset temperature is reached.

thermal turn ON: A method of thyristor turn ON that occurs when the heat level exceeds the limit of a thyristor (normally 110°C). See *thyristor.*

three-wire solid-state switch: A switch that has three connecting terminals or wires (exclusive of ground). See *terminal.*

throws: The number of different closed contact positions per pole that are available on a switch. See *contact* and *pole.*

thyristor: A solid-state switching device that switches current ON by a quick pulse of control current. See *device.*

time/division (time per division) control: An oscilloscope or scopemeter control that selects the width of the displayed waveform. See *oscilloscope, scopemeter,* and *waveform.*

thyristor

trace: A reference point/line that is visually displayed on the face of a scope screen. See *scope.*

transformer: A device that raises or lowers the voltage of alternating current of the source. See *device, voltage,* and *alternating current.*

transient voltages: Temporary, unwanted voltages in an electrical circuit.

transistor: A three-element device made of semiconductor material. See *device.*

transistor-transistor logic ICs (TTL ICs): A broad family of ICs that employ a two-transistor arrangement. See *transistor.*

transverse noise: Noise produced between the hot and neutral lines.

triac: A thyristor that is triggered into conduction in either direction. See *thyristor*.

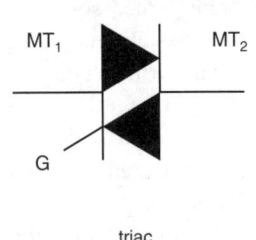

triac

troubleshooting: The systematic elimination of the various parts of a system or process to locate a malfunctioning part. See *system*, *malfunction*, and *process*.

troubleshooting procedure: A logical step-by-step process used to find a malfunction in a system or process as quickly and easily as possible. See *malfunction*, *system*, and *process*.

true power: Actual power used in an electrical circuit.

tungsten: A heavy high-melting ductile metal that is used for electrical purposes.

tungsten-halogen lamp: An incandescent lamp that is filled with a halogen gas (iodine or bromine). See *incandescent lamp*.

two-level registration: Determining the product level between the ON and OFF points of an operation using two sensors.

two-winding constant-wattage ballast: A ballast that uses a transformer that provides isolation between the primary and secondary circuits. See *ballast*.

two-wire solid-state switch: A switch that has two connecting terminals or wires (exclusive of ground). See *terminal*.

U

unit: An individual component that performs a specific task by itself. See *component*.

useful life: Period of time after the break-in period when most equipment operates properly. See *break-in period*.

V

valve: A device that controls fluid flow, direction, pressure, or flow rate. See *device*.

valve spool: The movable part of a valve that opens, shuts, or partially obstructs one or more passageways. See *valve*.

varistor: A resistor whose resistance is inversely proportional to the voltage applied to it. See *resistance* and *voltage*.

vertical chart recorder: A graphic pen and ink recorder that has measured values drawn on a continuous roll of graphic paper.

voltage: The amount of electrical pressure in a circuit.

voltage break-over turn ON: A method of thyristor turn ON that occurs when the voltage across the thyristor terminals exceeds the maximum voltage rating of the device. See *thyristor*, *voltage*, and *device*.

voltage dip: A momentarily low voltage.

voltage fluctuation: Any change in the voltage level above or below the specified normal range of the power supply.

voltage spike: An increase in voltage (typically several thousand volts) that lasts for a very short time (microseconds to milliseconds). See *voltage*.

voltage stabilizer (regulator): A device that provides precise voltage regulation to protect equipment from voltage dips and voltage surges. See *device*, *voltage dip*, and *voltage surge*.

voltage spike

voltage surge: A higher-than-normal voltage that temporarily exists on one or more power lines.

voltage tester: A device that indicates approximate voltage level and type (AC or DC) by the movement and vibration of a pointer on a scale. See *device*.

voltmeter: A meter that measures voltage.

voltage unbalance: The unbalance that occurs when the voltages at different motor terminals are not equal. See *voltage* and *terminal*.

voltage tester

volt-ohm-milliammeter (VOM): A meter that measures voltage, resistance, and current. See *voltage*, *resistance*, and *current*.

volts/division (volts per division) control: An oscilloscope and scopemeter control that selects the height of the displayed waveform. See *oscilloscope*, *scopemeter*, and *waveform*.

voltage variance: The difference in voltage between a voltage surge and a voltage dip. See *voltage*, *voltage surge*, and *voltage dip*.

W

wall-bulb temperature: The operating temperature of a fluorescent lamp's wall, that affects the mercury-vapor pressure inside the bulb. See *fluorescent lamp* and *bulb*.

watt density: The watts per square inch (W/sq in.) on the outer cover of a heating element.

wattmeter: An instrument that measures true power in a circuit. See *true power*.

waveform: The geometrical shape obtained by displaying a wave as a function of time when plotted over one period. See *period*.

way: The flow path through a valve. See *valve*.

wear-out period: Period of time after a piece of equipment's useful life terminates when normal equipment failure occurs. See *useful life*.

wet bulb temperature: The temperature of the air taken with a wet sock covering the temperature probe.

wiring (connection) diagram: A diagram that shows the connection of an installation or its component devices or parts. See *diagram*, *component*, and *device*.

wrap-around bar graph: On a digital meter, a bar graph that displays a fraction of the full range on the bar graph. See *bar graph*.

wye configuration: A transformer connection that has one end of each transformer coil connected together. See *transformer* and *coil*.

wye-connected 3φ motor: A motor that has one end of each of the three phases internally connected to the other phases. See *phase*.

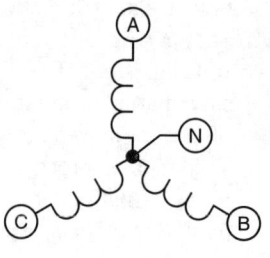

wye configuration

X

Y

Z

APPENDIX

page

Charts and Tables

Electrical Symbols	454
Fluid Power Symbols	458
Electrical Abbreviations	460
Three-Phase Voltage Values	461
Capacitor Ratings	461
Resistor Color Codes	461
Lamp Ratings	462
Lamp Identification	462
Ballast Loss	462
Recommended Light Levels	462
Recommended Ballast Output Voltage Limits	463
Recommended Short-Circuit Current Test Limits	463
Voltage Conversions	463
Motor Repair and Service Record	464
Semiannual Motor Maintenance Checklist	464
Annual Motor Maintenance Checklist	465
Heater Ambient Temperature Correction	465
Enclosure Types	466
Full-Load Currents — DC Motors	467
Full-Load Currents — 1ϕ, AC Motors	467
Full-Load Currents — 3ϕ, AC Motors	467
Typical Motor Efficiencies	467
Motor Frame Table	468
Motor Frame Letters	468
Motor Frame Dimensions	469

Formulas

Ohm's Law and Power Formula	470
Power Formulas — 1ϕ, 3ϕ	470
AC/DC Formulas	471
Horsepower Formulas	471
Voltage Drop Formulas — 1ϕ, 3ϕ	471
Capacitors	472
Sine Waves	472
Conductive Leakage Current	472
Temperature Conversions	472
Branch Circuit Voltage Drop	472
Flow Rate	472

CHARTS AND TABLES

ELECTRICAL SYMBOLS

CONTACTS

INSTANT OPERATING				TIMED CONTACTS - CONTACT ACTION RETARDED AFTER COIL IS:			
WITH BLOWOUT		WITHOUT BLOWOUT		ENERGIZED		DE-ENERGIZED	
NO	NC	NO	NC	NOTC	NCTO	NOTO	NCTC

OVERLOAD RELAYS

THERMAL	MAGNETIC

SUPPLEMENTARY CONTACT SYMBOLS

SPST NO		SPST NC		SPDT		TERMS
SINGLE BREAK	DOUBLE BREAK	SINGLE BREAK	DOUBLE BREAK	SINGLE BREAK	DOUBLE BREAK	SPST SINGLE-POLE, SINGLE-THROW

DPST, 2NO		DPST, 2NC		DPDT		
SINGLE BREAK	DOUBLE BREAK	SINGLE BREAK	DOUBLE BREAK	SINGLE BREAK	DOUBLE BREAK	

Terms:
- SPST — SINGLE-POLE, SINGLE-THROW
- SPDT — SINGLE-POLE, DOUBLE-THROW
- DPST — DOUBLE-POLE, SINGLE-THROW
- DPDT — DOUBLE-POLE, DOUBLE-THROW
- NO — NORMALLY OPEN
- NC — NORMALLY CLOSED

METER (INSTRUMENT)

INDICATE TYPE BY LETTER

TO INDICATE FUNCTION OF METER OR INSTRUMENT, PLACE SPECIFIED LETTER OR LETTERS WITHIN SYMBOL.

AM or A	AMMETER	VA	VOLTMETER
AH	AMPERE HOUR	VAR	VARMETER
μA	MICROAMMETER	VARH	VARHOUR METER
mA	MILLAMMETER	W	WATTMETER
PF	POWER FACTOR	WH	WATTHOUR METER
V	VOLTMETER		

PILOT LIGHTS

INDICATE COLOR BY LETTER

NON PUSH-TO-TEST	PUSH-TO-TEST

INDUCTORS

- IRON CORE
- AIR CORE

COILS

DUAL-VOLTAGE MAGNET COILS

HIGH-VOLTAGE	LOW-VOLTAGE
LINK	LINKS
1 2 3 4	1 2 3 4

BLOWOUT COIL

Appendix 457

ELECTRICAL ABBREVIATIONS

Abbreviation	Term	Abbreviation	Term
AC	Alternating current	MB	Magnetic brake
ALM	Alarm	MCS	Motor circuit switch
AM	Ammeter	MEM	Memory
ARM	Armature	MTR	Motor
AU	Automatic	MN	Manual
BAT	Battery (electric)	NEG	Negative
BR	Brake relay	NEUT	Neutral
CAP	Capacitor	NC	Normally closed
CB	Circuit breaker	NO	Normally open
CEMF	Counter electromotive force	OHM	Ohmmeter
CKT	Circuit	OL	Overload relay
CONT	Control	PB	Pushbutton
CR	Control relay	PH	Phase
CRM	Control relay master	PLS	Plugging switch
CT	Current transformer	POS	Positive
D	Down	PRI	Primary switch
DB	Dynamic braking contactor or relay	PS	Pressure switch
DC	Direct current	R	Reverse
DIO	Diode	REC	Rectifier
DISC	Disconnect switch	RES	Resistor
DP	Double-pole	RH	Rheostat
DPDT	Double-pole, double-throw	S	Switch
DPST	Double-pole, single-throw	SCR	Semiconductor-control rectifier
DS	Drum switch	SEC	Secondary
DT	Double-throw	1PH	Single-phase
EMF	Electromotive force	SOC	Socket
F	Forward	SOL	Solenoid
FLS	Flow switch	SP	Single-pole
FREQ	Frequency	SPDT	Single-pole, double-throw
FS	Float switch	SPST	Single-pole, single-throw
FTS	Foot switch	SS	Selector switch
FU	Fuse	SSW	Safety switch
GEN	Generator	T	Transformer
GRD	Ground	TB	Terminal board
IC	Integrated circuit	3PH	Three-phase
INTLK	Interlock	TD	Time delay
IOL	Instantaneous overload	THS	Thermostat switch
JB	Junction box	TR	Time delay relay
LS	Limit switch	U	Up
LT	Lamp	UCL	Unclamp

THREE-PHASE VOLTAGE VALUES

For 208 V × 1.732, use 360
For 230 V × 1.732, use 398
For 240 V × 1.732, use 416
For 440 V × 1.732, use 762
For 460 V × 1.732, use 797
For 480 V × 1.732, use 831

CAPACITOR RATINGS
110–125 VAC, 50/60 Hz, Starting Capacitors

Typical Ratings*	Dimensions** Diameter	Dimensions** Length	Model Number***
88–106	$1^{7}/_{16}$	$2^{3}/_{4}$	EC8815
108–130	$1^{7}/_{16}$	$2^{3}/_{4}$	EC10815
130–156	$1^{7}/_{16}$	$2^{3}/_{4}$	EC13015
145–174	$1^{7}/_{16}$	$2^{3}/_{4}$	EC14515
161–193	$1^{7}/_{16}$	$2^{3}/_{4}$	EC16115
189–227	$1^{7}/_{16}$	$2^{3}/_{4}$	EC18915A
216–259	$1^{7}/_{16}$	$3^{3}/_{8}$	EC21615
233–280	$1^{7}/_{16}$	$3^{3}/_{8}$	EC23315A
243–292	$1^{7}/_{16}$	$3^{3}/_{8}$	EC24315A
270–324	$1^{7}/_{16}$	$3^{3}/_{8}$	EC27015A
324–389	$1^{7}/_{16}$	$3^{3}/_{8}$	EC2R10324N
340–408	$1^{13}/_{16}$	$3^{3}/_{8}$	EC34015
378–454	$1^{13}/_{16}$	$3^{3}/_{8}$	EC37815
400–480	$1^{13}/_{16}$	$3^{3}/_{8}$	EC40015
430–516	$1^{13}/_{16}$	$3^{3}/_{8}$	EC43015A
460–553	$1^{13}/_{16}$	$4^{3}/_{8}$	EC5R10460N
540–648	$1^{13}/_{16}$	$4^{3}/_{8}$	EC54015B
590–708	$1^{13}/_{16}$	$4^{3}/_{8}$	EC59015A
708–850	$1^{13}/_{16}$	$4^{3}/_{8}$	EC70815
815–978	$1^{13}/_{16}$	$4^{3}/_{8}$	EC81515
1000–1200	$2^{1}/_{16}$	$4^{3}/_{8}$	EC100015A

220–250 VAC, 50/60 Hz, Starting Capacitors

Typical Ratings*	Diameter	Length	Model Number***
53–64	$1^{7}/_{16}$	$3^{3}/_{8}$	EC5335
64–77	$1^{7}/_{16}$	$3^{3}/_{8}$	EC6435
88–106	$1^{13}/_{16}$	$3^{3}/_{8}$	EC8835
108–130	$1^{13}/_{16}$	$3^{3}/_{8}$	EC10835A
124–149	$1^{13}/_{16}$	$4^{3}/_{8}$	EC12435
130–154	$1^{13}/_{16}$	$4^{3}/_{8}$	EC13035
145–174	$2^{1}/_{16}$	$3^{3}/_{8}$	EC6R22145N
161–193	$2^{1}/_{16}$	$3^{3}/_{8}$	EC6R2216N
216–259	$2^{1}/_{16}$	$4^{3}/_{8}$	EC21635A
233–280	$2^{1}/_{16}$	$4^{3}/_{8}$	EC23335A
270–324	$2^{1}/_{16}$	$4^{3}/_{8}$	EC27035A

* in μF
** in inches
*** Model numbers vary by manufacturer.

CAPACITOR RATINGS
270 VAC, 50/60 Hz, Running Capacitors

Typical Ratings*	Dimensions** Oval	Dimensions** Length	Model Number***
2		$2^{1}/_{8}$	VH5502
3		$2^{1}/_{8}$	VH5503
4	$1^{5}/_{16} \times 2^{5}/_{32}$	$2^{1}/_{8}$	VH5704
5		$2^{1}/_{8}$	VH5705
6		$2^{5}/_{8}$	VH5706
7.5		$2^{7}/_{8}$	VH9001
10	$1^{5}/_{16} \times 2^{5}/_{32}$	$2^{7}/_{8}$	VH9002
12.5		$3^{7}/_{8}$	VH9003
15	$1^{29}/_{32} \times 2^{29}/_{32}$	$2^{1}/_{8}$	VH9121
17.5		$2^{7}/_{8}$	VH9123
20		$2^{7}/_{8}$	VH5463
25	$1^{29}/_{32} \times 2^{29}/_{32}$	$3^{7}/_{8}$	VH9069
30		$3^{7}/_{8}$	VH5465
35	$1^{29}/_{32} \times 2^{29}/_{32}$	$3^{7}/_{8}$	VH9071
40		$3^{7}/_{8}$	VH9073
45	$1^{31}/_{32} \times 3^{21}/_{32}$	$3^{7}/_{8}$	VH9115
50		$3^{7}/_{8}$	VH9075

440 VAC, 50/60 Hz, Running Capacitors

Typical Ratings*	Oval	Length	Model Number***
10	$1^{5}/_{16} \times 2^{5}/_{32}$	$3^{7}/_{8}$	VH5300
15	$1^{29}/_{32} \times 2^{29}/_{32}$	$2^{7}/_{8}$	VH5304
17.5	$1^{29}/_{32} \times 2^{29}/_{32}$	$3^{7}/_{8}$	VH9141
20	$1^{29}/_{32} \times 2^{29}/_{32}$	$3^{7}/_{8}$	VH9082
25	$1^{29}/_{32} \times 2^{29}/_{32}$	$3^{7}/_{8}$	VH5310
30		$4^{3}/_{4}$	VH9086
35	$1^{29}/_{32} \times 2^{29}/_{32}$	$4^{3}/_{4}$	VH9088
40		$4^{3}/_{4}$	VH9641
45		$3^{7}/_{8}$	VH5351
50	$1^{31}/_{32} \times 3^{21}/_{32}$	$3^{7}/_{8}$	VH5320
55		$4^{3}/_{4}$	VH9084

* in μF
** in inches
*** Model numbers vary by manufacturer.

RESISTOR COLOR CODES

Color	Number 1st	Number 2nd	Multiplier	Tolerance (%)
Black (BK)	0	0	1	0
Brown (BR)	1	1	10	—
Red (R)	2	2	100	—
Orange (O)	3	3	1000	—
Yellow (Y)	4	4	10,000	—
Green (G)	5	5	100,000	—
Blue (BL)	6	6	1,000,000	—
Violet (V)	7	7	10,000,000	—
Gray (GY)	8	8	100,000,000	—
White (W)	9	9	1,000,000,000	—
Gold (Au)	—	—	0.1	5
Silver (Ag)	—	—	0.01	10
None	—	—	0	20

LAMP RATINGS

Lamp	Initial Lumen	Mean Lumen
40 W standard incandescent	480	N/A
100 W standard incandescent	1750	N/A
40 W standard fluorescent	3400	3100
100 W tungsten-halogen	1800	1675
100 W mercury-vapor	4000	3000
250 W mercury-vapor	12,000	9800
100 W high-pressure sodium	9500	8500
250 W high-pressure sodium	30,000	27,000
250 W metal-halide	20,000	17,000

BALLAST LOSS

Lamp	Lamp Rated Wattage (in W)	Ballast Loss (power loss %)
Low-pressure sodium	70	27
	100	25
	150	22
	250	20
	400	15
	1000	7
Mercury-vapor	40	32
	75	25
	100	22
	175	17
	250	16
	400	14
	700	7
	1000	7
Metal-halide	175	20
	250	17
	400	13
	1000	7
	1500	7
High-pressure sodium	50	30
	100	24
	150	22
	250	20
	400	16
	1000	7

LAMP IDENTIFICATION

Lamp	Letter	Number	Rating
Mercury-vapor	H	33	400 W
		34	1000 W/130 V
		35	700 W
		36	1000 W/265 V
		37	250 W
		38	100 W
		39	175 W
		43	75 W
Self-ballasted mercury-vapor	B	78	750 W/120 V
Metal-halide	M	47	1000 W
		48	1500 W
		57	175 W
		58	250 W
		59	400 W
High-pressure sodium	S	50	250 W
		51	400 W
		52	1000 W
		54	100 W
		55	150 W/55 V
		56	150 W/100 V
		62	70 W
		66	200 W
		67	310 W
		68	50 W
		76	35 W

RECOMMENDED LIGHT LEVELS

Area	Footcandles
Hospital operating table	2500
Factory assembly area	100-200
Accounting office	150
Major league baseball infield	150
Major league baseball outfield	100
School classroom	60-90
Home kitchen	50-70
Bank lobby	50
Active warehouse storage area	20-30
Inactive warehouse storage area	5
Auditorium	15
Elevator	10
Parking lot	5
Interstate roadway	1.4
Street	.9

RECOMMENDED BALLAST OUTPUT VOLTAGE LIMITS

Ballast	Lamp Size		rms Voltage (Volts)
	Wattage	ANSI Number	
Low-pressure sodium	18	L69	300-325
	35	L70	455-505
	55	L71	455-505
	90	L72	455-525
	135	L73	645-715
	180	L74	645-715
Mercury-vapor	50	H46	225-255
	75	H43	225-255
	100	H38	225-255
	175	H39	225-255
	250	H37	225-255
	400	H33	225-255
	700	H35	405-455
	1000	H36	405-455
Metal-halide	70	M85	210-250
	100	M90	250-300
	150	M81	220-260
	175	M57	285-320
	250	M80	230-270
	250	M58	285-320
	400	M59	285-320
	1000	M47	400-445
	1500	M48	400-445
High-pressure sodium	35	S76	110-130
	50	S68	110-130
	70	S62	110-130
	100	S54	110-130
	150	S55	110-130
	150	S56	200-250
	200	S66	200-230
	250	S50	175-225
	310	S67	155-190
	400	S51	175-225
	1000	S52	420-480

RECOMMENDED SHORT-CIRCUIT CURRENT TEST LIMITS

Ballast	Lamp Size		Short-Circuit Current (Amps)
	Wattage	ANSI Number	
Low-pressure sodium	18	L69	.30-.40
	35	L70	.52-.78
	55	L71	.52-.78
	90	L72	.8-1.2
	135	L73	.8-1.2
	180	L74	.8-1.2
Mercury-vapor	50	H46	.85-1.15
	75	H43	.95-1.70
	100	H38	1.10-2.00
	175	H39	2.0-3.6
	250	H37	3.0-3.8
	400	H33	4.4-7.9
	700	H35	3.9-5.85
	1000	H36	5.7-9.0
Metal-halide	70	M85	.85-1.30
	100	M90	1.15-1.76
	150	M81	1.75-2.60
	175	M57	1.5-1.90
	250	M80	2.9-4.3
	250	M58	2.2-2.85
	400	M59	3.5-4.5
	1000	M47	4.8-6.15
	1500	M48	7.4-9.6
High-pressure sodium	35	S76	.85-1.45
	50	S68	1.5-2.3
	70	S62	1.6-2.9
	100	S54	2.45-3.8
	150	S55	3.5-5.4
	150	S56	2.0-3.0
	200	S66	2.50-3.7
	250	S50	3.0-5.3
	310	S67	3.8-5.7
	400	S51	5.0-7.6
	1000	S52	5.5-8.1

VOLTAGE CONVERSIONS

To Convert	To	Multiply By
rms	Average	.9
rms	Peak	1.414
Average	rms	1.111
Average	Peak	1.567
Peak	rms	.707
Peak	Average	.637
Peak	Peak-to-peak	2

MOTOR REPAIR AND SERVICE RECORD

Motor File #: _____ Serial #: _____

Date Installed: _____ Motor Location: _____

MFR: _____ Type: _____ Frame: _____
HP: _____ Volts: _____ Amps: _____
RPM: _____ Filter Sizes: _____

Date	Operation	Mechanic

SEMIANNUAL MOTOR MAINTENANCE CHECKLIST

Motor File #: _____ Serial #: _____

Date Installed: _____ Motor Location: _____

MFR: _____ Type: _____ Frame: _____
HP: _____ Volts: _____ Amps: _____
RPM: _____ Date Serviced: _____

Step	Operation	Mechanic
1	Turn OFF and lock out all power to the motor and its control circuit.	
2	Clean motor exterior and all ventilation ducts.	
3	Check motor's wire raceway.	
4	Check and lubricate bearings as needed.	
5	Check drive mechanism.	
6	Check brushes and commutator.	
7	Check slip rings.	
8	Check motor terminations.	
9	Check capacitors.	
10	Check all mounting bolts.	
11	Check and record line-to-line resistance.	
12	Check and record megohmmeter resistance from L1 to ground.	
13	Check motor controls.	
14	Reconnect motor and control circuit power supplies.	
15	Check line-to-line voltage for balance and level.	
16	Check line current draw against nameplate rating.	
17	Check and record inboard and outboard bearing temperatures.	

ANNUAL MOTOR MAINTENANCE CHECKLIST

Motor File #: _____   Serial #: _____

Date Installed: _____   Motor Location: _____

_____

MFR: _____   Type: _____   Frame: _____

HP: _____   Volts: _____   Amps: _____

RPM: _____   Date Serviced: _____

Step	Operation	Mechanic
1	Turn OFF and lock out all power to the motor and its control circuit.	
2	Clean motor exterior and all ventilation ducts.	
3	Uncouple motor from load and disassemble.	
4	Clean inside of motor.	
5	Check centrifugal switch assemblies.	
6	Check rotors, armatures, and field windings.	
7	Check all peripheral equipment.	
8	Check bearings.	
9	Check brushes and commutator.	
10	Check slip rings.	
11	Reassemble motor and couple to load.	
12	Flush old bearing lubricant and replace.	
13	Check motor's wire raceway.	
14	Check drive mechanism.	
15	Check motor terminations.	
16	Check capacitors.	
17	Check all mounting bolts.	
18	Check and record line-to-line resistance.	
19	Check and record megohmmeter resistance from T1 to ground.	
20	Check and record insulation polarization index.	
21	Check motor controls.	
22	Reconnect motor and control circuit power supplies.	
23	Check line-to-line voltage for balance and level.	
24	Check line current draw against nameplate rating.	
25	Check and record inboard and outboard bearing temperatures.	

HEATER AMBIENT TEMPERATURE CORRECTION

ENCLOSURE TYPES

Type	Use	Service conditions	Tests	Comments
1	Indoor	No unusual	Rod entry, rust resistance	
3	Outdoor	Windblown dust, rain, sleet, and ice on enclosure	Rain, external icing, dust, and rust resistance	Do not provide protection against internal condensation or internal icing
3R	Outdoor	Falling rain and ice on enclosure	Rod entry, rain, external icing, and rust resistance	Do not provide protection against dust, internal condensation, or internal icing
4	Indoor/outdoor	Windblown dust and rain, splashing water, hose-directed water, and ice on enclosure	Hosedown, external icing, and rust resistance	Do not provide protection against internal condensation or internal icing
4X	Indoor/outdoor	Corrosion, windblown dust and rain, splashing water, hose-directed water, and ice on enclosure	Hosedown, external icing, and corrosion resistance	Do not provide protection against internal condensation or internal icing
6	Indoor/outdoor	Occasional temporary submersion at a limited depth		
6P	Indoor/outdoor	Prolonged submersion at a limited depth		
7	Indoor locations classified as Class I, Groups A, B, C, or D, as defined in the NEC®	Withstand and contain an internal explosion of specified gases, contain an explosion sufficiently so an explosive gas-air mixture in the atmosphere is not ignited	Explosion, hydrostatic, and temperature	Enclosed heat-generating devices shall not cause external surfaces to reach temperatures capable of igniting explosive gas-air mixtures in the atmosphere
7	Indoor locations classified as Class I, Groups A, B, C, or D, as defined in the NEC®	Withstand and contain an internal explosion of specified gases, contain an explosion sufficiently so an explosive gas-air mixture in the atmosphere is not ignited	Explosion, hydrostatic, and temperature	Enclosed heat-generating devices shall not cause external surfaces to reach temperatures capable of igniting explosive gas-air mixtures in the atmosphere.
9	Indoor locations classified as Class II, Groups E or G, as defined in the NEC®	Dust	Dust penetration, temperature, and gasket aging	Enclosed heat-generating devices shall not cause external surfaces to reach temperatures capable of igniting explosive gas-air mixtures in the atmosphere
12	Indoor	Dust, falling dirt, and dripping noncorrosive liquids	Drip, dust, and rust resistance	Do not provide protection against internal condensation
13	Indoor	Dust, spraying water, oil, and noncorrosive coolant	Oil explosion and rust resistance	do not provide protection against internal condensation

FULL-LOAD CURRENTS — DC MOTORS

Motor rating (HP)	Current (A) 120 V	Current (A) 240 V
¼	3.1	1.6
⅓	4.1	2.0
½	5.4	2.7
¾	7.6	3.8
1	9.5	4.7
1½	13.2	6.6
2	17	8.5
3	25	12.2
5	40	20
7½	48	29
10	76	38

FULL-LOAD CURRENTS — 1ɸ, AC MOTORS

Motor rating (HP)	Current (A) 115 V	Current (A) 230 V
⅙	4.4	2.2
¼	5.8	2.9
⅓	7.2	3.6
½	9.8	4.9
¾	13.8	6.9
1	16	8
1½	20	10
2	24	12
3	34	17
5	56	28
7½	80	40

FULL-LOAD CURRENTS — 3ɸ, AC INDUCTION MOTORS

Motor rating (HP)	208 V	230 V	460 V	575 V
¼	1.11	.96	.48	.38
⅓	1.34	1.18	.59	.47
½	2.2	2.0	1.0	.8
¾	3.1	2.8	1.4	1.1
1	4.0	3.6	1.8	1.4
1½	5.7	5.2	2.6	2.1
2	7.5	6.8	3.4	2.7
3	10.6	9.6	4.8	3.9
5	16.7	15.2	7.6	6.1
7½	24.0	22.0	11.0	9.0
10	31.0	28.0	14.0	11.0
15	46.0	42.0	21.0	17.0
20	59	54	27	22
25	75	68	34	27
30	88	80	40	32
40	114	104	52	41
50	143	130	65	52
60	169	154	77	62
75	211	192	96	77
100	273	248	124	99
125	343	312	156	125
150	396	360	180	144
200	—	480	240	192
250	—	602	301	242
300	—	—	362	288
350	—	—	413	337
400	—	—	477	382
500	—	—	590	472

TYPICAL MOTOR EFFICIENCIES

HP	Standard Motor (%)	Energy-Effiecient Motor (%)	HP	Standard Motor (%)	Energy-Efficient Motor (%)
1	76.5	84.0	30	88.1	93.1
1.5	78.5	85.5	40	89.3	93.6
2	79.9	86.5	50	90.4	93.7
3	80.8	88.5	75	90.8	95.0
5	83.1	88.6	100	91.6	95.4
7.5	83.8	90.2	125	91.8	95.8
10	85.0	90.3	150	92.3	96.0
15	86.5	91.7	200	93.3	96.1
20	87.5	92.4	250	93.6	96.2
25	88.0	93.0	300	93.8	96.5

MOTOR FRAME TABLE

Frame No. Series	Third/Fourth Digit of Frame No.							
	D	1	2	3	4	5	6	7
140	3.50	3.00	3.50	4.00	4.50	5.00	5.50	6.25
160	4.00	3.50	4.00	4.50	5.00	5.50	6.25	7.00
180	4.50	4.00	4.50	5.00	5.50	6.25	7.00	8.00
200	5.00	4.50	5.00	5.50	6.50	7.00	8.00	9.00
210	5.25	4.50	5.00	5.50	6.25	7.00	8.00	9.00
220	5.50	5.00	5.50	6.25	6.75	7.50	9.00	10.00
250	6.25	5.50	6.25	7.00	8.25	9.00	10.00	11.00
280	7.00	6.25	7.00	8.00	9.50	10.00	11.00	12.50
320	8.00	7.00	8.00	9.00	10.50	11.00	12.00	14.00
360	9.00	8.00	9.00	10.00	11.25	12.25	14.00	16.00
400	10.00	9.00	10.00	11.00	12.25	13.75	16.00	18.00
440	11.00	10.00	11.00	12.50	14.50	16.50	18.00	20.00
500	12.50	11.00	12.50	14.00	16.00	18.00	20.00	22.00
580	14.50	12.50	14.00	16.00	18.00	20.00	22.00	25.00
680	17.00	16.00	18.00	20.00	22.00	25.00	28.00	32.00

Frame No. Series	Third/Fourth Digit of Frame No.							
	D	8	9	10	11	12	13	14	15
140	3.50	7.00	8.00	9.00	10.00	11.00	12.50	14.00	16.00
160	4.00	8.00	9.00	10.00	11.00	12.50	14.00	16.00	18.00
180	4.50	9.00	10.00	11.00	12.50	14.00	16.00	18.00	20.00
200	5.00	10.00	11.00	—	—	—	—	—	—
210	5.25	10.00	11.00	12.50	14.00	16.00	18.00	20.00	22.00
220	5.50	11.00	12.50	—	—	—	—	—	—
250	6.25	12.50	14.00	16.00	18.00	20.00	22.00	25.00	28.00
280	7.00	14.00	16.00	18.00	20.00	22.00	25.00	28.00	32.00
320	8.00	16.00	18.00	20.00	22.00	25.00	28.00	32.00	36.00
360	9.00	18.00	20.00	22.00	25.00	28.00	32.00	36.00	40.00
400	10.00	20.00	22.00	25.00	28.00	32.00	36.00	40.00	45.00
440	11.00	22.00	25.00	28.00	32.00	36.00	40.00	45.00	50.00
500	12.50	25.00	28.00	32.00	36.00	40.00	45.00	50.00	56.00
580	14.50	28.00	32.00	36.00	40.00	45.00	50.00	56.00	63.00
680	17.00	36.00	40.00	45.00	50.00	56.00	63.00	71.00	80.00

MOTOR FRAME LETTERS

LETTER	DESIGNATION
G	Gasoline pump motor
K	Sump pump motor
M and N	Oil burner motor
S	Standard short shaft for direct connection
T	Standard dimensions established
U	Previously used as frame designation for which standard dimensions are established
Y	Special mounting dimensions required from manufacturer
Z	Standard mounting dimensions except shaft extension

MOTOR FRAME DIMENSIONS

Frame No.	Shaft U	Shaft V	Key W	Key T	Key L	A	B	D	E	F	BA
48	1/2	1 1/2*	flat	3/64	—	5 5/8*	3 1/2*	3	2 1/8	1 3/8	2 1/2
56	5/8	1 7/8*	3/16	3/16	1 3/8	6 1/2*	4 1/4*	3 1/2	2 7/16	1 1/2	2 3/4
143T	7/8	2	3/16	3/16	1 3/8	7	6	3 1/2	2 3/4	2	2 1/4
145T	7/8	2	3/16	3/16	1 3/8	7	6	3 1/2	2 3/4	2 1/2	2 1/4
182	7/8	2	3/16	3/16	1 3/8	9	6 1/2	4 1/2	3 3/4	2 1/4	2 3/4
182T	1 1/8	2 1/2	1/4	1/4	1 3/4	9	6 1/2	4 1/2	3 3/4	2 1/4	2 3/4
184	7/8	2	3/16	3/16	1 3/8	9	7 1/2	4 1/2	3 3/4	2 3/4	2 3/4
184T	1 1/8	2 1/2	1/4	1/4	1 3/4	9	7 1/2	4 1/2	3 3/4	2 3/4	2 3/4
203	3/4	2	3/16	3/16	1 3/8	10	7 1/2	5	4	2 3/4	3 1/8
204	3/4	2	3/16	3/16	1 3/8	10	8 1/2	5	4	3 1/4	3 1/8
213	1 1/8	2 3/4	1/4	1/4	2	10 1/2	7 1/2	5 1/4	4 1/4	2 3/4	3 1/2
213T	1 3/8	3 1/8	5/16	5/16	2 3/8	10 1/2	7 1/2	5 1/4	4 1/4	2 3/4	3 1/2
215	1 1/8	2 3/4	1/4	1/4	2	10 1/2	9	5 1/4	4 1/4	3 1/2	3 1/2
215T	1 3/8	3 1/8	5/16	5/16	2 3/8	10 1/2	9	5 1/4	4 1/4	3 1/2	3 1/2
224	1	2 3/4	1/4	1/4	2	11	8 3/4	5 1/2	4 1/2	3 3/8	3 1/2
225	1	2 3/4	1/4	1/4	2	11	9 1/2	5 1/2	4 1/2	3 3/4	3 1/2
254	1 1/8	3 1/8	1/4	1/4	2 3/8	12 1/2	10 3/4	6 1/4	5	4 1/8	4 1/4
254U	1 3/8	3 1/2	5/16	5/16	2 3/4	12 1/2	10 3/4	6 1/4	5	4 1/8	4 1/4
254T	1 5/8	3 3/4	3/8	3/8	2 3/4	12 1/2	10 3/4	6 1/4	5	4 1/8	4 1/4
256U	1 3/8	3 1/2	5/16	5/16	2 3/4	12 1/2	12 1/2	6 1/4	5	5	4 1/4
256T	1 5/8	3 3/4	3/8	3/8	2 3/4	12 1/2	12 1/2	6 1/4	5	5	4 1/4
284	1 1/4	3 1/2	1/4	1/4	2 3/4	14	12 1/2	7	5 1/2	4 3/4	4 3/4
284U	1 5/8	4 5/8	3/8	3/8	3 3/4	14	12 1/2	7	5 1/2	4 3/4	4 3/4
284T	1 7/8	4 3/8	1/2	1/2	3 1/4	14	12 1/2	7	5 1/2	4 3/4	4 3/4
284TS	1 5/8	3	3/8	3/8	1 7/8	14	12 1/2	7	5 1/2	4 3/4	4 3/4
286T	1 7/8	4 3/8	1/2	1/2	3 3/4	14	14	7	5 1/2	5	4 3/4
286TS	1 5/8	3	3/8	3/8	1 7/8	14	14	7	5 1/2	5	4 3/4
324	1 5/8	4 5/8	3/8	3/8	3 3/4	16	14	8	6 1/4	5 1/4	5 1/4
324U	1 7/8	5 3/8	1/2	1/2	4 1/4	16	14	8	6 1/4	5 1/4	5 1/4
324S	1 5/8	3	3/8	3/8	1 7/8	16	14	8	6 1/4	5 1/4	5 1/4
324T	2 1/8	5	1/2	1/2	3 7/8	16	14	8	6 1/4	5 1/4	5 1/4
324TS	1 7/8	3 1/2	1/2	1/2	2	16	14	8	6 1/4	5 1/4	5 1/4
326	1 5/8	4 5/8	3/8	3/8	3 3/4	16	15 1/2	8	6 1/4	6	5 1/4
326U	1 7/8	5 3/8	1/2	1/2	4 1/4	16	15 1/2	8	6 1/4	6	5 1/4
326S	1 5/8	3	3/8	3/8	1 7/8	16	15 1/2	8	6 1/4	6	5 1/4
326T	2 1/8	5	1/2	1/2	3 7/8	16	15 1/2	8	6 1/4	6	5 1/4
326TS	1 7/8	3 1/2	1/2	1/2	2	16	15 1/2	8	6 1/4	6	5 1/4
364	1 7/8	5 3/8	1/2	1/2	4 1/4	18	15 1/4	9	7	5 5/8	5 7/8
364S	1 5/8	3	3/8	3/8	1 7/8	18	15 1/4	9	7	5 5/8	5 7/8
364U	2 1/8	6 1/8	1/2	1/2	5	18	15 1/4	9	7	5 5/8	5 7/8
364US	1 7/8	3 1/2	1/2	1/2	2	18	15 1/4	9	7	5 5/8	5 7/8
364T	2 3/8	5 5/8	5/8	5/8	4 1/4	18	15 1/4	9	7	5 5/8	5 7/8
364TS	1 7/8	3 1/2	1/2	1/2	2	18	15 1/4	9	7	5 5/8	5 7/8
365	1 7/8	5 3/8	1/2	1/2	4 1/4	18	16 1/4	9	7	6 1/8	5 7/8
365S	1 5/8	3	3/8	3/8	1 7/8	18	16 1/4	9	7	6 1/8	5 7/8
365U	2 1/8	6 1/8	1/2	1/2	5	18	16 1/4	9	7	6 1/8	5 7/8
365US	1 7/8	3 1/2	1/2	1/2	1 7/8	18	16 1/4	9	7	6 1/8	5 7/8
365T	2 3/8	5 5/8	5/8	5/8	4 1/4	18	16 1/4	9	7	6 1/8	5 7/8
365TS	1 7/8	3 1/2	1/2	1/2	2	18	16 1/4	9	7	6 1/8	5 7/8
404	2 1/8	6 1/8	1/2	1/2	5	20	16 1/4	10	8	6 1/8	6 5/8
404S	1 7/8	3 1/2	1/2	1/2	2	20	16 1/4	10	8	6 1/8	6 5/8
404U	2 3/8	6 7/8	5/8	5/8	5 1/2	20	16 1/4	10	8	6 1/8	6 5/8
404US	2 1/8	4	1/2	1/2	2 3/4	20	16 1/4	10	8	6 1/8	6 5/8
404T	2 7/8	7	3/4	3/4	5 5/8	20	16 1/4	10	8	6 1/8	6 5/8
404TS	2 1/8	4	1/2	1/2	2 3/4	20	16 1/4	10	8	6 1/8	6 5/8
405	2 1/8	6 1/8	1/2	1/2	5	20	17 3/4	10	8	6 7/8	6 5/8
405S	1 7/8	3 1/2	1/2	1/2	2	20	17 3/4	10	8	6 7/8	6 5/8
405U	2 3/8	6 7/8	5/8	5/8	5 1/2	20	17 3/4	10	8	6 7/8	6 5/8
405US	2 1/8	4	1/2	1/2	2 3/4	20	17 3/4	10	8	6 7/8	6 5/8
405T	2 7/8	7	3/4	3/4	5 5/8	20	17 3/4	10	8	6 7/8	6 5/8
405TS	2 1/8	4	1/2	1/2	2 3/4	20	17 3/4	10	8	6 7/8	6 5/8
444	2 3/8	6 7/8	5/8	5/8	5 1/2	22	18 1/2	11	9	7 1/4	7 1/2
444S	2 1/8	4	1/2	1/2	2 3/4	22	18 1/2	11	9	7 1/4	7 1/2
444U	2 7/8	8 3/8	3/4	3/4	7	22	18 1/2	11	9	7 1/4	7 1/2
444US	2 1/8	4	1/2	1/2	2 3/4	22	18 1/2	11	9	7 1/4	7 1/2
444T	3 3/8	8 1/4	7/8	7/8	6 7/8	22	18 1/2	11	9	7 1/4	7 1/2
444TS	2 3/8	4 1/2	5/8	5/8	3	22	18 1/2	11	9	7 1/4	7 1/2
445	2 3/8	6 7/8	5/8	5/8	5 1/2	22	20 1/2	11	9	8 1/4	7 1/2
445S	2 1/8	4	1/2	1/2	2 3/4	22	20 1/2	11	9	8 1/4	7 1/2
445U	2 7/8	8 3/8	3/4	3/4	7	22	20 1/2	11	9	8 1/4	7 1/2
445US	2 1/8	4	1/2	1/2	2 3/4	22	20 1/2	11	9	8 1/4	7 1/2
445T	3 3/8	8 1/4	7/8	7/8	6 7/8	22	20 1/2	11	9	8 1/4	7 1/2
445TS	2 3/8	4 1/2	5/8	5/8	3	22	20 1/2	11	9	8 1/4	7 1/2
504U	2 7/8	8 3/8	3/4	3/4	7 1/4	25	21	12 1/2	10	8	8 1/2
504S	2 1/8	4	1/2	1/2	2 3/4	25	21	12 1/2	10	8	8 1/2
505	2 7/8	8 3/8	3/4	3/4	7 1/4	25	23	12 1/2	10	9	8 1/2
505S	2 1/8	4	1/2	1/2	2 3/4	25	23	12 1/2	10	9	8 1/2

* Not NEMA standard dimensions

FORMULAS

THREE-PHASE VOLTAGE VALUES
For 208 V × 1.732, use 360
For 230 V × 1.732, use 398
For 240 V × 1.732, use 416
For 440 V × 1.732, use 762
For 460 V × 1.732, use 797
For 480 V × 1.732, use 831

P = WATTS
I = AMPS
R = OHMS
E = VOLTS

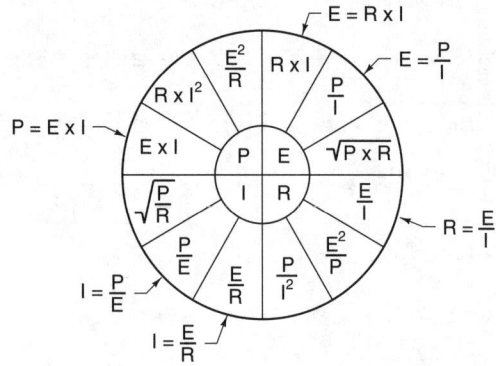

VALUES IN INNER CIRCLE ARE EQUAL TO VALUES IN CORRESPONDING OUTER CIRCLE

OHM'S LAW AND POWER FORMULA

Ohm's Law

Ohm's law is the relationship between the voltage, current, and resistance in an electrical circuit. Ohm's law states that current in a circuit is proportional to the voltage and inversely proportional to the resistance. It is written $I = E/R$, $R = E/I$, and $E = R \times I$.

Power Formula

The *power formula* is the relationship between the voltage, current, and power in an electrical circuit. The power formula states that the power in a circuit is equal to the voltage times the current. It is written $P = E \times I$, $E = P/I$, and $I = P/E$. Any value in these relationships is found using Ohm's Law and Power Formula.

POWER FORMULAS — 1φ, 3φ					
Phase	To Find	Use Formula	Example		
			Given	Find	Solution
1φ	I	$I = \dfrac{VA}{V}$	32,000 VA, 240 V	I	$I = \dfrac{VA}{V}$ $I = \dfrac{32{,}000 \text{ VA}}{240 \text{ V}}$ **I = 133 A**
1φ	VA	$VA = I \times V$	100 A, 240 V	VA	$VA = I \times V$ $VA = 100 \text{ A} \times 240 \text{ V}$ **VA = 24,000 VA**
1φ	V	$V = \dfrac{VA}{I}$	42,000 VA, 350 A	V	$V = \dfrac{VA}{I}$ $V = \dfrac{42{,}000 \text{ VA}}{350 \text{ A}}$ **V = 120 V**
3φ	I	$I = \dfrac{VA}{V \times \sqrt{3}}$	72,000 VA, 208 V	I	$I = \dfrac{VA}{V \times \sqrt{3}}$ $I = \dfrac{72{,}000 \text{ VA}}{360 \text{ V}}$ **I = 200 A**
3φ	VA	$VA = I \times V \times \sqrt{3}$	2 A, 240 V	VA	$VA = I \times V \times \sqrt{3}$ $VA = 2 \times 416$ **VA = 832 VA**

AC/DC FORMULAS

To Find	DC	AC 1ϕ, 115 or 220 V	AC 1ϕ, 208, 230, or 240 V	AC 3ϕ—All Voltages
I, HP known	$\dfrac{HP \times 746}{E \times E_{ff}}$	$\dfrac{HP \times 746}{E \times E_{ff} \times PF}$	$\dfrac{HP \times 746}{E \times E_{ff} \times PF}$	$\dfrac{HP \times 746}{1.73 \times E \times E_{ff} \times PF}$
I, kW known	$\dfrac{kW \times 1000}{E}$	$\dfrac{kW \times 1000}{E \times PF}$	$\dfrac{kW \times 1000}{E \times PF}$	$\dfrac{kW \times 1000}{1.73 \times E \times PF}$
I, kVA known		$\dfrac{kVA \times 1000}{E}$	$\dfrac{kVA \times 1000}{E}$	$\dfrac{kVA \times 1000}{1.763 \times E}$
kW	$\dfrac{I \times E}{1000}$	$\dfrac{I \times E \times PF}{1000}$	$\dfrac{I \times E \times PF}{1000}$	$\dfrac{I \times E \times 1.73 \times PF}{1000}$
kVA		$\dfrac{I \times E}{1000}$	$\dfrac{I \times E}{1000}$	$\dfrac{I \times E \times 1.73}{1000}$
HP (output)	$\dfrac{I \times E \times E_{ff}}{746}$	$\dfrac{I \times E \times E_{ff} \times PF}{746}$	$\dfrac{I \times E \times E_{ff} \times PF}{746}$	$\dfrac{I \times E \times 1.73 \times E_{ff} \times PF}{746}$

E_{ff} = efficiency

HORSEPOWER FORMULAS

To Find	Use Formula	Example Given	Example Find	Example Solution
HP	$HP = \dfrac{I \times E \times E_{ff}}{746}$	240 V, 20 A, 85% E_{ff}	HP	$HP = \dfrac{I \times E \times E_{ff}}{746}$ $HP = \dfrac{240 \text{ V} \times 20 \text{ A} \times 85\%}{746}$ $HP = \mathbf{5.5}$
I	$I = \dfrac{HP \times 746}{E \times E_{ff} \times PF}$	10 HP, 240 V, 90% E_{ff}, 88% PF	I	$I = \dfrac{HP \times 746}{E \times E_{ff} \times PF}$ $I = \dfrac{10 \text{ HP} \times 746}{240 \text{ V} \times 90\% \times 88\%}$ $I = \mathbf{39 \text{ A}}$

VOLTAGE DROP FORMULAS —1ϕ, 3ϕ

Phase	To Find	Use Formula	Example Given	Example Find	Example Solution
1ϕ	VD	$VD = \dfrac{2 \times R \times L \times I}{CM}$	240 V, 40 A, 60′ L, 16,510 CM, 12 R	VD	$VD = \dfrac{2 \times R \times L \times I}{CM}$ $VD = \dfrac{2 \times 12 \times 40 \times 60}{16,510}$ $VD = \mathbf{3.5}$
3ϕ	VD	$VD = \dfrac{2 \times R \times L \times I}{CM}$	208 V, 110 A, 75′ L, 66,360 CM, 12 R, .866 multiplier	VD	$VD = \dfrac{2 \times R \times L \times I}{CM}$ $VD = \dfrac{2 \times 12 \times 75 \times 110}{66,360}$ $VD = 2.98 \times .866$ $VD = \mathbf{2.58}$

* $\dfrac{\sqrt{3}}{2} = .866$

CAPACITORS

Connected in Series		Connected in Parallel	Connected in Series/Parallel
Two Capacitors	**Three or More Capacitors**		
$C_T = \dfrac{C_1 \times C_2}{C_1 + C_2}$ where C_T = total capacitance (in μF) C_1 = capacitance of capacitor 1 (in μF) C_2 = capacitance of capacitor 2 (in μF)	$\dfrac{1}{C_T} = \dfrac{1}{C_1} + \dfrac{1}{C_2} + ...$	$C_T = C_1 + C_2 + ...$	1. Calculate the capacitance of the parallel branch. $C_T = C_1 + C_2 + ...$ 2. Calculate the capacitance of the series combination. $C_T = \dfrac{C_1 \times C_2}{C_1 + C_2}$

SINE WAVES

Frequency	Period	Peak-to-Peak Value
$f = \dfrac{1}{T}$ where f = frequency (in hertz) 1 = constant T = period (in seconds)	$T = \dfrac{1}{f}$ where T = period (in seconds) 1 = constant f = frequency (in hertz)	$V_{p-p} = 2 \times V_{max}$ where 2 = *constant* V_{p-p} = peak-to-peak value V_{max} = peak value

Average Value	rms Value
$V_{avg} = V_{max} \times .637$ where V_{avg} = average value (in volts) V_{max} = peak value (in volts) .637 = constant	$V_{rms} = V_{max} \times .707$ where V_{rms} = rms value (in volts) V_{max} = peak value (in volts) .707 = constant

CONDUCTIVE LEAKAGE CURRENT

$I_L = \dfrac{V_A}{R_I}$

where
I_L = leakage current (in microamperes)
V_A = applied voltage (in volts)
R_I = insulation resistance (in megohms)

TEMPERATURE CONVERSIONS

Convert °C to °F	Convert °F to °C
°F = (1.8 × °C) + 32	°C = $\dfrac{(°F - 32)}{1.8}$

FLOW RATE

$Q = \dfrac{N \times V_d}{231}$

where
Q = flow rate (in gpm)
N = pump drive speed (in rpm)
V_d = pump displacement (in cu in./rev)
231 = constant

BRANCH CIRCUIT VOLTAGE DROP

$\%V_D = \dfrac{V_{NL} - V_{FL}}{V_{FL}} \times 100$

where
$\%V_D$ = percent voltage drop (in volts)
V_{NL} = no-load voltage drop (in volts)
V_{FL} = full-load voltage drop (in volts)
100 = constant

INDEX

A

abbreviations, 25, *55*
acceptable AC voltage ranges, 91
AC motors, *221-228*
  troubleshooting, *229-235*
  troubleshooting guide, 235-238
AC sine wave, *30*
actuator, 331
adjustable motor base, *199*
alignment chart, 40-*41*
allowable motor starting time, 184
alternating current, 26
alternating current motor, *211*
alternator, *381*
alternation, *30*
ambient temperature, 29
amplitude, 32
analog display, *77*
angular misalignment, *201*
anode, *373*
apparent power, 27
application drawing, *41*
arcing, *150*
arc suppressor, 151
arc tube, *297*
armature, *209*
average value, 32

B

ballast, *293*
  sound rating, 293
  ventilation, *294*
bar graph, *40*, *79*
basic logic functions, *117*
battery, *58*, *74*
belt cleaner, *200*
bench testing, 120
block diagram, *42*
board, *4*
branch circuit voltage drop, 135
break-in period, 9
breaks, 149
brown-out, 245
brushes, *209*
  troubleshooting, *212*-213

C

candela, 284
capacitance, 28
capacitive circuit, 34
capacitive leakage current, *115*
capacitor, 115
capacitor connection, 60-62
capacitor motor, *224-225*
  troubleshooting, *232-234*
capacitor-start motor, *224*
capacitor-run motor, *224*
capacitor start-and-run motor, *225*
cathode, *373*
cell, 58
centrifugal force, 223
centrifugal switch, *222-223*
charts, *40-41*
circuit breaker, 141
  troubleshooting, 141-*142*
circular chart recorder, *430*
clamping diode, *390*
clamp-on ammeter, *94*
  AC measurement, 94-95
codes, 15-*16*
color rendering, 299
common noise, 247
commutator, *209*
  troubleshooting, *213*
  troubleshooting commutator windings, 215
complimentary metal-oxide semiconductor ICs, 118
component, *4*, 74
compound motor, *211*
conduction angle, *384*
conductive leakage current, *114*
conductor, 27
  terminations, *115*
  testing, 97
constant-wattage autotransformer ballast, *300*-301
contact protection circuit, *151*
contact tachometer, *111*-112
  speed measurement, 112
contact thermometer, *109*
  temperature measurement, *109*-110
contactor, 156
  troubleshooting, 170
  troubleshooting guide, 171-172
continuity checker, 96
continuity test, *96*
conveyor adjustment, *436*
current, 26
  alternating, 26
  direct, 26
  normal, *13*
  fault, *13*
  gain, *376*
cycle, *30*

D

danger tag, *433*
data file, 409
DC motors, *209-211*
  performance characteristics, 184
  troubleshooting, *211-215*
  troubleshooting guide, 216
DC voltage variations, 183
delta configuration, *251*
delta-connected, 3φ motor, 226-227
diac, *380*
diagrams, 34
  block, *42*
  interconnection, *37*
  line, *35-36*
  operational, 41-*42*
  schematic, *36*
  terminal, *38*
  wiring, *37*
dielectric absorption test, *187*
dielectric material, 232
digital display, 78-*79*
digital logic probe, 116-*117*
  use, 118-*119*
diode, *373*
  current and voltage, *373*
direct current, 26
direct drive method, *435*
direct light, *283*
disable command, 403-*404*
distorted waveforms, *139*
distribution panel protection, 249
double-acting solenoid valve, *333*
drawings, 34-41
dry bulb temperature, *127*
dual-voltage, 3φ motor, 227
dual-voltage, delta-connected, 3φ motor, 228
dual-voltage, wye-connected, 3φ motor, 227-228

E

effective light beam, 362
effective value, 33
electric discharge lamp, 297
electrical balance, 258

electrical noise, *247*, 415
electrical prefixes, 25
electrical safety, 13
electrical shock, *12*
electrical symbols, *55*
electrolyte, 232
electromagnetic overload relay, 156
electromechanical relay, *149*
   contact failure, 152
   contact life, 150
   contact material, 151
   mechanical life, 150
   troubleshooting, 165-*166*
emergencies, 11
enclosed-coil heating element, *338*
enclosure, 433-*434*
equipment typical life expectancy curve, *9*
erratic relay operation, 169
error code, 400-403

F

field, *209*
filament, *289*
fire extinguisher classes, *14*
floating input, 118
flow chart, *6*, 40-*41*
flow switch, *363*
fluorescent lamp, 291-297
   bases, *292*
   bulb discoloration, *313*
   construction, *292*
   interference, *295*
   seating, *313*
   troubleshooting, *313-316*
   troubleshooting guide, 320-321
flush-mounted proximity sensor, *362*
footcandle, 283
force command, 403-*404*
forward bias, 387
frequency, *31*
   variation, 183
fuses, 140
   standard sizes, *51*
   troubleshooting, 140-*141*

G

gate turn ON, *379*
ghost voltage, 80
graphs, *40-41*
   bar, *40*
   line, *40*
grounded circuit, 214
grounding, 15
   methods, 15

H

half-waving, 170
hardware, 2
hard-wired unit, 3
harmonic analyzer, 139
harmonic distortion, *138*-139
   measurement, *139*-140
hazardous location, 433-434
   class, 433
   division, 434
   group, 434
head end, *435*
heating element, 337-340
   specifications, 339
   terminal protection, *340*
   troubleshooting, 340-342
heat sink, 149, 153, 391
   mounting, 154, *392*
   selection, 153
high-intensity discharge lamp, 297-302
   ballast loss, 302
   bases, 299
   identification, 302
   measuring ballast output current, *319*-320
   measuring ballast output voltage, *318*-319
   measuring line voltage, *317*
   operating characteristics, 299
   selection, 302
   troubleshooting guide, 322
high-pressure sodium lamp, *298*-299
   ballast, 301
high-reactance autotransformer ballast, *300*-301
horizontal pulley drive, *200*

I

illumination, 283
improper phase sequence, *182*
incandescent lamp, 289-292
   bases, *289*
   bulb positioning, *311*
   bulb shapes, *290*
   characteristics, *309*
   troubleshooting, *309-312*
   troubleshooting guide, 320
indirect light, *283*
inductance, 28, 390
inductive circuit, 34
inductive reactance, 222
infrared meter, *110*
   temperature measurement, 110-*111*
initial lumen, 283
in-line ammeter, 92
   DC measurement, *93*
in-phase, *33*

input module, *400*
input shielding, 415
instant-start circuit, *296*
   testing, *315*
insulation spot test, *186*
insulation step voltage test, *188*
insulator, 27
intensity, 121
interconnection diagram, *37*
interface, 2
interference, 361
interlock, 333-*334*
isolated grounded receptacle, *250*
isometric drawing, *38*

J

jumper link, *405*

K

L

lamp efficiency, 287
leakage current, 114
let-through voltage, 259
level switch, 364
light meter, *284*
linear scale, *77*
linear speed, *112*
line diagram, *35-36*
line graph, *40*
load, 271
locked rotor torque, 224
lockout, *431*
   devices, *432*
low-pressure sodium lamp, *297*-298
   ballast, *300*
lumen, 283
lumen per watt, 287

M

malfunction, 1, 8, 11
mean lumen, 283
measurement precautions, 73
mechanical switch, 355
   testing, 355-*356*
megohmmeter, *113*, 185
   resistance measurement, 115-*116*
   tests, *185-188*
mercury-vapor lamp, *298*
metal-halide, 298
   ballast, 301
metal-halide lamp, *298*
meter abbreviations, *73*
meter loading effect, *136*-137

meter symbols, 73-*74*
meter terminology, *75-77*
minimum holding current, 357
minimum/maximum readings, 135-*136*
module, *4*
motor
  AC, *221*
    altitude deratings, 198
    belt tension, *199*
    cleaning, *427*-428
    couplings, *201*
      ratings, 201
      damage, 202
  DC, *209*
  defect, *202*
  heat problems, *197*-198
  insulation class, 197
  maintenance, 427-429
  misalignment and vibration, 201
  mounting and positioning, *199*, *429*
  repair and service record, *8*
  safety, 14
  starter, *156*
    troubleshooting, *170*-171
    troubleshooting guide, 171-172
multimeter, *87*
  diode test, *388*
  settings, 87
  use, *87*-88

N

neon light, *88*
non-flush-mounted proximity sensor, *362*
nonlinear scale, *77*
normally-closed valve, *332*
normally-open valve, *332*
NPN transistor, *374*

O

off-site troubleshooting, *2-3*
off-state leakage current, 150
ohmmeter, 95
  diode test, *387*
  resistance measurement, 95-*96*
Ohm's law, *26*
one-level registration, *364*
one-time fuse, 140
on-site troubleshooting, 2
open circuit, 214
operational diagram, 41-*42*
oscilloscope, *120*
output module, *400*
overcurrent, 271
overcycling, *184*
overload, *198*
  selecting, 158
  tripping, 157

trip time, *157*
overload relay, 156

P

parallel circuit, 59
parallel connection, 55-*56*
parallel load connections, *57*
parallel misalignment, *201*
PC board repair, *119*
peak value, *32*
peak-to-peak value, *32*
period, 30
permanent-magnet motor, *211*
phase shift, 33
phase unbalance, *179*
  derating factor, 179
photo tachometer, *111-112*
  speed measurement, *112*
photocell, 59
pictorial drawing, 38
pie chart, 40-*41*
piping system, *8*
plug-in protection devices, *250*
PNP transistor, *374*
polarity, *58*
poles, 149
position, 331
power, 27
  apparent, 27
  calculation, 27
  circuit, 271
  factor, 28
  formula, *26*-27
  supply, *399*
  true, 27
prefixes, 25
preheat circuit, *296*
  testing, *314*
pressure switch, 363
preventive maintenance, 4, 11, 427
primary division, *78*
prints, *10*
process, 1
processor, *400*
  modes, *400*
programmable controller, 4, *399*
  flow chart, *414*
  preventive maintenance, 415
  startup procedure, 404-*413*
    testing the system, 412-*413*
    troubleshooting input devices, *409*-410
    troubleshooting input modules, 407-*408*
    troubleshooting output devices, *411*-412
    troubleshooting output modules, *410*-411

programming terminal, *400*
  handheld, *3*
protecting switch contacts, *361*
proximity switch ratings, *130*
psychrometric chart, *127*
pull-up resistor, *118*
pure DC power, 183

Q

R

rapid-start circuit, *297*
  testing, *316*
RC circuit, 170
reactor ballast, *300*-301
recording meter, *430*
relay, *149*
  contact arrangements, *150*
  contact abbreviations, 149
renewable fuse, 140
resistance, 27
resistive circuit, 33
reverse bias, 387
roller drive method, *435*
root-mean-square value, 33
rotational speed, *112*
rotor, *221*
rough service lamp, 311

S

schematic diagram, *36*
scope, 120
  display, *121*
  horizontal positioning, 121
  measurement — AC voltage, 123-*124*
  measurement — DC voltage, 125-*126*
  measurement — frequency, 124-*125*
  trace, *121*
  vertical positioning, 121
scopemeter, *120*
SCR, *377*
  DC voltage control, *377-378*
  testing, *390*
secondary division, *78*
sectional view drawing, *39*
  section lines, *39*
series circuit, 59
series connection, 55-*56*
series load connections, *57*
series motor, *210*
series/parallel circuit, 60
series/parallel connection, 55
series/parallel load connections, 57-*58*
service bulletin, *7*
shaded-pole motor, *211*-212

troubleshooting, *229*
troubleshooting guide, 235
short circuit, 214, *271*
shunt motor, *210*
single-phase motor, 221
single phasing, *181*
single-voltage, 3ϕ motor, 226
slip, 129
snubber circuit, *416*
snub roller, *435-436*
software, 2
solenoid, *331*
  specifications, 334
  voltage measurement, *333*
solenoid-operated valve, *331-332*
  troubleshooting, *336-337*
solid-state relay, *152*
  current problems, 154
  temperature problems, 152-153
  troubleshooting, *167-170*
  voltage drop, *155*-156
  voltage problems, 154
split-phase motor, *222*
  troubleshooting, *229-230*
  troubleshooting guide, 236-237
standards, *15-16*
standard receptacle grounding, *250*
startup procedure, 10, 404-406
static turn ON, *380*
stator, *211*
status indicator, 400-402
strobe tachometer, *111*, *113*
  speed measurement, *113*
subdivision, *78*
substitution, 3
surface leakage current, *115*
sulfidation, 151
surge suppressor, *248*
sweep, 121
switch connections, *56*
system, 1
  electrical/electronic, *1-2*
  protection, 248-*249*
  analysis, 2
systems analyst, 2

T

tabular chart, *40*
tachometer, *111*
tagout, *431*

tail end, *435*
take-up pulley, *435-436*
taps, 252
temperature, 29
  ambient, 29
  conversion, 29-30
terminal diagram, *38*
test light, *88*
  use, 88-89
thermal overload relay, 156
thermal resistance, 153
thermal switch, *223*
  testing, *231*
thermal turn ON, 380
three-phase motors, *225*
  troubleshooting, *234-235*
  troubleshooting guide, 237-238
three-wire solid-state switch, 359-361
  testing, *360*-361
throws, 149
thyristor, *377*
  triggering methods, *379*
time/division control, *122-123*
trace, 121
transformer deratings, 257
transformer load cycle, *251*
transformer sizing and balancing, *253-258*
transformer tap connections, 252
transient voltages, 167, 246
transistor, *374*
  testing, *389*
transistor-transistor logic ICs, 118
transverse noise, 247
triac, *377*
  AC voltage control, *378*
  testing, *391*
troubleshooter responsibilities, 10
troubleshooting, 1
  levels, *1*
  methods, 4
troubleshooting procedure, 6-8
troubleshooting circuit faults, *214*-215
true power, 27
tungsten-halogen lamp, *291*
two-level registration, *364*
two-winding constant-wattage ballast, *300*-301
two-wire solid-state switch, 356-359
  testing, *359*

U

unit, 3
useful life, 9

V

varistor, 151
vertical chart recorder, *430*
vertical pulley drive, *200*
voltage, 26
voltage break-over turn ON, *380*
voltage conversions, 33
voltage dip, *245*
voltage fluctuation, 245
voltage spike, *247*
voltage stabilizer, *246*
voltage surge, *182-183*, 245
voltage tester, *89*
  use, *90*
voltage unbalance, 180-*181*
voltage variation, 183, 245
voltmeter, *90*
  AC voltage measurement, *92*
  DC voltage measurement, *91*
  load connection, *137*
  switch connection, *138*
volt-ohm-milliammeter, 87
volts/division control, *122*

W

wall-bulb temperature, 294
watt density, 338-339
wattmeter, *97*
  power measurement, 97-*98*
way, 331
wear-out period, 9
wet bulb temperature, *127*
wiring diagram, *37*
wrap-around bar graph, 79-*80*
wye configuration, *251*
wye-connected, 3ϕ motor, *226*

X

Y

Z